# Cellular and Molecular Biology of Hemopoietic Stem Cell Differentiation

Tak W. Mak and Ernest A. McCulloch, Editors

The Ontario Cancer Institute and University of Toronto, Toronto, Ontario, Canada



Proceedings of a Symposium held
at Honey Harbor, Ontario, Canada,
September 20–24, 1981

Alan R. Liss, Inc., New York

Address all Inquiries to the Publisher
Alan R. Liss, Inc., 150 Fifth Avenue, New York, NY 10011



Printed in the United States of America.



**Library of Congress Cataloging in Publication Data**
Symposium on Cellular and Molecular Biology of
Hemopoietic Stem Cell Differentiation (1981:
Honey Harbor, Ont.)
Cellular and molecular biology of hemopoietic
stem cell differentiation.

Includes index.
1. Hematopoietic system—Diseases—Congresses.
2. Hematopoietic stem cells—Congresses.
3. Cell differentiation—Congresses. 4. Hemato-
poietic system—Cancer—Congresses. I. Mak, Tak W.,
1945- . II. McCulloch, Ernest A., 1926- .
III. Title. [DNLM: 1. Hematopoietic stem cells—
Cytology—Congresses. 2. Cell differentiation—
Congresses. 3. Molecular biology—Congresses.
WH 380 S988c 1981]
RC644.5.S95 1981 616.99'44107 82-15348
ISBN 0-8451-0215-X AACR2

These proceedings are published both as *Journal of Cellular Physiology*, Supplement 1, 1982, and as *Cellular and Molecular Biology of Hemopoietic Stem Cell Differentiation.*

# Contents

# Contributors

*The boldface number in brackets following each contributor's address indicates the opening page of that author's article or articles.*

JOHN W. ADAMSON, University of Washington School of Medicine and Veterans Administration Medical Center, Seattle, Washington 98195 **[79]**

ARTHUR AXELRAD, University of Toronto, Toronto, Ontario M5S 1A8, Canada **[165]**

MENASHE BAR-ELI, University of California at Los Angeles, Los Angeles, California 90024 **[213]**

HARTMUT BEUG, German Cancer Research Center, 6900 Heidelberg, Federal Republic of Germany **[195]**

KARL G. BLUME, City of Hope National Medical Center, Duarte, California 91010 **[99]**

W. PAUL BOWMAN, St. Jude Children's Research Hospital, Memphis, Tennessee 38101 **[127]**

RAYMOND S. BUSH, The Ontario Cancer Institute, Toronto, Ontario M4X 1K9, Canada **[1]**

L.J-A. CHANG, University of Toronto and The Ontario Cancer Institute, Toronto, Ontario M4X 1K9, Canada **[103]**

MARTIN J. CLINE, University of California at Los Angeles, Los Angeles, California 90024 **[213]**

GEOFFREY M. COOPER, Harvard Medical School, Boston, Massachusetts 02115 **[209]**

FRANKLIN COSTANTINI, University of Oxford, Oxford OX1 3PS, United Kingdon **[219]**

HELENA CROIZAT, University of Toronto, Toronto, Ontario M5S 1A8, Canada **[165]**

T.M. DEXTER, Christie Hospital and Holt Radium Institute, Manchester M20 9BX, United Kingdom **[87]**

GABI DOEDERLEIN, German Cancer Research Center, 6900 Heidelberg, Federal Republic of Germany **[195]**

LOIS W. DOW, St. Jude Children's Research Hospital, Memphis, Tennessee 38101 **[127]**

DENISE ESKINAZI, University of Toronto, Toronto, Ontario M5S 1A8, Canada **[165]**

PHILIP J. FIALKOW, University of Washington, Seattle, Washington 98195 **[37, 79]**

STEPHEN J. FORMAN, City of Hope National Medical Center, Duarte, California 91010 **[99]**

CHRISTA FREUDENSTEIN, German Cancer Research Center, 6900 Heidelberg, Federal Republic of Germany **[195]**

EMILIA FRINDEL, Institut Gustave-Roussy, 94800 Villejuif, France **[13]**

HENK VAN DER GAAG, University of Toronto, Toronto, Ontario M5S 1A8, Canada **[165]**

EUGENE GOLDWASSER, University of Chicago, Chicago, Illinois 60637 **[133]**

THOMAS GRAF, German Cancer Research Center, 6900 Heidelberg, Federal Republic of Germany **[195]**

MELVYN F. GREAVES, Imperial Cancer Research Fund, London WC2A 3PX, United Kingdom **[113]**

L.J. GUILBERT, Albert Einstein College of Medicine, Bronx, New York 10461 **[65]**

N.N. ISCOVE, Basel Institute for Immunology, CH-4005, Basel, Switzerland **[65]**

C.A. IZAGUIRRE, University of Toronto and The Ontario Cancer Institute, Toronto, Ontario M4X 1K9, Canada **[45,103]**

N. JAMAL, The Ontario Cancer Institute and University of Toronto, Toronto, Ontario M4X 1K9, Canada **[45]**

G.R. JOHNSON, Walter and Eliza Hall Institute of Medical Research, Royal Melbourne Hospital, Melbourne, 3050 Victoria, Australia **[23]**

G.M. KELLER, The Ontario Cancer Institute and University of Toronto, Toronto, Ontario M4X 1K9, Canada **[23, 31]**

ELIZABETH LACY, University of Oxford, Oxford OX1 3PS, United Kingdom **[219]**

A. THOMAS LOOK, St. Jude Children's Research Hospital, Memphis, Tennessee 38101 **[127]**

TAK W. MAK, The Ontario Cancer Institute and University of Toronto, Toronto, Ontario M4X 1K9, Canada **[185]**

ALVIN M. MAUER, St. Jude Children's Research Hospital, Memphis, Tennessee 38101 **[127]**

NILS MAURITZSON, University of California at Los Angeles, Los Angeles, California 90024 **[213]**

E.A. McCULLOCH, University of Toronto and The Ontario Cancer Institute, Toronto, Ontario M4X 1K9, Canada **[103]**

KAREN E. MERCOLA, University of California at Los Angeles, Los Angeles, California 90024 **[213]**

H.A. MESSNER, The Ontario Cancer Institute and University of Toronto, Toronto, Ontario M4X 1K9, Canada **[45]**

D. METCALF, Walter and Eliza Hall Institute of Medical Research, Royal Melbourne Hospital, Melbourne, 3050 Victoria, Australia **[175]**

MALCOLM A.S. MOORE, Sloan-Kettering Institute for Cancer Research, New York, New York 10021 **[53]**

N.A. NICOLA, Walter and Eliza Hall Institute of Medical Research, Royal Melbourne Hospital, Melbourne, 3050 Victoria, Australia **[23]**

THALIA PAPAYANNOPOULOU, University of Washington, Seattle, Washington 98195 **[145]**

R.A. PHILLIPS, The Ontario Cancer Institute and University of Toronto, Toronto, Ontario M4X 1K9, Canada **[31]**

JERRY S. POWELL, University of Washington School of Medicine and Veterans Administration Medical Center, Seattle, Washington 98195 **[79]**

C.A. ROITSCH, Basel Institute for Immunology, CH-4005, Basel, Switzerland **[65]**

LEO SACHS, Weizmann Institute of Science, Rehovot 76100, Israel **[151]**

R. SCHOFIELD, Christie Hospital and Holt Radium Institute, Manchester M20 9BX, United Kingdom **[95]**

TSUNEFUMI SHIBUYA, The Ontario Cancer Institute and University of Toronto, Toronto, Ontario M4X 1K9, Canada **[185]**

DENNIS J. SLAMON, University of California at Los Angeles, Los Angeles, California 90024 **[213]**

L.J. SMITH, University of Toronto and The Ontario Cancer Institute, Toronto, Ontario M4X 1K9, Canada **[103]**

O. SMITHIES, University of Wisconsin – Madison, Madison, Wisconsin 53706 **[137]**

GEORGE STAMATOYANNOPOULOS, University of Washington, Seattle, Washington 98195 **[145]**

HOWARD D. STANG, University of California at Los Angeles, Los Angeles, California 90024 **[213]**

SOLOMON STEWART, University of Toronto, Toronto, Ontario M5S 1A8, Canada **[165]**

J.E. TILL, The Ontario Cancer Institute and University of Toronto, Toronto, Ontario M4X 1K9, Canada **[3]**

MAURICE TUBIANA, Institut Gustave-Roussy, 94800 Villejuif, France **[13]**

DUSHANDHAN VAITHILINGAM, University of Toronto, Toronto, Ontario M5S 1A8, Canada **[165]**

DOROTHY WILLIAMS, St. Jude Children's Research Hospital, Memphis, Tennessee 38101 **[127]**

N. WILLIAMS, University of Melbourne, Parkville, Victoria 3052, Australia **[65]**

# Participants

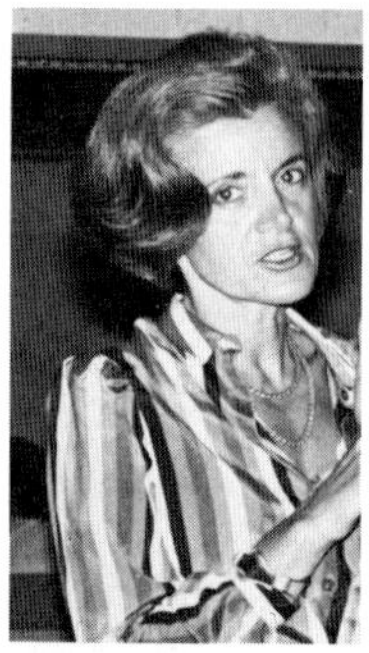

**Top:** N. Teich / O. Smithies / T. Papayannopoulou / P. Harrison, H. Eisen / M. Cline

**Middle:** L. Sachs / J. Singer, P. Fialkow / D. Metcalf, T. Hoang

**Bottom:** A. Axelrad, E. Goldwasser / G. Cooper, M. Lane / Y. Niho, J. Till, R. Bush

# Participants

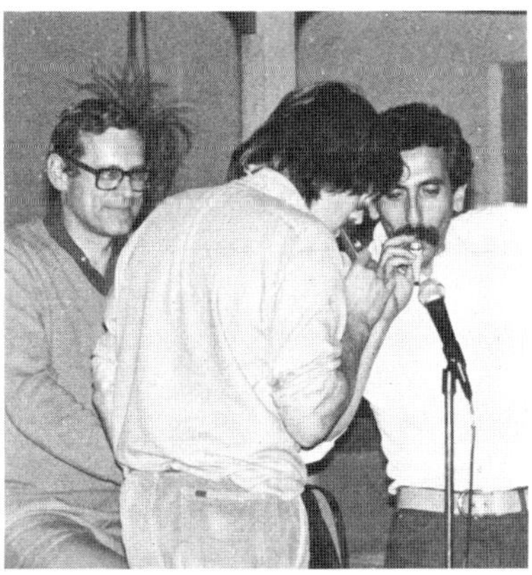

**Top:** J. Adamson, E. McCulloch, R. Bush / M. Moore / G. Johnson, N. Iscove

**Middle:** J. Till, B. Acres, A. Bernstein / M. Greaves / H. Weintraub, D. Housman

**Bottom:** R. Phillips, S. Oliphant / M. Dexter / M. Aye, T. Mak, H. Messner

# Preface

A major challenge for modern cancer research is to achieve integration of findings at the clinical, cellular and molecular levels. In the last decade cellular studies have been dominant, with emphasis on kinetics, growth control, cytogenetics and sensitivity to chemotherapeutic agents. Clinical studies have rested heavily on empiricism, guided only modestly by the insights provided by cell biology. At the molecular level, where many believe mechanisms may be specified most precisely, work has been hindered by the marked heterogeneity of cellular populations.

The Symposium on Cellular and Molecular Biology of Hemopoietic Stem Cell Differentiation, the proceedings of which are published here, was held at a time when clinical and molecular studies were challenging the primacy of cell biology. Its major objective was to bring together distinguished investigators representing several disciplines in the hope of encouraging interactions. Hemopoiesis and the malignant diseases of hemopoietic stem cells were chosen to provide a common theme, a role that they have played frequently in cancer research.

The contribution of cell biology was seen to continue to be of great importance. Many speakers presented data showing that colony assays in culture are available for the progenitors of normal and malignant hemopoiesis; these include pluripotent stem cells, the committed progenitors of granulopoiesis, erythropoiesis, megakaryocytopoiesis, B and T lymphocytes, as well as the blast cells of acute leukemia. These methods have permitted the determination of progenitor properties and the purification of substances that stimulate their growth in culture.

Characterization of the phenotypes of normal and leukemic cells has been greatly facilitated by the development of monoclonal antibodies. This work has provided a major link with clinical research by permitting ALL patients to be classified by the probability of achieving remission and perhaps cure. Clinical studies based on these insights achieved a new level of resolution. The possiblity that marrow transplantation might prove curative for certain leukemia patients was explored; not only were clinical results presented, but also mechanisms of engraftment and graft-vs-host disease were discussed in light of recent advances in immunobiology.

Perhaps the greatest advances were seen in the papers reporting studies at the molecular level. By using the new techniques of DNA transfer and gene cloning, many of the problems of cellular heterogeneity have been circumvented; indeed, mechanisms have been reported that underlie the generation of biologically important variation. The cloning of DNA directly or indirectly affected by the primary lesions responsible for malignancy promises to signal the beginning of a new era. It will be a fruitful era to the extent that integration with cellular and clinical research is maintained.

The Symposium was organized, arranged and managed by a committee consisting of Drs. A. Axelrad, A. Bernstein, R. Buick, T. Mak, S. Oliphant and R. Phillips with the help of the secretaries, postdoctoral fellows and students at The Ontario Cancer Institute. Everyone is grateful for their efforts. We also thank Dr. L. Siminovich for his summary of the meeting. Much credit is also due P. West and D. Quon for their help in the coordination of the manuscripts and to The Ontario Cancer Institute and the National Cancer Institute of Canada for sponsoring the meeting. Not least, we thank the contributors, not only for the quality of their papers, but also for submitting manuscripts promptly. Because of this cooperation and the hard work of S. Hannan and Alan Liss and his staff, the Proceedings appear just six months after the meeting was held.

*T.W. Mak*
*Ernest A. McCulloch*
Toronto, 1982

# The Symposium

In Canada, we have just experienced the effect that one determined and remarkable man can have on every citizen of the country. This man, young in years but mature in spirit, lost his leg, and finally his life, to cancer. After the diagnosis was made, and treatment had been given, he managed to bring an excitement to, and high expectations from, his life. This had a marked affect on all Canadians. To focus the public's awareness of the need for increased support of cancer research, he decided to run across Canada from sea to sea. This is a distance of approximately 5,000 miles. He reached the western boundary of Ontario, roughly 2,500 miles after he started from the edge of the Atlantic at St. John's, New Foundland. He had two goals. One was to complete his run, but this was not to be; the second was to raise for cancer research in Canada one dollar for every person in the country. This goal he did reach. But of course he did much more than that; he brought a renewed sense of need, hope, endeavour, and of excitement for research.

This meeting has a sense of excitement similar to that raised by Terry Fox, and it is in keeping with the expectations that have been present in the study of hemopoietic cells over the last twenty years. It was that long ago when the first paper was published reporting the quantitative assessment of the bone marrow cells that could give rise to a variety of mature blood cells. That paper marks a point where there was a clear change from a purely morphological approach in hematological research to a cell biology one. There was a high level of excitement with the introduction of quantitative cell biology, and the extensive activity generated by that excitement has lasted. A great deal of knowledge has been obtained about the cellular characteristics of the various primitive cells in the marrow of both healthy and diseased individuals. The development of in vitro culture methods has contributed significantly to this knowledge.

We are now at another point of transition, giving a renewed sense of excitement and expectation. Methods are now available in molecular genetics to allow the examination of questions that heretofore were clearly unanswerable. Techniques can be applied that should provide a view of the inner control mechanisms of both normal and malignant hematological cells. This view should give similar insight into the same control mechanisms of cells from other tissues. Human cells can now be cultured with relative ease, so that many fundamental questions may be answered using human material. This increases the expectation that the increased knowledge may quickly be transferred to the evaluation and management of patients, and that such applications will benefit them by an improved outcome.

This meeting, then, is timely, because it brings together people from cellular biology, molecular genetics, and clinical practice, with the expectation that all three are ready for close interaction. Members of the staff of The Ontario Cancer Institute have played important parts in all three fields, and they propose to continue doing so.

This international forum gives us an opportunity to discuss where we are and to explore possibilities for the future. On behalf of The Ontario Cancer Institute, and the National Cancer Institute of Canada, I thank the organizers of this meeting, and the participants. The presentations published here will stand in the literature to summarize this outstanding meeting.

*Raymond S. Bush, M.D.*
Director, The Ontario Cancer Institute

0021-9541/82/SUPPL. 1-0001$01.00 

JOURNAL OF CELLULAR PHYSIOLOGY SUPPLEMENT 1:3–11 (1982)

# Stem Cells in Differentiation and Neoplasia

J. E. TILL*
*Division of Biological Research, The Ontario Cancer Institute and School of Graduate Studies, University of Toronto, Toronto, Ontario M4X 1K9, Canada*

The stem cell concept continues to be very useful both for studies on cellular differentiation and for studies on neoplasia. At present the concept is being enlarged to include consideration of a new aspect: stem cell genetics, at the molecular level. Some questions are still to be answered. What is the basis for genetic regulation of "commitment," that is, the transition from pluripotent stem cells to committed progenitor cells? Does neoplasia involve changes in the genetic material of hemopoietic stem cells? To illustrate the ways in which genetic approaches may be used to provide new answers to these important questions, a minimal model of commitment at the molecular level is presented. The model is based on the assumption that movable genetic elements may play a hitherto-unsuspected role in the early stages of stem cell differentiation. It provides a simple explanation for the apparent close association between loss of potential for stem cell renewal and commitment. The possibility that analogous movable genetic elements may be involved in both the development of neoplastic cells and the differentiation of normal cells is also discussed. The need for caution in interpreting evidence for apparent differences in the genetic material of normal and neoplastic cells is stressed.

The purpose of this introductory paper is to present a personal view of where we are now and where we are going, in relation to the cellular biology of hemopoietic stem cells. My intent is to focus attention on some issues that I believe are central to the further development of the field. Let me begin with a little bit of history.

The concept of hemopoietic stem cells has proved to be a durable one. So far, it has evolved through two main phases, which I shall call the morphological phase and the functional phase. The first phase was characterized by a reliance on microscopy as the primary technique of experimental analysis. The main area of controversy about stem cells during this era was whether or not they can be pluripotent in their potential for differentiation. This question was answered with an unequivocal "yes" only at the beginning of the second phase, after functional approaches became available (Till and McCulloch, 1961; Becker et al., 1963; Wu et al., 1967). An example of a simplified model for the differentiation of functionally recognizable hemopoietic stem cells is shown in Figure 1. The committed progenitor cell types shown in Figure 1 are also recognized using functional assays (for a recent review, see Till and McCulloch, 1980). In models of this type, it is usually assumed that commitment is irreversible. It should of course, be remembered that processes involved in the regulation of cell differentiation must involve multiple stages, with many such steps intervening between the primordial progenitor cell (the fertilized ovum) and the fully differentiated cells of the blood-forming system.

The function of stem cells that gives them fundamental importance and that distinguishes them from progenitor cells or any of the other more differentiated cell types (Fig. 1) is their property of "independence" (McCulloch et al., 1965; Till and McCulloch, 1980). Stem cells may be regarded as the only cells with the potential to regenerate a functional blood-forming system after severe depletion, such as that which follows exposure to total-body radiation at lethal dose levels. For a stem cell to be capable of independence, it must be able to give rise to new stem cells ("self-renewal"). Thus, an essential feature of stem cells is that they inherit the capacity for stem cell renewal. Loss of potential for self-renewal seems to occur concomitantly with the irre-

*To whom correspondence/reprint requests should be addressed.

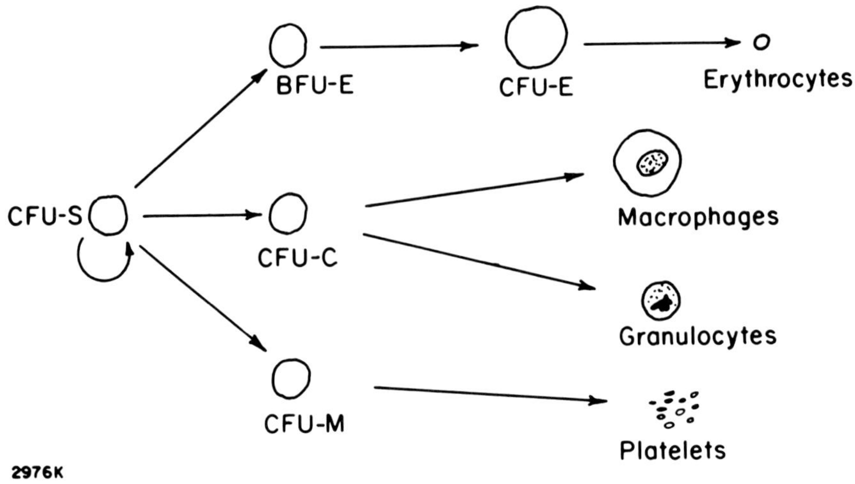


Fig. 1. Simplified illustration of the differentiation of myelopoietic stem cells. The types of cells detected on the basis of assays of cell function are shown as stem cells [CFU-S, colony-forming unit(s) in spleen] and committed progenitor cells [BFU-E, burst-forming unit(s), erythroid; CFU-C, also termed CFU-GM, granulocyte, macrophage CFU; CFU-M, megacaryocyte CFU and CFU-E, erythroid CFU]. The essential property of stem cells, capacity for self-renewal, is represented by the semicircular arrow on the left.

versible commitment of the stem cell to a more restricted potential for differentiation. These two processes, commitment and loss of potential for stem cell renewal, seem to be closely associated and probably bear a cause-effect relationship to each other (Till et al., 1964).

The crucial nature of stem cell renewal raises the question: Is there a genetic basis for inheritance by cells of the potential for stem cell renewal and for the loss of this potential that accompanies an irreversible commitment to a particular pathway of differentiation? We can only begin to answer this question on the basis of present experimental evidence. It is my belief that attempts to answer such questions will lead to the evolution of stem cell concepts into a third phase, which I shall call the genetic phase. This view is based on a conviction that the essential difference between stem cells and their more differentiated progeny is at the level of genetic information. Until we can understand the regulation of flow of genetic information in stem cells and their progeny, we will not be able to understand rigorously the renewal and differentiation of stem cells.

## THE CONSTANCY HYPOTHESIS

One key genetic question that must be addressed is whether or not the loss of capacity for stem cell renewal is accompanied by any detectable change at the level of the cell genome. The supposed identity of the genetic material in different somatic cells has been called the "constancy hypothesis" (Davidson, 1968). My question, rephrased, is does the constancy hypothesis hold for stem cells and their descendants? If it does, then stem cells do not differ from their descendants at the genotypic level, and differences between them involve only their ability to express certain genes belonging to a constant, unchanging endowment of DNA present in the same form in somatic cells and in the germ line.

I will not attempt to review here all the evidence for and against the constancy hypothesis. Some of the best-known evidence in favor of the hypothesis comes from nuclear transplantation experiments (King and Briggs, 1956; Gordon and Uehlinger, 1966; Illmensee and Hoppe, 1981). These experiments have been interpreted to mean that, for the nuclei of at least some somatic cells, no irreversible changes leading to loss of totipotency have occurred at the level of the genome. This conclusion can not only be questioned at a technical level, (McKinnell, 1978), but also, it is conceptually unsatisfying, because it fails to provide an answer to the question why somatic cells appear to differ from each other in so stable a manner, if no genetic changes are involved (Campbell, 1981).

Evidence in favor of the lability of genetic material, rather than its constancy, have come from studies of movable genetic elements in a variety of organisms. Some of this evidence is summarized in Table 1. For at least some of these examples, there is clear evidence that DNA rearrangements play a demonstrable role in controlling gene expression or cellular differentiation. These include the classical studies on controlling elements in maize (McClintock, 1952), phase variation in Salmonella (Silverman et al., 1981; Simon et al., 1980), the determination of mating type in yeast (Herskowitz et al., 1980; Hicks et al., 1977), the differentiation of B-lymphocytes leading to the generation of diversity of immunoglobulin genes (Hozumi and Tonegawa, 1976; Bernard et al., 1978; Seidman et al., 1979; Early et al., 1980; Sakano et al., 1981; Davis et al., 1980), and the acquisition of drug resistance by mammalian cells in vitro (Schimke, et al., 1978).

Additional evidence that the genome of all cells, including somatic mammalian cells, may be more labile than we once thought has come from work on a variety of procaryotic and eucaryotic movable genetic elements (Calos and Miller, 1980; Green, 1980; Williamson et al., 1981; Pays et al., 1981; Hoeijmakers et al., 1980; Campbell, 1981; Yarmolinsky, 1981), including intriguing studies on retroviruses and the activation of cellular transforming genes by these viruses (see, for example, Temin, 1980; McClements et al., 1981; Neel et al., 1981; Payne et al., 1981; Fung et al., 1981).

The extensive work that has been done on the generation of diversity of immunoglobins has provided the only strong evidence that the constancy hypothesis can be violated during the differentiation of normal cells forming part of the hemopoietic system. It was originally suggested by Dreyer and Bennett (1965) that the generation of immunoglobulin diversity might involve violation of the principle of "one gene, one polypeptide chain." They proposed that rearrangement of immunoglobulin gene segments might precede immunoglobulin synthesis. Subsequently, convincing evidence has been obtained for rearrangement of segments of structural genes responsible for synthesis of immunoglobulin light chain genes (Fig. 2) (Hozumi and Tonegawa, 1976; Bernard et al., 1978; Seidman et al., 1979) and heavy chain genes (Early et al., 1980; Davis et al., 1980; Sakano et al., 1981). Evidence for rearrangement of DNA sequences during T-cell differentiation is, at present, more controversial (Kurosawa et al., 1981; Cayre et al., 1981).

TABLE 1. Some examples of evidence for rearrangement of genetic material

| Organisms | Type of rearrangement |
|---|---|
| Maize | Movable controlling elements |
| Bacteria | Insertion sequences and transposons |
| Yeast | Movable elements, e.g., Ty 1 |
| Fruit fly | Movable elements, e.g., copia |
| Trypanosomes | Rearrangement of surface antigen genes |
| Mammalian cells | Rearrangement of immunoglobulin genes |
| Mammalian cells | Retrovirus insertion |
| Mammalian cells | Gene duplication, e.g., resistance to methotrexate |

Studies of other structural genes of hemopoietic cells, such as globin genes (Maniatis et al., 1980; Proudfoot et al., 1980) have not provided evidence for rearrangement, but the possibility remains that rearrangement of genetic material may play a more general role in the early stages of the differentiation of mammalian cells. For example, perhaps gene rearrangements occur frequently during the early stages of differentiation, but mainly affect genes with regulatory functions. Hence, the ultimate tests of constancy hypothesis may have to await the molecular cloning of regulatory genes controlling early events in cellular differentiation. Genetic changes responsible for disorders of human hemoglobin synthesis, such as thalassemia (Leder, 1978; Bank et al., 1980), have been explored from this point of view, but these changes appear to affect structural genes, not regulatory genes (Williamson, 1981).

If the constancy hypothesis is, in fact, not tenable during differentiation, reproducible differences in DNA sequences should be found between stem cells and their more differentiated progeny, and in selected experimental systems, appropriately designed gene transfers might permit a differentiated cell to be converted back into a stem cell. This issue is important enough from both a conceptual and a practical viewpoint to merit a great deal of attention.

## MODELS OF COMMITMENT

A particularly important stage in the regulation of cellular differentiation is commitment, involving both loss of the unique capacity of stem cells for self-renewal, and restriction of their potentiality for further differentiation. This stage is represented in Figure 1 by the transition from stem cell to progenitor cell, e.g., from CFU-S to BFU-E. What

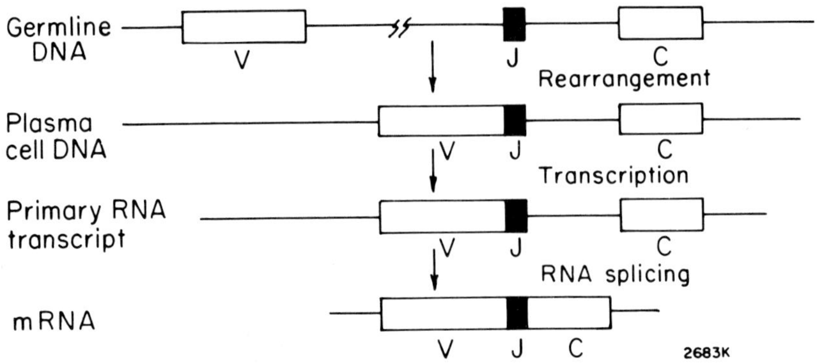


Fig. 2. Stages in the expression of immunoglobulin genes. V, J, C: variable, joining, and constant region gene segments, respectively.

is the basis for genetic regulation of this transition?

Two main classes of models for the regulation of this stage of differentiation have been proposed. They will be referred to here as stochastic and deterministic models. Stochastic models are based on the assumption that some sort of random (or pseudorandom) event is involved in commitment. The clonal selection model (Burnet, 1959) for the commitment of B-lymphocytes to the synthesis of a particular immunoglobulin is the best-known example of a stochastic model. Commitment, in this case, involves a change in the sequence of bases in DNA, either because of the rearrangement of genetic material as outlined in Figure 2 or because of classical gene mutations (Cohn et al., 1980). Environmentally derived stimuli (antigens) act on cells that have already undergone a random genetic change; there is selection of the appropriate B-lymphocyte progenitor and stimulation, by exposure to antigen, of its proliferation and further differentiation.

The HER (hemopoiesis engendered at random) model for the regulation of stem cell renewal and commitment (Till et al., 1964; Korn et al., 1973; Till and McCulloch, 1980; Till, 1981) is another example of a stochastic model. In this model, (Fig. 3) it is assumed that a random event leads to loss of capacity for stem cell renewal and to commitment to a particular pathway of differentiation. The nature of this event at the molecular level was not specified; it could be either a random change in DNA sequence (transposition or mutation) or the initiation of expression of preexisting DNA sequences in randomly chosen stem cells. Environmentally derived stimuli are assumed to act at the stem cell level only to influence the probability of stem cell renewal or commitment. Consequently, only populations of many stem cells would be expected to show predictable responses to environmental stimuli; individual stem cells should not. An appropriate analogy is the decay of radioactive nuclei; on the average, decay occurs according to a predictable half-life, but the decay of any individual nucleus cannot be predicted accurately because of its stochastic nature.

The second class of models for the regulation of the transition from stem cells to committed progenitor cells may be called deterministic. In such models, an appropriate external stimulus acts directly on stem cells to initiate the transition. Antigen-directed stimulation of a stem cell to induce its transition to a B-lymphocyte progenitor committed to synthesis of the antibody specific for that antigen is a now-discredited example of a deterministic model. Another example is the HIM (hemopoietic inductive microenvironment) model for the commitment of stem cells to differentiation along a particular pathway (Curry and Trentin, 1967). Here, the microenvironment surrounding the stem cells is assumed to exert an inductive effect on stem cells; those exposed to erythropoietic HIM would differentiate along the erythropoietic pathway, while those exposed to granulopoietic HIM would initiate granulopoiesis. The detailed mechanism for induction at the molecular level was not specified. Direct interaction of the hormone erythropoietin with stem cells, to induce their differentiation along the erythropoietic pathway, represents a more specific example of this kind of deterministic model (Goldwasser, 1975).

If deterministic models are correct, one might expect to observe rather uniform responses of stem cells to environmental stimuli,

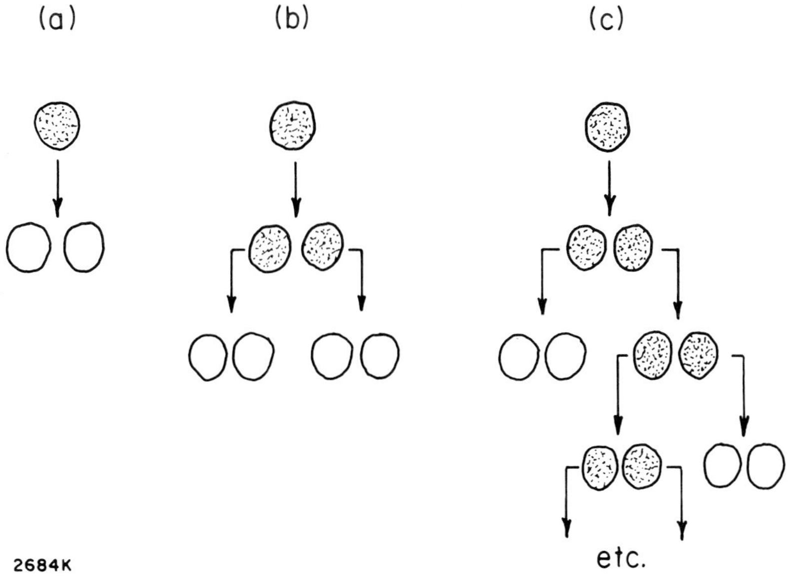


Fig. 3. Stochastic model for the regulation of stem cell renewal and commitment at the cellular level. Shaded circles: stem cells still capable of self-renewal. Open circles: progenitor cells committed to a particular pathway of differentiation. It is assumed that a randomly occurring event results in both loss of potential for stem cell renewal and in commitment. In (a) and (b), extinction of stem cells has occurred because of commitment events during the 1st and 2nd cell generations, respectively. In (c), stem cells persist as a consequence of retention of potential for stem cell renewal.

under conditions where these stimuli can be subjected to experimental control. Studies on hemopoietic cells in culture, where experimental control of environmental stimuli is much greater than can be achieved in vivo, have indicated that regulation of stem cell renewal is lax in vitro (Humphries et al., 1981), as it is in vivo (Siminovitch et al., 1963; Till et al., 1964). These results favor a stochastic over a deterministic interpretation.

If it is assumed that a stochastic model of commitment is correct, that the constancy hypothesis is violated during commitment, and that commitment and loss of potential for stem cell renewal are a consequence of the same event, then the simple model for commitment shown in Figure 4 seems an attractive one. This model is based on the view that transposition of genetic material results in loss of potential for stem cell renewal, followed by initiation of a particular differentiation program. Such a transposition could result in an apparently irreversible commitment to this pathway of differentiation if the forward and reverse transpositions differ markedly in their probability of occurrence because of, for example, differential recognition of the sites involved in the forward and reverse reactions (Campbell, 1981). In the model shown in Figure 4, violation of the constancy hypothesis is assumed to be based on transposition, rather than on deletion of genetic material as occurs during the rearrangement of immunoglobulin genes. Immunoglobulin genes may be atypical, because they must generate an exceptional level of diversity, in comparison with genes involved in most other pathways of differentiation.

## EXPERIMENTAL APPROACHES

To obtain definitive information about the genetic regulation of commitment, more direct experimental approaches are needed. There are three prerequisites: 1) experimental circumstances that permit stem cell renewal and commitment to occur under well controlled conditions, 2) convenient methods for analysis of the genotype of somatic mammalian cells, and 3) efficient means for obtaining relatively pure populations of stem cells. The first of these three prerequisites has largely been met. Techniques of cell culture are available that permit stem cell proliferation and renewal over periods of many weeks (Dexter and Lajtha, 1976; Dexter and Testa, 1980). Cell culture methods are also available for studying the development of clones containing both differentiated

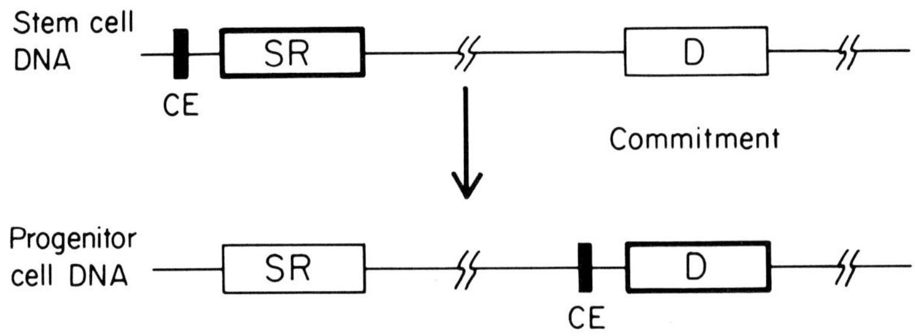


Fig. 4. A minimal model for the regulation of stem cell renewal and commitment at the molecular level. CE: a movable genetic controlling element responsible for initiation of the expression of contiguous genes. SR: a gene (or gene complex) responsible for regulation of stem cell renewal. D: a gene complex responsible for regulation of the programmed expression of phenotypic markers involved in the differentiation of a committed progenitor cell. It is assumed that transposition of the controlling element results in the turning on of the D gene complex and the turning off of the SR gene(s), as indicated by the heavy lines in the upper left and lower right.

cells and new stem cells (Johnson and Metcalf, 1977; Metcalf et al., 1979; Humphries et al., 1981).

Approaches to the analysis of the genome of somatic cells are developing rapidly. A particularly important area of methodological development is that concerned with means for the transfer of functional genetic material from one cell to another (Graham et al., 1980; Wigler et al., 1979; Shih et al., 1981; Cooper et al., 1980; Klobutcher and Ruddle, 1981; Mercola et al., 1980).

The main impediment to genetic analysis of stem cells appears to be the absence of convenient and reliable ways to obtain purified populations of stem cells. Techniques of purification based on differences between the physical properties of stem cells and their descendants have been under exploration for some time (Worton et al., 1969; Haskill et al., 1970; Zeiller et al., 1972). The heterogeneity of stem cells in regard to their physical properties, coupled with an apparent lack of major differences in physical properties between stem cells and committed progenitor cells, have inhibited the progress of this approach to purification. Perhaps the most promising possibility is that differences between stem cells and progenitor cells in cell surface markers might be used for purposes of purification (van den Engh and Golub, 1974; Krogsrud and Price, 1979; Till, 1976; Berridge and Okech 1980; Morstyn et al., 1981; Goldschneider et al., 1980; Visser et al., 1981).

In the absence of effective methods for the purification of normal stem cells, cell lines established in cell culture, or populations of cells derived from hemopoietic malignancies have been examined, on the assumption that they may provide relatively homogeneous populations of cells arrested at an early stage of differentiation (see, for example, Boss et al., 1980). This assumption needs to be validated for any cell population used for such a purpose. It may be incorrect (Till and McCulloch, 1980).

## GENETICS OF NEOPLASIA

The use of established cell lines or populations of cells derived from tumors to study the genetic basis for differentiation raises another important question. What is the genetic basis for the neoplastic transformation of hemopoietic cells? For example, the view that neoplasia involves changes in the genome of somatic cells is widely (Ames, 1976; Sager, 1979; Crains, 1981), but not universally (Rubin, 1980), accepted. Crains (1981) has suggested that transpositions of genetic material, rather than classical gene mutations, may be the major type of genetic change associated with neoplasia.

A human hematological malignancy, chronic myeloid leukemia, appears to arise in pluripotent stem cells (Fialkow et al., 1980; Martin et al., 1980). Gross changes in chromosome structure are regularly seen in the clinically detectable stages of chronic myeloid leukemia (Nowell and Hungerford, 1961) and also in acute myeloid leukemia (Rowley, 1980). However, detailed comparisons of DNA sequences in normal stem cells and preneoplastic stem cells in the initial stages of development of leukemia are not yet technically feasible. If Cairns' (1981) model is applicable to hemopoietic cells, one might expect to find evidence for transposition of genetic material in preleukemic or leukemic stem cells, in comparison with normal stem cells.

A well-known pitfall in the interpretation of apparent differences in phenotype between normal and neoplastic cells is the use of inappropriate comparisons. Phenotypic differences may be a reflection of the influence of confounding variables, such as differences in cell cycle state or stage of differentiation, rather than a direct consequence of the neoplastic transformation. For example, studies of the phenotypes of cells derived from patients with leukemia have not yet provided unequivocal evidence for leukemia-specific antigens. Instead, the antigens detected on leukemic cells may be a reflection of the level of maturation expressed by these cells (Greaves, 1981). When comparisons are made of the genomes of normal and neoplastic stem cells, attention must be paid to the possibility of an analogous pitfall. This possibility can be illustrated with a hypothetical example.

Suppose that the early stages in stem cell differentiation and the initial steps in the formation of neoplastic cells both involve similar processes of genetic change, such as transpositions of genetic material in related regulatory genes. Suppose also, for simplicity (Fig. 4), that the genetic transpositions in normal stem cells lead to loss of potential for stem cell renewal and a consequent commitment to differentiation. This in turn leads, after a few cell generations, to an irreversible loss of capacity for proliferation. In contrast, suppose that the genetic transpositions in the related genes associated with the initiation of preneoplastic cells also affect the regulation of the same function, stem cell renewal, but in a different way. In this case, the transpositions are assumed to lead to a retention, rather than a loss, of potential for stem cell renewal. These two different effects on regulation of stem cell renewal would be expected to lead to markedly different consequences. In the case of normal stem cells, there would be immediate selection against further propagation of that particular genetic change within the stem cell population, since it would result inevitably in loss of potential for stem renewal. However, for neoplastic stem cells, there would be selection in favor of propagation of analogous genetic changes within the stem cell population, since they lead to preservation of stem cell renewal. As a result of this selection, a direct comparison of populations of normal and neoplastic stem cells could yield an erroneous interpretation of the kinds of genetic changes which each can experience. Transpositions of genetic material would be found routinely in neoplastic stem cells, and only very rarely, if at all, in normal stem cells, because their occurrence in normal stem cells would result in loss of stem cell function.

The lesson to be learned from this hypothetical example is that just as analyses of the phenotypes of neoplastic cells may tell more about the phenotypes of minority populations of normal cells than about differences between normal and neoplastic cells, so analysis of genetic changes in neoplastic cells might give information about genetic events in a minority population of normal cells, the stem cells. Perhaps one should not simply assume that all genetic changes observed in neoplastic cells are related to the events responsible for neoplastic transformation. From this point of view, which is based on the assumption (Fig. 4) that violations of the constancy hypothesis may be a normal occurrence during the early stages of differentiation of hemopoietic cells, transpositions of genetic material need not be confined to neoplastic cells (Bernstein and Till, 1981). It may be their effects on the regulation of cell function, rather than their mere occurrence, that differ between normal and neoplastic cells.

## ACKNOWLEDGMENTS

This work was supported in part by the Ontario Cancer Treatment and Research Foundation and the National Cancer Institute of Canada.

## LITERATURE CITED

Ames, B.N. (1976) Carcinogenicity tests. Science, *191*:241–245.

Bank, A., J.G. Mears, and F. Ramirez (1980) Disorders of human hemoglobin. Science, *207*:486–493.

Becker, A.J., E.A. McCulloch, and J.E. Till (1963) Cytological demonstration of the clonal nature of spleen colonies derived from transplanted mouse marrow cells. Nature, *197*:452–454.

Bernard, O., N. Hozumi, and S. Tonegawa (1978) Sequences of mouse immunoglobulin light chain genes before and after somatic changes. Cell, *15*:1133–1144.

Bernstein, A., and J.E. Till (1981) Differentiation and neoplasia (in preparation).

Berridge, M.V., and N. Okech (1980) Surface antigens of murine hemopoietic stem cells. II. Rosetting method used to compare the cross reacting of antisera against mouse brain, sperm, and testis cells with bone marrow stem cells and progenitor cells. J. Immunol., *124*:732–738.

Boss, M.A., D. Delia, J.B. Robinson, and M.F. Greaves (1980) Differentiation-linked expression of cell surface markers on HL-60 leukemic cells. Blood, *56:*910–916.

Burnet, F.M. (1959) The Clonal Selection Theory of Acquired Immunity. Nashville: Vanderbilt University Press, pp. 139–140.

Cairns, J. (1981) The origin of human cancers. Nature, *289:*353–357.

Calos, M.P., and J.H. Miller (1980) Transposable elements. Cell, *20:*579–595.

Campbell, A. (1981) Some general questions about movable elements and their implications. Cold Spring Harbor Symp. Quant. Biol., *45:*1–9.

Cayre, Y., M.A. Palladino, K.B. Marcu, and J. Stavnezer (1981) Expression of an antigen receptor on T cells does not require recombination at the immunoglobulin $J_H$-$C\mu$ locus. Proc. Natl. Acad. Sci. U.S.A., *78:*3814–3818.

Cohn, M., R. Langman, and W. Geckeler (1980) Diversity 1980. In Fougereau M., Dausset J., eds.: Immunology 80, Progress in Immunology IV, Vol. 1. London: Academic Press, pp. 153–201.

Cooper, G.M., S. Okenquist, and L. Silverman (1980) Transforming activity of DNA of chemically transformed and normal cells. Nature, *284:*418–421.

Curry, J.L., and J.J. Trentin (1967) Hemopoietic spleen colony studies. I. Growth and differentiation. Dev. Biol., *15:*395–413.

Davidson, E.H. (1968) Gene Activity in Early Development. New York: Academic Press, pp. 5–7.

Davis, M.M., K. Calame, P.W. Early, D.L. Livant, R. Joho, I.L. Weissman, and L. Hood (1980) An immunoglobulin heavy chain gene is formed by at least two recombinational events. Nature, *283:*733–739.

Dexter, T.M., and L.G. Lajtha (1976) Proliferation of haemopoietic stem cells and development of potentially leukaemic cells in vitro. In Clemmesen J., Yohn D.S., eds.: Comparative Leukemia Research. Basel: Karger, pp. 1–5.

Dexter, T.M., and N.G. Testa (1980) In vitro methods in haemopoiesis and lymphopoiesis. J. Immunol. Methods, *38:*177–190.

Dreyer, W.J., and J.C. Bennett (1965) The molecular basis of antibody formation: A paradox. Proc. Natl. Acad. Sci. U.S.A., *54:*864–869.

Early, P., H. Huang, M. Davis, K. Calame, and L. Hood. (1980) An immunoglobulin heavy chain variable region gene is generated from three segments of DNA: $V_H$, D, and $J_H$. Cell, *19:*981–992.

Fialkow, P.J., R.J. Jacobson, J.W. Singer, R.A. Sacher, R.W. McGuffin, and J.R. Neefe (1980) Philadelphia chromosome (Ph1)-negative chronic myelogenous leukemia (CML): A clonal disease with origin in a multipotent stem cell. Blood, *56:*70–73.

Fung, Y.-K.T., A.M. Fadly, L.B. Crittenden, and H.-J. Kung (1981) On the mechanism of retrovirus-induced avian lymphoid leukosis: Deletion and integration of the proviruses. Proc. Natl. Acad. Sci. U.S.A., *78:*3418–3422.

Goldschneider, I., D. Metcalf, F. Battye, and T. Mandel (1980) Analysis of rat hemopoietic cells on the fluorescence-activated cell sorter. I. Isolation of pluripotent hemopoietic stem cells and granulocyte-macrophage progenitor cells. J. Exp. Med., *152:*419–437.

Goldwasser, E. (1975) Erythropoietin and the differentiation of red blood cells. Fed. Proc., *34:*2285–2292.

Graham, F.L., S. Bachetti, R. McKinnon, C.P. Stanners, B. Cordell, and H.M. Goodman (1980) Transformation of mammalian cells with DNA using the calcium technique. In Baserga, R., Croce C., Rovera, G. eds.: Progress in Clinical and Biological Research: Introduction of Macromolecules into Viable Mammalian Cells. New York: Alan Liss, pp. 3–25.

Greaves, M. (1982) 'Target' cells, cellular phenotypes and lineage fidelity in human leukaemia. J. Cell. Physiol. Suppl. 1.

Green, M.M. (1980) Transposable elements in Drosophila and other Diptera. Annu. Rev. Genet., *14:*109–120.

Gurdon, J.B., and V. Uehlinger (1966) "Fertile" intestine nuclei. Nature, *210:*1240–1241.

Haskill, J.S., T.A. McNeill, and M.A.S. Moore (1970) Density distribution analysis of in vivo and in vitro colony forming cells in bone marrow. J. Cell. Physiol., *75:*167–180.

Herskowitz, I., L. Blair, D. Forbes, J. Hicks, Y. Kassir, P. Kushner, J. Rine, G. Sprague Jr., and J. Strathern (1980) Control of cell type in the yeast Saccharomyces cerevisiae and a hypothesis for development in higher eukaryotes. In Leighton T., Loomis, W.F. eds.: The Molecular Genetics of Development. New York: Academic Press, pp. 79–118.

Hicks, J.B., J. Strathern, and I. Herskowitz (1977) The cassette model of mating type interconversion. In Bukhari, A.I., Shapiro, J.A., Adhya, S.L. eds.: DNA Insertion Elements, Plasmids, and Episomes. Cold Spring Harbor, New York: Cold Spring Harbor Laboratory, pp. 457–462.

Hoeijmakers, J.H.J., A.C.C. Frasch, A. Bernards, P. Borst, and G.A.M. Cross (1980) Novel expression-linked copies of the genes for variant surface antigens in trypanosomes. Nature, *284:*78–80.

Hozumi, N., and S. Tonegawa (1976) Evidence for somatic rearrangement of immunoglobin genes coding for variable and constant regions. Proc. Natl. Acad. Sci. U.S.A., *73:*3628–3632.

Humphries, R.K., A.C. Eaves, and C.J. Eaves (1981) Self-renewal of hemopoietic stem cells during mixed colony formation in vitro. Proc. Natl. Acad. Sci. U.S.A., *78:*3629–3633.

Illmensee, K., and P.C. Hoppe (1981) Nuclear transplantation in Mus musculus: Developmental potential of nuclei from preimplantation embryos. Cell, *23:*9–18.

Johnson, G.R., and D. Metcalf (1977) Pure and mixed erythroid colony formation in vitro stimulated by spleen conditioned medium with no detectable erythropoietin. Proc. Natl. Acad. Sci. U.S.A., *74:*3879–3882.

King, T.J., and R. Briggs (1956) Serial transplantation of embryonic nuclei. Cold Spring Harbor Symp. Quant. Biol., *21:*271–290.

Klobutcher, L.A., and F.H. Ruddle (1981) Chromosome mediated gene transfer. Annu. Rev. Biochem., *50:*533–554.

Korn, A.P., R.M. Henkelman, F.P. Ottensmeyer, and J.E. Till (1973) Investigations of a stochastic model of haemopoiesis. Exp. Hematol., *1:*362–375.

Krogsrud, R.L. and G.B. Price (1979) Hemopoietic stem cells bear sperm-associated antigens. Exp. Hematol., *7:*377–382.

Kurosawa, Y., H. von Boehmer, W. Haas, H. Sakano, A. Trauneker, and S. Tonegawa (1981) Identification of D segments of immunoglobulin heavy-chain genes and their rearrangement in T lymphocytes. Nature, *290:*565–570.

Leder, P. (1978) Discontinuous genes. New Engl. J. Med., *289:*1079–1081.

Maniatis, T., E.F. Fritch, J. Lauer, and R.M. Lawn (1980) The molecular genetics of human hemoglobins. Annu. Rev. Genet. *14:*145–178.

Martin, P.J., V. Najfeld, J.A. Hausen, G.K. Penfold, R.J. Jacobson, and P.J. Fialkow (1980) Involvement of the B-lymphoid system in chronic myelogeneous leukemia. Nature, *287:*49–50.

McClements, W.L., R. Dhar, D.G. Blair, L. Enquist, M. Oskarsson, and G.F. Vande Woude (1981) The long terminal repeat of Moloney sarcoma provirus. Cold Spring Harbor Symp. Quant. Biol., *45:*699–706.

McClintock, B. (1952) Chromosome organization and genic expression. Cold Spring Harbor Symp. Quant. Biol., *16:*13–47.

McCulloch, E.A., J.E. Till, and L. Siminovitch (1965) The role of independent and dependent stem cells in the control of hemopoietic and immunologic responses. In V. Defendi, ed.: Methodological Approaches to the Study of Leukemia. Philadelphia: Wistar Institute Press, pp. 61–68.

McKinnell, R.G. (1978) Cloning: Nuclear Transplantation in Amphibia. Minneapolis: University of Minnesota Press, pp. 134–147.

Mercola, K.E., H. Stang, J. Browne, W. Salser, and M.J. Cline (1980) Insertion of a new gene of viral origin into bone marrow cells of mice. Science, *208:*1033–1035.

Metcalf, D., G.R. Johnson, and T.E. Mandel (1979) Colony formation in agar by multipotent hemopoietic cells. J. Cell. Physiol., *98:*401–420.

Morstyn, G., N.A. Nicola, and D. Metcalf (1980) Purification of hemopoietic progenitor cells from human marrow using a fucose-binding lectin and cell sorting. Blood, *56:*798–805.

Neel, B.G., W.S. Hayward, H.L. Robinson, J. Fang, and S.M. Astrin (1981) Avian leukosis virus-induced tumors have common proviral integration sites and synthesize discrete new RNAs: Oncogenesis by promoter insertion. Cell, *23:*323–334.

Nowell, P.C., and D.A. Hungerford (1961) Chromosome studies in human leukemia. II. Chronic granulocytic leukemia. J. Natl. Cancer Inst., *27:*1013–1035.

Payne, G.S., S.A. Courtneidge, L.B. Crittenden, A.M. Fadly, J.M. Bishop, and H.E. Varmus (1981) Analysis of avian leukosis virus DNA and RNA in bursal tumors: Viral gene expression is not required for maintenance of the tumor state. Cell, *23:*311–322.

Pays, E., N. van Meirvenne, D. LeRay, and M. Steinert (1981) Gene duplication and transposition linked to antigenic variation in *Trypanosoma brucei*. Proc. Natl. Acad. Sci. U.S.A., *78:*2673–2677

Proudfoot, N.J., M.H.M. Shander, J.L. Manley, M.L. Gefter, and T. Maniatis (1980) Structure and in vitro transcription of human globin genes. Science, *209:*1329–1336.

Rowley, J.D. (1980) Chromosome abnormalities in human leukemia. Annu. Rev. Genet., *14:*17–39.

Rubin, H. (1980) Is somatic mutation the major mechanism of malignant transformation? J. Natl. Cancer Inst., *64:*995–1000.

Sager, R. (1979) Transposable elements and chromosomal rearrangements in cancer: A possible link. Nature, *282:*447–448.

Sakano, H., Y. Kurosawa, M. Weigert, and S. Tonegawa (1981) DNA segment (D) of immunoglobulin heavy chain genes. Nature, *290:*562–565.

Schimke, R.T., R.J. Kaufman, F.W. Alt, and R.F. Kellems (1978) Gene amplification and drug resistance in cultured murine cells. Science, *202:*1051–1055.

Seidman, J.G., E.E. Max, and P. Leder (1979) A K-immunoglobulin gene is formed by site specific recombination without further somatic mutation. Nature, *280:*370–375.

Shih, C., L.C. Padhy, M. Murray, and R.W. Weinberg (1981) Transforming genes of carcinomas and neuroblastomas introduced into mouse fibroblasts. Nature, *290:*261–264.

Silverman, M., J. Zieg, G. Mandel, and M. Simon (1981) Analysis of the functional components of the phase variation system. Cold Spring Harbor Symp. Quant. Biol., *45:*17–26.

Siminovitch, L., E.A. McCulloch, and J.E. Till (1963) The distribution of colony-forming cells among spleen colonies. J. Cell. Comp. Physiol., *62:*327–336.

Simon, M., J. Zieg, M. Silverman, G. Mandel, and R. Doolittle (1980) Phase variation: Evolution of a controlling element. Science, *209:*1370–1374.

Temin, H. (1980) Origin of retrovirus from cellular moveable genetic elements. Cell, *21:*599–600.

Till, J.E. (1976) Regulation of hemopoietic stem cells. In Cairnie, A.B., Lala P.K., Osmond, D.G. eds.: Stem Cells in Renewing Cell Populations. New York: Academic Press, pp. 143–145.

Till, J.E. (1981) Cellular diversity in the blood-forming system. Am. Sci., *69:*522–527.

Till, J.E., and E.A. McCulloch (1961) A direct measurement of the radiation sensitivity of normal mouse bone marrow cells. Radia. Res., *14:*213–222.

Till, J.E., and E.A. McCulloch (1980) Hemopoietic stem cell differentiation. Biochim. Biophys. Acta, *605:*431–459.

Till, J.E., E.A. McCulloch, and L. Siminovitch (1964) A stochastic model of stem cell proliferation, based on the growth of spleen colony-forming cells. Proc. Natl. Acad. Sci. U.S.A., *51:*29–36.

Van den Engh, G.J. and E.S. Golub (1974) Antigenic differences between hemopoietic stem cells and myeloid progenitors. J. Exp. Med., *139:*1621–1627.

Visser, J.W.M., S.J.L. Bol, and G. van den Engh (1981) Characterization and enrichment of murine hemopoietic stem cells by fluorescence activated cell sorting. Exp. Hematol., *9:*644–655.

Wigler, M., R. Sweet, G.K. Sim, B. Wold, A. Pellicer, E. Lacy, T. Maniatis, S. Silverstein, and R. Axel (1979) Transformation of mammalian cells with genes from procaryotes and eucaryotes. Cell, *16:*777–785.

Williamson, R. (1981) Thalassemia from theory to practice. Nature, *292:*405–406.

Williamson, V.M., E.T. Young, and M. Ciriacy (1981) Transposable elements associated with constitutive expression of the yeast alcohol dehydrogenase II. Cell, *23:*605–617.

Worton, R.G., E.A. McCulloch, and J.E. Till (1969) Physical separation of hemopoietic stem cells from cells forming colonies in culture. J. Cell. Physiol., *74:*171–182.

Wu, A.M., J.E. Till, L. Siminovitch, and E.A. McCulloch (1967) A cytological study of the capacity for differentiation of normal hemopoietic colony-forming cells. J. Cell. Physiol., *69:*177–184.

Yarmolinsky, M.B. (1981) Summary. Cold Spring Harbor Symp. Quant. Biol., *45:*1009–1015.

Zeiller, K., J.C.F. Schubert, F. Walther, and K. Hannig (1972) Electrophoretic distribution analysis of in vivo colony forming cells in mouse bone marrow. Hoppe Seylers Z. Physiol. Chem., *353:*95–104.

JOURNAL OF CELLULAR PHYSIOLOGY SUPPLEMENT 1:13–21 (1982)

# Regulation of Pluripotent Stem Cell Proliferation and Differentiation: The Role of Long-Range Humoral Factors

MAURICE TUBIANA* AND EMILIA FRINDEL
*Institut de Radiobiologie Clinique, Institut Gustave-Roussy, 94800 Villejuif, France*

The proliferative status of the hemopoietic pluripotential stem cells (CFU-S) is controlled by inhibitors and stimulators, which have been studied by an in vivo-in vitro technique. Inhibitors protect CFU-S during iterative administration of cycle specific drugs. Among stimulators are long-range humoral factors that are released by bone marrow following irradiation or drug administration. After the same treatment, bone marrow also releases a long-range humoral factor that increases the rate of differentiation of CFU-S, probably in order to compensate for the depletion of the maturing compartment. This differentiation is qualitatively different from normal differentiation. When the bone marrow of mice treated with Ara-C is transplanted to lethally irradiated mice, the total number of nodules remains constant; however the number of erythroid (E) colonies in the spleen is significantly increased, while the number of granulocytic (G) colonies is significantly decreased, and the number of mixed colonies is slightly decreased. Similar observations for E and G colonies are made when normal bone marrow is injected into lethally irradiated mice following in vitro incubation with humoral factors released by cytosine-arabinoside (Ara-C)-treated mice. In both cases most splenic colonies contain CFU-S and GM-CFC, even when they appear histologically E colonies. After irradiation or iterative administration of Ara-C the E/G ratio is decreased. The factors involved, pluripoietins, are released by both bone marrow and spleen and are found in the serum of treated mice. The mechanism by which they act is unknown; however two hypotheses can be made: a) Modulation of differentiation potential towards only one of the cell lineages. During 6 to 7 days after Ara-C administration the determination of CFU-S is modified. However this restriction to erythroid determination is temporary. This hypothesis is compatible with the "hemopoietic inductive model," but microenvironment is not playing any role. b) Specific inhibition or stimulation of the committed stem cells to which the CFU-S having received the "message" gives birth. This information lasts during one week but disappears when the cells are plated in vitro. Whatever the mechanism, the primary events occur at CFU-S level and the information is transmitted to the descendants, showing that humoral factors can channel the differentiation pathways.

The pioneer work of Till and McCulloch (1961) has provided a unique opportunity for the understanding of a cell-renewal system and for the study of the regulation of differentiation program, which remains one of the central issues in modern biology (Till and Mc Culloch, 1980).

Besides, the work of Till and McCulloch had an important clinical impact. Several diseases of hemopoiesis originate in pluripotent stem cells. Furthermore, hemopoietic tissues are the main critical tissues in most cancer treatments. During the past decades the increased knowledge of the kinetics of cell proliferation in bone marrow and of the role of hemopoietic stem cells has made possible a more meaning-

*To whom correspondence/reprint requests should be addressed.

ful investigation of the acute and late effects of irradiation and chemotherapy both in experimental animals and in patients (Tubiana et al., 1979).

The origin of our interest at Villejuif in this topic was double.

A clinical observation: We observed a little over 20 years ago that during the course of radiotherapy, the fall in blood granulocyte number was greater than could be ascribed to the proportion of bone marrow that was irradiated (Abbatucci et al., 1958). We thought that the only interpretation of this observation was an abscopal effect on unirradiated hemopoietic tissues (Tubiana and Frindel, 1962).

This hypothesis led us to undertake experiments in which mice were submitted to single or repeated sessions of partial body irradiation. We observed in the protected bone marrow a decreased cellularity as well as a reduction in the uptake of radioactive iron (Tubiana et al., 1971). In order to better understand this effect we concentrated on the study of pluripotent stem cells (CFU-S) because they play a crucial role in hemopoiesis (Croizat et al., 1970a, 1976).

Experimental studies: In the early 1960s, we were interested in the phenomenon of cell synchronization and its possible exploitation in human therapy (Frindel et al., 1963). We studied the time course changes of the radiosensitivity of the CFU-S after a priming dose of irradiation (Frindel et al., 1966) as well as bone marrow cell kinetics (Frindel et al., 1967).

CFU-S are quiescent in normal mice. After various types of aggression such as drugs (Vassort et al., 1971; Van Putten, 1973), irradiation (Croizat et al., 1970a), antigen (Croizat et al., 1970b, Frindel et al., 1976a), bleeding (Rencricca et al., 1970), etc., quiescent stem cells can be recruited into cycle. However, the mechanism of the stimulation was controversial and short range microenvironmental factors were usually evoked (Wolff and Trentin, 1968, Luriya and Friedenstein, 1977). We have recently demonstrated that long range humoral factors play an important role in the stimulation (Frindel et al., 1976b) and inhibition (Frindel and Guigon, 1977; Guigon and Frindel, 1978) of CFU-S proliferation kinetics.

The last few years have made it clear that committed progenitor cells are sensitive to regulatory mechanisms that are specific for each one of the cell lineages considered (Krantz et al., 1963, Bradley and Metcalf, 1966). However, very little is known about the regulation of CFU-S differentiation towards the progenitor cells. Random differentiation (Till et al., 1964) and the microenvironment (Wolff and Trentin 1968) have been suggested as probable mechanisms. We now have data suggesting that long-range humoral factors may also play a role in channeling the differentiation into the erythrocytic or granulopoietic pathways.

## STIMULATION OF QUIESCENT CFU-S

Two series of experiments were undertaken in which the origin or the nature of the signal could be identified. The proliferative status of the stem cells of mouse bone marrow was studied after a) partial body irradiation and b) selective killing of maturing hemopoietic cells. This series of experiments provided data on the feedback mechanisms at 3 levels: 1) between different hemopoietic areas, 2) between the differentiated cell and the stem cell compartments, and 3) within the stem cell compartment itself.

Subtotal irradiation has been used (Tsuya et al., 1961; Kurnick and Nokay, 1964; Croizat et bal., 1970a; Gidali and Lajtha, 1972; Lahiri et al., 1972; Croizat et al., 1976) for studying the indirect effect of irradiation on an undisturbed area of bone marrow. In the experiments we carried out, the mice were irradiated with one leg shielded (that is about one tenth of the hemopoietic tissues). Soon after exposure the number of CFU-S falls considerably in the shielded areas. The size of the reduction in number increases with the dose. Thus, 15 minutes after irradiation the number of CFU-S in the shielded leg, expressed as the percentage of the preirradiation number, is 90% for a dose of 150 rads, 70% for 500 rads, 63% for 1,000 rads, and 55% for 1,500 rads (Croizat et al., 1976).

An indirect killing effect being unlikely, this decrease can be the consequence of either an increased rate of migration and/or an increased rate of differentiation. Migration from unirradiated hemopoietic sites, with consequent seeding in the irradiated bone marrow or spleen, has been demonstrated for CFU-S (Hanks, 1964). However its magnitude is small (Croizat et al., 1980) and it does not increase with dose. Differentiation is therefore in this case the most likely explanation for this loss of CFU-S. In the case of external irradiation, the depletion of the CFU-S compartment is followed by the recruitment of most of the previously quiescent CFU-S, and 8 hours after irradiation the CFU-S in the protected leg are triggered into proliferation (Croizat et al., 1970, 1976).

In a normal mouse most of the CFU-S are quiescent and an in vivo administration of radioactive thymidine kills only a very small, and undetectable, percentage of them by a suicide effect. However, the committed progenitor cells and the recognizable differentiated cells are actively proliferating and a third to a half of them die owing to incorporation of radioactive thymidine in their DNA. During the hours following an intravenous injection of thymidine two effects are observed: first, a decrease in the number of CFU-S, and second, as following partial body irradiation the entry into S phase of at least half of the CFU-S (Vassort et al., 1973). Some drugs acting on cells in S phase, such as hydroxyurea (Vassort et al., 1971) or cytosine arabinoside (Frindel et al., 1980). also induce differentiation of CFU-S and stimulate quiescent CFU-S.

These results indicate that the damage to the compartment of differentiated cells causes a differentiation of stem cells by a feedback mechanism. The consequence of this accelerated differentiation is a depletion of the stem cell compartment. This leads to recruitment into cycle of the quiescent cells.

These two sets of data suggest the presence of humoral factors. One of them, probably released by committed progenitors or more differentiated cells, increases the rate of differentiation of CFU-S, probably in order to compensate for the depletion of the maturing compartments. Another humoral factor has a stimulatory effect on CFU-S recruitment. A number of papers suggest that humoral factors are released by cells after various treatments (Tubiana et al., 1974). The influences of these factors are difficult to study in vivo since neither the active fractions nor the metabolism of such factors are known as yet. We therefore devised a simple technique to study the eventual influence of diffusible factors on CFU-S kinetics (Frindel et al., 1976b, 1978).

The system is composed of two cellular compartments separated by a Millipore filter that permits the diffusion of molecules but isolates the two cell populations from direct contact. The treated fragments are deposited on the filter. The bone marrow is used in the form of a plug and six to eight plugs are placed on each filter, which floats on the surface of tissue culture medium containing the potentially responsive cells that are present in normal bone marrow. These cells are harvested from mouse femoras and tibias by flushing with culture medium using a syringe and needle. Then $10^7$ nucleated bone marrow cells are suspended in 4 ml of medium containing 15% fetal calf serum (FCS). Thereafter the cells are placed in Petri dishes (Falcon Plastics, Los Angeles, CA, diameter 60 mm, 3002) and incubated at 37°C in a continuous flow incubator (95% air and 5% $CO_2$). The responder cells and the treated cells are always of the same strain.

After 20–24 hours of incubation, the responder bone marrow cells from three Petri dishes are harvested and pooled in order to assess the proliferation kinetics of their CFU-S. The suicide technique is used as for in vivo experiments and $1.4 \times 10^5$ cells injected per mouse.

The recipient mice are killed 9 days after injection of bone marrow cells. The spleens are removed and nodules counted after 24 hours of fixation in Bouin's solution.

This technique has shown that a stimulatory factor is secreted almost immediately after irradiation (Croizat and Frindel, 1980)—2 hours after Ara-C treatment, and a few days after antigen stimulation (Frindel et al., 1980). After irradiation the amount of factor secreted is a function of the dose; when the latter is too large there is no secretion, showing that the factor is probably secreted by viable cells. The duration of the release is short, only a few hours. This long-range humoral factor (or factors) is found in the blood after administration of a cytotoxic drug. It is secreted not only by the bone marrow but also by the spleen (Sainteny et al., in press). In vivo it induces a temporary proliferation of CFU-S. When these factors have been purified it might be possible to use them to enhance the recovery of bone marrow during or after a course of radiotherapy or chemotherapy. They might also be used to stimulate the proliferation of CFU-S after a bone marrow graft.

## INHIBITION OF CFU-S PROLIFERATION

Since phase specific drugs trigger CFU-S into cycle, rendering them vulnerable to subsequent doses of chemotherapy, it seemed of interest to try to protect the bone marrow by keeping the CFU-S out of cycle.

We have shown the existence of a specific inhibitor of CFU proliferation that is able to block the entry of the stem cells into a new cycle without damaging them. This factor can be extracted from fetal calf bone marrow (Frindel and Guigon 1977; Guigon and Frindel, 1978; Guigon et al., 1980).

It exerts a specific and statistically significant protection of CFU-S when administered

TABLE 1. Protective effect of inhibitors on CFU-S kinetics, CFU-S survival, and mouse survival after various treatment protocols

| | | Biological tests[1] | | | | | | |
|---|---|---|---|---|---|---|---|---|
| | Inhibitor | Percent CFU-S in S phase | | Percent CFU-S survival | | Percent mouse survival | | |
| Experiment | Fetal calf marrow extract | Without | With inhibitor | Without | With inhibitor | Without | With inhibitor | References |
| In vivo | | | | | | | | |
| Irradiation | Dialysate | 44 ± 5 | 3 ± 12 | | | ND | ND | Frindel and Guigon (1977) |
| | | | | 14 ± 1.81 | 29.7 ± 3.5 | | | Guigon et al. (1980) |
| Ara-C | Fraction | 28.5 ± 2.2 | 7.3 ± 2.3 | ND | ND | 6 ± 6 | 79.2 ± 0.7 | Guigon and Frindel (1981a) |
| | | | | | | | | Guigon and Frindel (1981b) |
| In vitro-in vivo | | | | | | | | |
| Irradiation | Dialysate | 28 ± 4.3 | 0 | ND | ND | ND | ND | Guigon and Frindel (1978) |
| Ara-C | Dialysate | 30 ± 1.7 | 0 | ND | ND | ND | ND | Guigon and Frindel (1978) |

[1]All differences are statistically significant.

to mice during sequential treatments with cytosine-arabinoside (Ara-C). Its protective effect is greater when it is given 2 hours prior to injection of Ara-C. The inhibitor seems to be specific for CFU-S and has no effect on EMT6 tumor cells in vitro (Guigon and Frindel, 1981b). A purified fraction was obtained by chromatography, which greatly increases the survival of mice receiving lethal doses of Ara-C (Guigon et al., 1981a; Guigon et al., in press) (Table 1).

However the relationships between inhibition of CFU-S entry into cycle, survival of CFU-S, and survival of mice after lethal doses of Ara-C deserve further investigations.

Convincing evidence concerning regulatory factors of CFU-S kinetics have now also been reported from other laboratories (Lord et al., 1976, 1977; Wright and Lord, 1977, 1978, 1979; Cork et al., 1981; Riches et al., 1981). It thus appears that CFU-S proliferation kinetics can be modulated. In the steady state, inhibitors and stimulators are in equilibrium. The disturbance of the steady state induces or inhibits the production of one of the regulators according to the needs of the organism to restore homeostasis.

## DIFFERENTIATION PATHWAYS OF CFU-S

After drug treatment, the first event is a decrease of cycling committed stem cells followed by a decrease of CFU-S, due to differentiation (Vassort et al., 1971; Frindel et al., 1980). After subtotal irradiation, the first event is a decrease in CFU-S number of the shielded bone marrow (Croizat et al., 1976). This decrease is mostly due to differentiation and led us to conclude that humoral factors could influence the rate of CFU-S differentiation (Tubiana et al., 1974). In order to investigate whether the accelerated differentiation was also qualitatively altered (Frindel et al., 1979a), we analyzed the spleen colonies generated by control or treated bone marrow.

The first series of results showed that bone marrow from Ara-C treated mice injected into syngeneic lethally irradiated recipients generated colonies with an E/G ratio significantly above control values (Frindel, 1979a). The total number of spleen colonies remained constant; the number of E colonies was increased while the number of G colonies was decreased, as well as the number of mixed colonies. Since the recipient mice were treated in an identical

TABLE 2. Histologic characteristics of spleen nodules: Number of spleen colonies of the different hematological types and E/G ratios of these colonies generated by CFU-S from normal and Ara-C-treated mice

| Experiment | Number of spleen | Number of nodules | E Number | E Percent | G Number | G Percent | UND. Number | UND. Percent | Mega Number | Mega Percent | Mixed Number | Mixed Percent | E/G |
|---|---|---|---|---|---|---|---|---|---|---|---|---|---|
| In vivo control | 4 | 85 | 39 | 45.9 | 22 | 25.9 | 4 | 4.7 | 2 | 2.4 | 18 | 21.2 | 1.8 |
| Ara-C | 4 | 67 | 44 | 65.7 | 9 | 13.4 | 1 | 1.5 | 3 | 4.5 | 10 | 14.9 | 4.9 |

TABLE 3. Influence of humoral factors on histologic characteristics of spleen nodules: Number of spleen colonies of different hematological types and E/G ratios of these colonies generated by the normal responder cell population exposed to factors released by bone marrow of normal or Ara-C-treated mice

| Experiment | Number of spleen | Number of nodules | E Number | E Percent | G Number | G Percent | UND. Number | UND. Percent | Mega Number | Mega Percent | Mixed Number | Mixed Percent | E/G |
|---|---|---|---|---|---|---|---|---|---|---|---|---|---|
| In vivo-in vitro 9 days | | | | | | | | | | | | | |
| Plug from normal mice | 16 | 414 | 184 | 44.3 | 109 | 26.2 | 25 | 6.0 | 24 | 5.9 | 72 | 17.5 | 1.9 |
| Plug from Ara-C treated mice | 16 | 414 | 284 | 68.4 | 40 | 9.8 | 22 | 5.3 | 27 | 6.6 | 41 | 9.9 | 7.6 |

manner in both the control and the treated groups, the microenvironment of the spleen did not play a role in these results. However, these in vivo experiments could not exclude a direct action of the drug on the CFU-S without the involvement of humoral factors (Table 2).

We therefore used the in vivo-in vitro technique (Table 3), in which any modification of the responder cells means that a factor has diffused, via the cell-proof filter, from the treated bone marrow to the recipient marrow and has brought to it information that influenced the outcome of the CFU-S and the lineage of their progeny. Control experiments ruled out the possibility of diffusion of Ara-C from the treated bone marrow to the recipient cells (Frindel, 1979b). Analysis with mathematical models (Mary et al., submitted for publication) of the data obtained with the control group (bone marrow incubated with normal bone marrow on the filter) and the treated group (normal bone marrow incubated with bone marrow of animal treated with Ara-C on the filter) show that in the latter case about half of the nodules that should have been granulocytic were erythrocytic and about half of the nodules which should normally be mixed E + G nodules were purely erythrocytic. These results might be owing to an effect on CFU-S and/or to a stimulation, or an inhibition, of the proliferation of the progenitors (GM-CFC or BFU-E) if these cells are able to seed in the spleen and give rise to colonies.

In order to answer to this question we used anti-CFU-S serum. Incubation with anti-CFU serum, with or without complement, significantly reduced spleen colonies whereas the number of GM-CFC and BFU-E were unchanged. This suggests that the spleen colonies examined were mostly generated by CFU-S (Frindel et al., 1980). However as the specificity of anti-CFU serum is questioned, we also undertook another experiment in which individual spleen colonies were dispersed and retransplanted and the number of CFU-S per single colony were assayed as well as the histologic characteristics of the spleen nodules in the secondary hosts (Duménil and Frindel, in press).

When bone marrow of control and treated mice were injected into lethally irradiated mice (in vivo experiments), 80% and 67% of the nodules from normal and Ara-C treated mice, respectively, gave rise to spleen colonies

TABLE 4. Stem cell content of spleen nodules: Analyses of stem cell content of each individual spleen colony generated by CFU-S from normal and Ara-C-treated mice[1]

| | Untreated control mice | Number[2] | Ara-C-treated mice (2 hr after 20 mg) | Number[2] |
|---|---|---|---|---|
| Primary nodules | | | | |
| Number of CFU-S per leg | 2400 ± 187 | | 2425 ± 265 | |
| E/G ratio | 1.8 ± 0.2 | 1344 | 4.2 ± 0.3 | 901 |
| Percent of nodules giving rise to secondary spleen nodules (CFU-S) | 80.2 ± 8.0 | 58 | 67.2 ± 15.0 | 56 |
| Mean number of CFU-S per nodule | 6.5 ± 1.8 | | 6.2 ± 1.4 | |
| Percent of nodules in which GM-CFC are present | 72.1 ± 8.7 | 78 | 83.4 ± 7.0 | 69 |
| Mean number of GM-CFC per nodule | 42.4 ± 13.0 | | 74.0 ± 3.40 | |
| Secondary nodules | | | | |
| E/G ratio | 3.5 ± 0.7 | 44 | 2.0 ± 0.6 | 43 |
| Percent of nodules giving rise to | | | | |
| erythroid colonies | 23.6 ± 4.8 | | 24.6 ± 5.9 | |
| granuloid colonies | 0.3 ± 0.3 | | 13.5 ± 7.5 | |
| mixed colonies | 70.8 ± 10.4 | | 58.7 ± 16.3 | |

[1]In vivo experiments (Duménil and Frindel, 1981).
[2]Number of colonies analyzed.

in secondary hosts; these spleen colonies were of various histologic types (Table 4). These results demonstrate the presence of CFU-S in the nodules and therefore prove that a CFU-S was at their origin. In the in vivo-in vitro experiments in which normal bone marrow responder cells were submitted to effects of a factor prior to their injection, over 70% of primary spleen nodules generated by this marrow gave rise to secondary colonies. In particular the nodules from the bone marrow incubated with Ara-C-treated bone marrow generated colonies of more than one type (Table 5). These results confirm that the increase of the E/G ratios (Frindel et al., 1980) is owing to an effect on the CFU-S. Furthermore over 70% of the nodules contain GM-CFC, even when the E/G ratio is highly elevated after Ara-C treatment (Tables 4 and 5).

In summary, it was found that 8 days after injection of bone marrow incubated in vitro with a diffusible factor secreted by bone marrow treated by Ara-C, the number of erythrocytic colonies increased and the number of granulocytic or mixed colonies decreased; however in practically all the nodules GM-CFC are present. Two mechanisms could explain these findings. a) The CFU-S proliferate and differentiate normally but the proliferation of the GM-CFC is blocked and these progenitors do not differentiate. This reversible maturation arrest hypothesis is compatible with the "hemopoiesis-engendered-randomly model" (Korn et al., 1973; Till and McCulloch, 1980). b) During 6 to 7 days the determination of the CFU-S is modified as the result of an inductive humoral factor. CFU-S proliferate but differentiate only towards the erythrocytic line. However this restriction to erythroid differentiation is temporary and after a few days the CFU-S still present in the nodule can give birth to GM-CFC. This hypothesis is compatible with the "hemopoietic inductive model," but in this case microenvironment does not play any role.

It should be underlined that in both cases the primary event occurs at the CFU-S level and the information is transmitted to the descendants. In the first hypothesis the information, which blocks GM-CFC proliferation, lasts during at least one week but disappears when the cells are plated in vitro. In the second hypothesis the information that influences CFU-S determination towards BFU-E instead of GM-CFC is transmitted to the descendants during at least ten generations; thereafter the differentiation proceeds in the normal way.

Modern molecular biology provides many possible mechanisms for such phenomena. A reversible block of the proliferation and differentiation of certain types of cells has been observed during embryogenesis and is well documented. However, no irrefutable data have as yet demonstrated a modification of the stochastic differentiation of the CFU-S; however there is no theoretical reason to exclude this possibility (Till and McCulloch, 1980; Levenson and Housman, 1981), and our Ara-C data suggest that CFU-S differentiation can indeed be channeled. Metcalf (1980) and Hor-

TABLE 5. Stem cell content of spleen nodules: Analyses of stem cell content of each individual spleen colony generated by CFU-S from the normal responder cell population exposed to factors released by marrow from normal and Ara-C-treated mice[1]

| | Bone marrow incubated with bone marrow plugs from untreated mice | Number[2] | Bone marrow incubated with bone marrow plugs from Ara-C-treated mice | Number[2] |
|---|---|---|---|---|
| Primary nodules | | | | |
| Number of CFU-S per $10^5$ responder cells | 9.5 ± 1.3 | | 9.8 ± 1.5 | |
| E/G ratio | 2.2 ± 0.3 | 1368 | 5.1 ± 0.4 | 1663 |
| Percent of nodules giving rise to secondary nodules (CFU-S) | 71.2 ± 2.5 | | 86.6 ± 3.3 | 38 |
| Mean number of CFU-S per nodule | 6.7 ± 1.4 | | 12.5 ± 2.5 | |
| Percent of nodules in which GM-CFC are present | 97.0 ± 3.0 | 36 | 100 | 37 |
| Mean number of GM-CFC per nodule | 51.0 ± 21.2 | | 110.9 ± 40 | |
| Secondary nodules | | | | |
| E/G ratio | 2.6 ± 1.0 | 26 | 2.4 ± 1.1 | 31 |
| Percent of nodules giving rise to | | | | |
| only erythroid colonies | 35.9 ± 2.6 | | 0 | |
| only granuloid colonies | 5.5 ± 5.03 | | 0 | |
| more than 1 type of colonies | 58.5 ± 4.2 | | 100 | |

[1]In vivo-in vitro experiments (Duménil and Frindel, 1981).
[2]Number of colonies analyzed.

iuchi et al. (1979) have shown that "manipulation" is possible at the more mature committed stem cell level; differentiation towards the granulocytic or macrophage cell lineage can be modulated either by changing the concentration of CSF (Metcalf, 1980) or by selecting purified fractions of the regulator (Horiuchi et al., 1979). Furthermore recent data from Gallichio and Chen (1980) also suggest that lithium can channel differentiation into the granulocytic pathway and favor granulopoiesis over erythropoiesis. However all of these experiments, as well as those of Hellman et al. (1969), can be interpreted by any of the two previous hypotheses.

Further experiments are in progress that try to elucidate the mechanism involved. A few points may be stressed. Irradiated bone marrow releases a diffusible factor (or factors), which instead of increasing the number of erythrocytic colonies provokes an increase in the number of granulocytic or megakaryocytic colonies. This is consistent with the data reported by Hellman et al. (1969). The same observation is made with bone marrow of mice having received multiple doses of Ara-C. The differences observed in the E/G ratios in the different protocols might be related to differences in concentration or in the nature of the factors: the secretion of these putative pluripoietins may be either increased because of continuous stimulation or, on the contrary, decreased because of increased cell death of the secreting cells.

Though the exact nature of the cells secreting these factors is not yet known, experiments using fractionated bone marrow indicate that after one injection of Ara-C the factor is secreted by the fraction enriched in CFU-S (Frindel, 1979b). It is too early to determine the exact role played by CFU-S in the secretion of "pluripoietins"; however our present results suggest that there may exist a correlation between the determination of differentiation and CFU-S survival. Two hours after one injection of 20 mg of Ara-C, there is no CFU-S death and the E/G ratio is above normal values. After multiple doses of Ara-C, CFU-S kill is important and the E/G ratio decreases. The E/G ratio is also decreased after 5 Grays of X irradiation when the CFU-S survival is very low. Thus there might be a relationship between, on the one hand, a low E/G ratio and, on the other hand, a decreased number of CFU-S and a preferential orientation of their differentiation towards granulopoiesis (Guigon et al., 1981).

In conclusion, one can envisage the possibility that the hemopoietic system functions as the endocrine system, with feedback mechanisms, hormones (factors), and regulatory cells. The data of these experiments suggest that various insults to the hematopoietic tissue trigger a cascade of events that are responsible for the modifications of CFU-S differentiation pathways. These events are, at least in part, under the control of humoral factors, the putative pluripoietins, and are probably one of the mechanisms involved in the regulation of pluripotent stem cells in their efforts to restore normal hematopoietic processes. The exact nature of the medullar cells secreting these factors is still unknown, as are the mechanisms of action, which may be multiple: changes in membrane receptors, inhibition or stimulation of differentiation proteins, phenotypic mechanisms, etc. Whatever the mechanism, the available data already suggest that by the use of humoral factors it will be possible to channel the differentiation pathways towards erythropoiesis or granulopoiesis.

## LITERATURE CITED


Abbatucci, J., A. Dutreix, P. Arichaux, J.P. Peyroux, J. Surmont, and M. Tubiana, (1958) Etude des modifications hematologiques consécutives à certaines techniques radiothérapiques. Ann. Radiol., *1*:365–379.

Bradley, T.R., and D. Metcalf (1966) The growth of mouse bone marrow cells in vitro. Aust. J. Exp. Biol. Med. Sci., *44*:297–300.

Cork, M., I. Anderson, D.B. Thomas, and A. Richez (1981) Regulation of the growth fraction of CFU-S by an inhibitor produced by bone marrow. Leuk. Res., *5*:101–105.

Croizat, H., E. Frindel, and M. Tubiana (1970a) Proliferative activity of the stem cell in the bone marrow of mice after single and multiple irradiation. Int. J. Radiat. Biol., *18*:347–358.

Croizat, H., E. Frindel, M. Tubiana, and J.C. Salomon (1970b) Antigenic stimulation of DNA synthesis in haematopoietic stem cells of anemic mice. Nature, *228*:1187–1188.

Croizat, H., E. Frindel, and M. Tubiana (1976) Abscopal effect of irradiation on haemopoietic stem cells of shielded bone marrow. Role of irradiation. Int. J. Radiat. Biol., *30*:347–358.

Croizat, H., E. Frindel, and M. Tubiana (1980) The effect of partial body irradiation on haemopoietic stem cell migration. Cell Tissue Kinet., *13*:309–317.

Croizat, H., and E. Frindel (1980) Study of a CFU-S stimulating factor liberated by bone marrow cells after total and partial body irradiation. Exp. Hematol., *8*:185–191.

Duménil D., and E. Frindel Modification of pluripotent stem cell differentiation after Ara-C treatment: Clonal analyses of CFU-S progeny. Leuk Res., in press.

Frindel, E., R.F. Kallman, and H.S. Kaplan (1963) DNA synthesis and mitotic activity in bone marrow, intestine, and mammary tumors of mice after X irradiation. Radiat. Res., *19*:205.

Frindel, E., F. Charruyer, M. Tubiana, H.S. Kaplan, and C. Alpen (1966) Radiation effects on DNA synthesis and cell division in the bone marrow of the mouse. Int. J. Radiat. Biol., *11*:435–443.

Frindel, E., M. Tubiana, and F. Vassort (1967) The generation cycle of mouse bone marrow. Nature, *214*:1017–1018.

Frindel, E., E. Leuchars, and A.J.S. Davies (1976a) Thymus dependency of bone marrow stem cell proliferation in response to certain antigens. Exp. Hematol. *4*:275–284.

Frindel, E., H. Croizat, and F. Vassort (1976b) Stimulating factors liberated by treated bone marrow: In vitro effect on CFU kinetics. Exp. Hematol., *4*:56–61.

Frindel, E., and M. Guigon (1977) Inhibition of CFU entry into cycle by a bone marrow extract. Exp. Hematol., *5*:74–76.

Frindel, E., M. Guigon, D. Duménil, and M.P. Fache (1978) Stimulating factors and cell recruitment in murine bone marrow stem cells and EMT6 tumors. Cell Tissue Kinet., *11*:393–403.

Frindel, E. (1979a) Humoral regulation of pluripotent stem cell differentiation. Biomedicine, *31*:113–117.

Frindel, E. (1979b) Regulation of bone marrow stem cell kinetics. In Le Douarin, N. ed.: Cell Lineage, Stem Cell Determination. North-Holland, Amsterdam, pp. 227–239.

Frindel, E., D. Duménil, and F. Sainteny (1980) Role of pluripoietins in murine bone marrow stem cell differentiation. Leuk. Res., *4*:287–299.

Gallichio, V.S., and M.G. Chen (1980) Modulation of murine pluripotential stem cell proliferation in vivo by lithium carbonate. Blood, *56*:1150–1152.

Gidali, J., and L.G. Lajtha (1972) Regualtion of haemopoietic stem cell turnover in partially irradiated mice. Cell Tissue Kinet., *5*:147–157.

Guigon, M., and E. Frindel (1978) Inhibition of CFU-s entry into cycle after irradiation and drug treatment. Biomedicine, *29*:176–178.

Guigon, M., J. Enouf, and E. Frindel (1980) Effects of CFU-s inhibitors on murine bone marrow during Ara-C treatment. Leuk. Res., *4*:385–391.

Guigon, M., and E. Frindel (1981a) Inhibiteurs de la prolifération des cellules souches hematopoiétiques. Perspectives d'utilisation en chimiothérapie. Bull. Cancer, *68*:150–153.

Guigon, M., and E. Frindel (1981b) Inhibition of CFU-s recruitment and survival of mice during fractionated chemotherapy. Exp. Hematol., 9 suppl.9:75.

Guigon, M., D. Duménil, J. Enouf, and E. Frindel (1981) CFU-S differentiation after multiple doses of cytosine-arabinoside: Effect of inhibitors. Leuk. Res. *5*:367–372.

Hanks, G.E. (1964) In vivo migration of colony forming units from shielded bone marrow in the irradiated mouse. Nature, *203*:1393–1395.

Hellman, S., H.E. Grate, and J.T. Chaffey (1969) Effects of radiation on the capacity of the stem cell compartment to differentiate into granulocytic and erythrocytic progeny. Blood, *39*:141–156.

Horiuchi, M., K. Nagata, and Y. Ichikawa (1979) Determination of differentiation of bipotent hemopoietic progenitor cells in vitro. Exp. Cell Res., *123*:247–252.

Korn, A.P., R.M. Henkelman, F.P. Ottensmeyer, and J.E. Till (1973) Investigations of a stochastic model of haemopoiesis. Exp. Hematol. *1*:362–375.

Krantz, S.B., O. Gallien-Lartigue, and E. Goldwasser (1963) The effect of erythropoietin upon heme synthesis by marrow cells in vitro. J. Biol. Chem., *238*:4085–4090.

Kurnick, N.B., and N. Kokay (1964) Abscopal effects and bone marrow repopulation in man and mouse. Ann. N.Y. Acad. Sci, *114*:528–537.

Lahiri, S.K., and L.M. Van Putten (1972) Localisation of the Go phase in the cell cycle of the mouse haemopoietic spleen colony forming cells. Cell Tissue Kinet., *5*:365–369.

Levenson, R., and D. Housman (1981) Commitment: How do cells make the decision to differentiate? Cell, *25*:5–6.

Lord, B.I., K.J. Mori, E.G. Wright, and L.G. Lajtha (1976) An inhibitor of stem cell proliferation in normal bone marrow. Brit. J. Haematol., *34*:441–445.

Lord, B.I., K.J. Mori, and E.G. Wright (1977) A stimulator of stem cell proliferation in regenerating bone marrow. Biomedicine, *27*:223–226.

Lord, B.I., E.G. Wright, and L.G. Lajtha (1979) Actions of the haematopoietic stem cell proliferation inhibitor. Biochem. Pharmacol., *28:*1843–1848.

Luriya, E.A., and A.Y.A. Fridenshtein (1977) Recent experience in monolayer and organ cultures of haematopoietic tissue. In Luriya ed.: Haemopoietic and Lymphoid Tissue in Culture. Moscow: pp. 151–164 (translated from Russian)—Consultant Burlou, New York.

Mary, J.Y., F. Sainteny, D. Duménil, and E. Frindel Analysis of spleen colony numbers according to their histological type after Ara-C treatment: An attempt to clarify CFU-s differentiation patterns. (submitted for publication).

Metcalf, D. (1980) Clonal analysis of proliferation and differentiation of paired daughter cells: Action of granulocyte-macrophage colony-stimulating factor on granulocyte-macrophage precursors. Proc. Natl. Acad. Sci. U.S.A. *77:*5327–5330.

Rencricca, N.J., V. Rizzoli, D. Howard, P. Duffy, and F. Stohlman (1970) Stem cell migration and proliferation during severe anemia. Blood, *36:*764–771.

Riches, A.D., D.B. Thomas, and M.J. Cork (1981) Model for studying the regulation of hematopoietic stem cell proliferation. Exp. Hematol., *1981:*59–68.

Sainteny, F., D. Duménil, and E. Frindel (1981) Regulation of splenic CFU-S kinetics after cytosine-arabinoside treatment in mice. I. In vivo studies. Leukemia Res. *5:*373–378.

Till, J.E., and E.A. McCulloch (1961) A direct measurement of the radiation sensitivity of normal mouse bone marrow cells. Radiat. Res., *14:*213–222.

Till, J.E., E.A. McCulloch, and L. Siminovitch (1964) A stochastic model of stem cell proliferation, based on the growth of spleen colony forming cells. Proc. Natl. Acad. Sci. U.S.A., *51:*29–36.

Till, J.E., and E.A. McCulloch (1980) Hemopoietic stem cell differentiation. Bioch. Biophys. Acta, *605:*431–459.

Tsuya, A., V.P. Bond, T.M. Fliedner , and L.E. Feinendegen (1961) Cellularity and desoxyribonucleic acid synthesis in bone marrow after total and partial body irradiation. Radiat. Res., *14:*618–632.

Tubiana, M., E. Frindel, H. Croizat, and F. Vassort (1974) Study of some of the factors influencing the proliferation and differentiation of the multipotential hemopoietic stem cells. In Clarkson, B, Baserga, R. eds.: Control of Proliferation in Animal Cells. Cold Spring Harbor, New York: Cold Spring Harbor Laboratory, pp. 933–944.

Tubiana, M., E. Frindel, and H. Croizat (1971) Hematological effects of partial body irradiation: Clinical and experimental data. In Biological Aspects of Radiation Protection. Tokyo, Igaku Shin: pp. 53–62.

Tubiana, M., and E. Frindel (1962) Effets sur les tissus hématopoiétiques et les cellules sanguines d'irradiation subtotale par séances quotidiennes répétées pendant un mois. C. R. Acad. Sci. (Paris), *254:*183–185.

Tubiana, M., E. Frindel, H. Croizat, and C. Parmentier (1979) Effects of radiation on bone marrow. Pathol. Biol. (Paris), *27:*326–334.

Van Putten, L.M. (1973) Recrutement: Une arme à double tranchant dans la chimiothérapie du cancer. Bull. Cancer, *60:*131–142.

Vassort, F., E. Frindel, and M. Tubiana (1971) Effects of hydroxyurea on the kinetics of colony forming units of bone marrow in the mouse. Cell Tissue Kinet. *4:*423–431.

Vassort, F., M. Wintherholer, E. Frindel, and M. Tubiana (1973) Kinetic parameters of bone marrow stem cells using in vivo suicide by tritiated thymidine or by hydroxyurea. Blood, *41:*789–796.

Wolf, N.S., and J.J. Trentin (1968) Hemopoietic colony studies. J. Exp. Med., *127:*205–227.

Wright, E.G., and B.I. Lord (1977) Regulation of CFU-S proliferation by locally produced endogenous factors. Biomedicine, *27:*215–218.

Wright, E.G., and B.I. Lord (1978) Production of stem cell proliferation stimulators and inhibitors by haemopoietic cell suspensions. Biomedicine, *28:*156–160.

Wright, E.G., and B.I. Lord (1979) Production of stem cell proliferation regulators by fractionated haemopoietic cell suspensions. Leuk. Res., *3:*15–22.

JOURNAL OF CELLULAR PHYSIOLOGY SUPPLEMENT 1:23–30 (1982)

# Differentiation and "Renewal" of Multipotential Cells In Vitro

G. R. JOHNSON,* G. M. KELLER, AND N. A. NICOLA
*Cancer Research Unit, Walter and Eliza Hall Institute of Medical Research, Royal Melbourne Hospital, Melbourne, 3050 Victoria, Australia (G.R.J., N.A.N.); The Ontario Cancer Institute and University of Toronto, Toronto, Ontario M4X 1K9, Canada (G.M.K.)*

Cells obtained from in vitro colonies expressing multiple hemopoietic lineages (mixed-erythroid colonies) have been shown to produce secondary mixed-erythroid colonies. When individual mixed-erythroid colonies are studied, considerable variation has been observed in the capacity to produce secondary mixed-erythroid colonies. That this heterogeneity in self-renewal capacity may be an intrinsic property of the primary mixed-erythroid colonies has been shown by the ability to fractionate fetal liver mixed-erythroid colony-forming cells with high and low levels of secondary colony formation. In addition mixed-erythroid colony-forming cells obtained from spleens consistently produce higher numbers of secondary mixed-erythroid colonies (up to 250 secondary mixed-erythroid colonies per primary mixed-erythroid colony) than bone marrow-derived mixed-erythroid colonies (no secondary mixed-erythroid colonies obtained) suggesting that external factors may predetermine the ability of primary mixed-erythroid colony-forming cells to self-renew in vitro. The ability of supernatants obtained from long-term marrow cultures to enhance secondary colony formation by bone marrow-derived mixed-erythroid colony-forming cells raises the possibility of defining the nature of the proposed external factors that determine stem cell self-renewal.

Since the discovery of a procedure to quantify hemopoietic pluripotential cells (Till and McCulloch, 1961), many experiments have been performed to determine the factors regulating the differentiation of these cells (CFU-S). Colony formation by CFU-S can produce differentiated progeny of several hemopoietic lineages (Wu et al., 1967; Trentin, 1970). However, the role of external or internal regulatory factors in determining the decision to produce differentiated progeny with finite proliferative potentials remains unknown.

As well as possessing the capacity for producing cells of different hemopoietic lineages, CFU-S may also have the capacity of producing progeny with equivalent proliferative ability to the parent cell. This concept of CFU-S self-renewal was obtained from experiments in which it was noted that primary spleen colonies contained cells capable of producing secondary spleen colonies apparently identical to the parent colony (Siminovitch et al., 1963; Metcalf and Moore, 1971). As with differentiation of CFU-S, the factors controlling CFU-S self-renewal remain obscure.

As an alternative to the CFU-S assay, an in vitro procedure has been developed that allows single cells to produce colonies containing differentiated progeny of several hemopoietic lineages (Johnson and Metcalf, 1977; Hara and Ogawa, 1978; Humphries et al., 1979). In addition to producing differentiated progeny, the cells capable of producing multipotential clones in vitro are also able to produce cells capable of secondary multipotential clone formation, as well as CFU-S (Metcalf et al., 1978; Humphries et al., 1979). This in vitro assay appears therefore to detect cells identical to CFU-S (Humphries et al., 1979; Johnson, 1980).

Owing to the possibility of varying the in vitro conditions in which single cells may differentiate, the in vitro multipotential cell assay provides an alternative means to analyze the factors determining stem cell self-renewal and differentiation. It has for instance been observed that in vitro multipotential clones

*To whom correspondence/reprint requests should be addressed.

always contain erythroid progeny, even when cultured under a variety of experimental conditions. The cells producing these clones have therefore been termed mixed-erythroid colony-forming cells (Johnson and Metcalf, 1977) and a theory of hemopoietic stem cell differentiation that accounts for these observations has been proposed (Johnson, 1981).

The self-renewal capacity of mixed-erythroid colonies has been examined recently (Humphries et al., 1981). Large variations were observed in the numbers of daughter stem cells detectable in individual primary clones, an observation consistent with previously published data (Metcalf et al., 1978). It has been proposed therefore that in vitro the decision to self-replicate or not is ultimately determined at the single cell level by a mechanism involving a random-event component that is intrinsic to the stem cell itself (Humphries et al., 1981).

The present work was performed to determine some of the parameters affecting in vitro stem cell self-renewal. Data will be presented that indicates that in vitro mixed-erythroid colony-forming cells with differing self-renewal capacities can be physically separated from one another. In addition experiments with long-term culture supernatants indicate that the dicision to self-renew can possibly be altered by external regulatory factors.

## MATERIALS AND METHODS

### Mice

Mice used were adult CBA/J obtained from the Jackson Laboratory, Bar Harbor, ME, or CBA/CaH, C57BL/6, or BALB/c animals bred and maintained at the Walter and Eliza Hall Institute animal facility.

### Cultures

All in vitro colony assays were performed in either methylcellulose or agar cultures using CBA cells.

The medium and method of preparation of agar cultures for the growth of mixed-erythroid colonies has been described in detail elsewhere (Johnson, 1980). Briefly, equal volumes of double strength Dulbello's Modified Eagle's Medium (DME) containing 40% fetal calf serum and 0.6% agar (Difco, Bactoagar) were mixed; cells were then added and 1 ml aliquots were pipetted into 35 mm petri dishes containing 0.1 ml of lectin-stimulated spleen cell conditioned media (SCM). The cells in agar medium were mixed with the SCM by agitation and after gelling the cultures were incubated for 7 days (unless otherwise stated) at 37°C in a fully humidified atmosphere containing 10% $CO_2$ in air.

Methylcellulose cultures were established using Iscove's Modified Dulbecco's Medium (Gibco, Catalogue Number 430 2200), 0.9% methylcellulose (Methocel MC 4000 c P, Fluka, Buchs, Switzerland), 20% fetal calf serum, 7.5 $\times$ $10^{-5}$M 2-mercaptoethanol, and 10% SCM. One milliliter volumes of the methylcellulose-medium mixture containing appropriate numbers of cells were cultured for 7 days in 35 mm petri dishes at 37°C in a fully humidified atmosphere containing 5% $CO_2$ in air.

Cultures were scored for colonies using an Olympus dissecting microscope (35–40 $\times$) with indirect illumination. Mixed-erythroid and other colony types were identified by procedures outlined previously (Johnson, 1980).

For colony replating experiments, individual colonies were removed with a sterile, finely drawn out Pasteur pipette and placed into 0.25 ml of DME. Cells were dispersed by gentle pipetting and 0.2 ml cultured in duplicate agar medium cultures stimulated by SCM as described above.

### Conditioned media

All cultures were stimulated by predetermined optimal concentrations of SCM. SCM was prepared from either C57BL/6 or BALB/c spleen cells as described previously by Metcalf and Johnson (1978). Bone marrow culture conditioned media were obtained during the weekly media replacement of CBA/J bone marrow long-term cultures. These cultures were established using the method of Dexter et al. (1977) except that the contents of a single femur were flushed into a 25 $cm^2$ plastic tissue culture flask (Falcon 3013) containing 10 ml of medium. The medium, DME, contained 20% horse serum and hydrocortisone ($10^{-7}$ M) and the initial cell innoculum was the only source of cells added to the cultures. One half of the medium was removed and replaced with fresh medium at weekly intervals. Medium obtained from 3–10 week old cultures was pooled and concentrated 10-fold over an Amicon UM-10 membrane. Concentrated conditioned medium was added in 0.1 ml volumes to each 1 ml agar medium culture.

### Fluorescence-activated cell sorting (FACS)

Fractionated 13–14 day CBA/CaH fetal liver cell populations were obtained after FACS separation using procedures published previously

(Nicola et al., 1981). Briefly, cells were separated on the basis of low-angle light scatter, lack of binding to an antimouse neutrophil antiserum (αN), and binding to fluorescein-conjugated pokeweed mitogen (PWM). With these parameters two populations were obtained; one population (1R) contained the majority of CFU-S (> 100-fold enrichment, overall recovery 60–70%). The second population (1L) contained the majority of in vitro colony-forming cells (pure- and mixed-erythroid and granulocyte-macrophage) with enrichments varying between 40–90-fold and 65–93% overall recoveries (Nicola et al., 1981).

## RESULTS

### Secondary colony formation by adult spleen or bone marrow mixed-erythroid colonies

Previous experiments with fetal liver cells have demonstrated that primary mixed-erythroid colonies contain cells capable of secondary mixed-erythroid colony formation (Metcalf et al., 1978). However, with these cells the number and size of the secondary colonies obtained was small. These results may reflect intrinsic properties of fetal liver-derived cells or, alternatively, be a measure of the degree to which cell proliferation and demands for cell differentiation are occurring within the rapidly developing fetal liver.

To obtain data to contrast with those previously obtained, mixed-erythroid colonies were grown in cultures of normal CBA spleen or bone marrow cells. Primary cultures were established in methylcellulose since preliminary experiments showed that single cell suspensions of colony cells were more easily attained in this medium than with colonies obtained from agar cultures.

Mixed-erythroid colonies were recognized by their characteristic shape due to hemoglobin-containing cells and peripheral halo of neutrophils and macrophages (Johnson, 1980). Sequential mixed-erythroid colonies were individually harvested and recultured from day 7 and day 14 primary bone marrow cultures. The results of these experiments are shown in Table 1. No secondary mixed-erythroid colonies were obtained when primary bone marrow-derived mixed-erythroid colony cells were recultured and only low numbers of pure-erythroid colonies were observed in secondary cultures. Granulocyte and macrophage secondary colonies were obtained from 8 primary day 7 bone marrow mixed-erythroid colonies, but by day 14 of culture only 2 primary colonies produced secondary granulocyte and macrophage colonies (Table 1). A similar reduction between days 7 and 14 was observed in the proportion of bone marrow mixed-erythroid colonies capable of producing clusters in secondary cultures.

In contrast to the data obtained with bone marrow cultures, the proportion of spleen primary mixed-erythroid colonies capable of secondary colony formation was markedly increased. Similarly, an increase in secondary colony formation was observed when day 14 spleen mixed-erythroid colonies were compared to day 7 spleen mixed-erythroid colonies (Table 1). The increase in secondary colony formation occurred for all 3 categories of colonies assayed. Thus at 7 days, 3 primary colonies (6%) produced secondary pure erythroid colonies compared with 14 (19%) day 14 primary colonies. Similarly, only 2 (4%) day 7 mixed-erythroid colonies produced secondary mixed-erythroid colonies, but by day 14 the proportion of primary colonies capable of producing secondary mixed-erythroid colonies had increased to 22%. Similar changes occurred with secondary granulocyte and macrophage colony formation (Table 1).

The increased proportion of primary colonies positive for secondary colony formation was paralled by an increase with time in the total numbers of secondary colonies obtained from

TABLE 1. Secondary colony formation by mixed-erythroid colonies obtained from spleen or bone marrow cultures[1]

| Cells cultured | Primary culture age (days) | Total number of colonies transferred | Percent of primary colonies producing | | | |
|---|---|---|---|---|---|---|
| | | | Clusters | Pure-erythroid colonies | Mixed-erythroid colonies | Granulocyte macrophage colonies |
| Bone marrow | 7 | 48 | 67 | 4 | 0 | 17 |
| | 14 | 53 | 25 | 2 | 0 | 4 |
| Spleen | 7 | 51 | 59 | 0 | 4 | 8 |
| | 14 | 73 | 64 | 19 | 22 | 51 |

[1]Individual, sequential mixed-erythroid colonies, dispersed and recultured after 7 or 14 days in methylcellulose cultures. Colony cells obtained from individual primary colonies cultured for 7 days in duplicate secondary agar cultures. All cultures (1 ml) contained 20% fetal calf serum and were stimulated by 0.1 ml SCM. Primary cultures contained either $5 \times 10^4$ bone marrow cells or $1 \times 10^6$ spleen cells.

individual primary colonies. After 7 days of primary culture the mean numbers of secondary pure-erythroid, mixed-erythroid, and granulocyte-macrophage colonies per positive primary colony were 8 ± 7, 2 ± 1, and 15 ± 18, respectively. The maximum number of secondary pure-erythroid, mixed-erythroid, or granulocyte-macrophage secondary colonies obtained from a single day 7 primary colony being 23, 3, and 331, respectively. When the day 14 primary mixed-erythroid colony data were examined the mean numbers of secondary colonies per positive primary colony were pure-erythroid colonies 50 ± 82, mixed-erythroid 42 ± 66, and granulocyte-macrophage 203 ± 273. The maximum number of secondary colonies obtained from a single day 14 primary colony being 325 pure-erythroid, 250 mixed-erythroid, and 900 granulocyte-macrophage.

The increased secondary colony formation observed with primary spleen mixed-erythroid colonies was more marked in primary cultures containing $1 \times 10^6$ cells than in cultures with $1 \times 10^5$ cells. These observations suggested that accessory cells or their products may be responsible for altering the degree of secondary colony formation.

### Effect of long-term culture supernatants upon secondary colony formation by mixed-erythroid colonies

Since the long-term bone marrow culture system devised by Dexter et al. (1977) is capable of maintaining CFU-S populations for many months, the effects of supernatants obtained from these cultures upon mixed-erythroid colony formation was examined. Supernatants obtained from 3–10-week-old cultures were pooled and concentrated 10-fold and added to agar cultures of bone marrow cells stimulated by SCM. Bone marrow was chosen for these experiments since mixed-erythroid colonies obtained from these cultures stimulated by SCM were relatively poor at producing secondary colonies (see previous section).

Mixed-erythroid colonies grown in the presence of SCM and long-term culture supernatant were picked from day 7 primary cultures and colony cells were recultured for a further 7 days in the presence of SCM alone. Control experiments were performed with primary mixed-erythroid colonies obtained from cultures stimulated with SCM. The data obtained from these experiments is shown in Table 2. The addition of long-term marrow supernatant increased the total number of secondary colonies obtained from individual primary mixed-erythroid colonies such that the mean number of secondary pure- and mixed-erythroid colonies increased 16-fold (from 0.4 per primary colony with SCM alone to 6.4 per primary colony in the presence of supernatant) and secondary granulocyte-macrophage colonies increased 3-fold (Table 2).

To determine whether the enhanced secondary colony formation observed, following the addition of long-term culture supernatants, was specific for primary mixed-erythroid colonies, a second series of experiments were performed in which primary granulocyte-macrophage colonies were recultured. These data are shown in Table 3. As expected no secondary pure- or mixed-erythroid colonies were obtained from the 55 primary granulocyte-macrophage colonies recultured (Table 3). The addition of long-term culture supernatant to primary cultures did however increase the total numbers of secondary granulocyte-macrophage colonies obtained in cultures stimulated by SCM (Table 3). Compared with control primary cultures stimulated with SCM alone, the addition of supernatant increased secondary granulocyte-macrophage colony formation 7-fold (Table 3).

The ability of long-term marrow supernatants to enhance secondary colony formation

TABLE 2. Effect of long-term culture supernatants upon secondary colony formation by mixed-erythroid colony cells[1]

| Number of experiments | Primary stimulus | Number of colonies recultured | Total number of secondary colonies: Erythroid/mixed-erythroid | Total number of secondary colonies: Granulocyte-macrophage |
|---|---|---|---|---|
| 3 | SCM + Super | 25 | 160 (6.4)[2] | 463 (18.5) |
| 3 | SCM | 28 | 10 (0.4) | 170 (6) |

[1]Individual, sequential primary colonies, dispersed and recultured after 7 days in agar cultures. Secondary colony numbers determined from duplicate agar cultures of cells from individual primary colonies after 7 days' incubation. All cultures (1 ml) contained 20% fetal calf serum and were stimulated by 0.1 ml of SCM. Primary cultures contained $5 \times 10^4$ bone marrow cells and, where indicated, 0.1 ml of 10-fold concentrated long-term bone marrow culture supernatant (see Materials and Methods).

[2]( ) = Mean number of colonies per primary mixed-erythroid colony.

TABLE 3. Effect of long-term culture supernatants upon secondary colony formation by granulocyte-macrophage colony cells[1]

| Number of experiments | Primary stimulus | Number of colonies recultured | Total number of secondary colonies | |
|---|---|---|---|---|
| | | | Erythroid/mixed-erythroid | Granulocyte-macrophage |
| 3 | SCM + Super | 27 | 0 | 609 (23)[2] |
| 3 | SCM | 28 | 0 | 87 (3.1) |

[1]Data obtained from the same cultures used in Table 2.
[2]( ) = Mean number of secondary colonies per primary colony.

could have occurred if its addition to primary cultures selectively enhanced primary colony formation by cells intrinsically capable of increased secondary colony formation. Evidence against this proposal was obtained by examination of primary day 7 cultures for the distribution of the various colony types in the presence of SCM alone or SCM plus long-term culture supernatant. In cultures containing 50,000 cells and stimulated by SCM alone, means of 93 ± 30 granulocyte-macrophage, 20 ± 15 erythroid and mixed-erythroid, and 2 ± 1 megakaryocyte colonies were obtained. No significant differences were observed in cultures stimulated by SCM plus supernatant where means of 109 ± 19 granulocyte-macrophage, 14 ± 9 pure- and mixed-erythroid, and 2 ± 2 megakaryocyte colonies per culture were obtained.

### Ability of cells fractionated by FACS sorting to produce secondary colonies

The heterogeneity in the ability of primary mixed-erythroid colonies to produce secondary colonies could occur if subpopulations of mixed-erythroid colony-forming cells existed with differing intrinsic abilities for secondary colony formation. To examine this proposal cells obtained from 13–14 day fetal livers were fractionated into two populations (designated 1L and 1R), shown previously to be highly enriched for both CFU-S and in vitro colony-forming cells (Nicola et al., 1981) (Fig. 1).

The 1L fraction enriched for CFC was PWM-binding$^{++}$ (channels 140–190), αN-binding$^{-}$ channels 10–90), and high intensity of forward light scatter (channels 110–200). The 1R fraction enriched for CFU-S was PWM-binding$^{+}$ (channels 80–115), αN-binding$^{-}$ channels 10–80), and high intensity of forward light scatter (channels 110–200) (Fig. 1). The 1L fraction contained a morphologically homogeneous population of blast cells with an in vitro colony-forming efficiency of at least 30% and total clone-forming efficiency of 70–90%. The 1R fraction had a colony-forming efficiency of 1% and a CFU-S content of 2–4%.

Methylcellulose cultures containing either 200 1L cells or 2,000 1R cells were stimulated with SCM and at 7 days colonies were removed for secondary cell culture. Preliminary experiments were performed in which 200 sequential clones (colonies and clusters) were individually recultured from primary cultures of either 1L or 1R cells. No secondary colonies were obtained from primary clusters or macrophage colonies so that in subsequent experiments sequential granulocyte, granulocyte-macrophage, pure-erythroid, and mixed-erythroid colonies were recultured (colony type being determined by gross morphological appearance in the culture dish and having been previously verified by colony removal and staining). Table 4 shows the numbers of secondary colonies obtained following the sequential removal and reculturing of 1L and 1R cultured cells.

Only 1 primary colony (2%) (mixed-erythroid colony) from the 1L fraction produced secondary colonies all of which contained granulocytes and macrophages. By comparison, 11 of 38 (29%) primary 1R colonies produced secondary colonies, including 218 pure-erythroid and 17 mixed-erythroid colonies (Table 4). Of the 11 primary 1R-derived colonies, 6 produced secondary colonies containing either neutrophils, macrophages, erythroid, or various mixtures of these cells, confirming the multipotentiality of the cell initiating the primary colony.

In a second series of experiments 10 sequential mixed-erythroid colonies were recultured from cultures of 1L cells of which 2 produced secondary colonies. By comparison, 7 mixed-erythroid colonies were recultured from primary 1R cultures, with secondary pure-erythroid, mixed-erythroid, and granulocyte-mac-

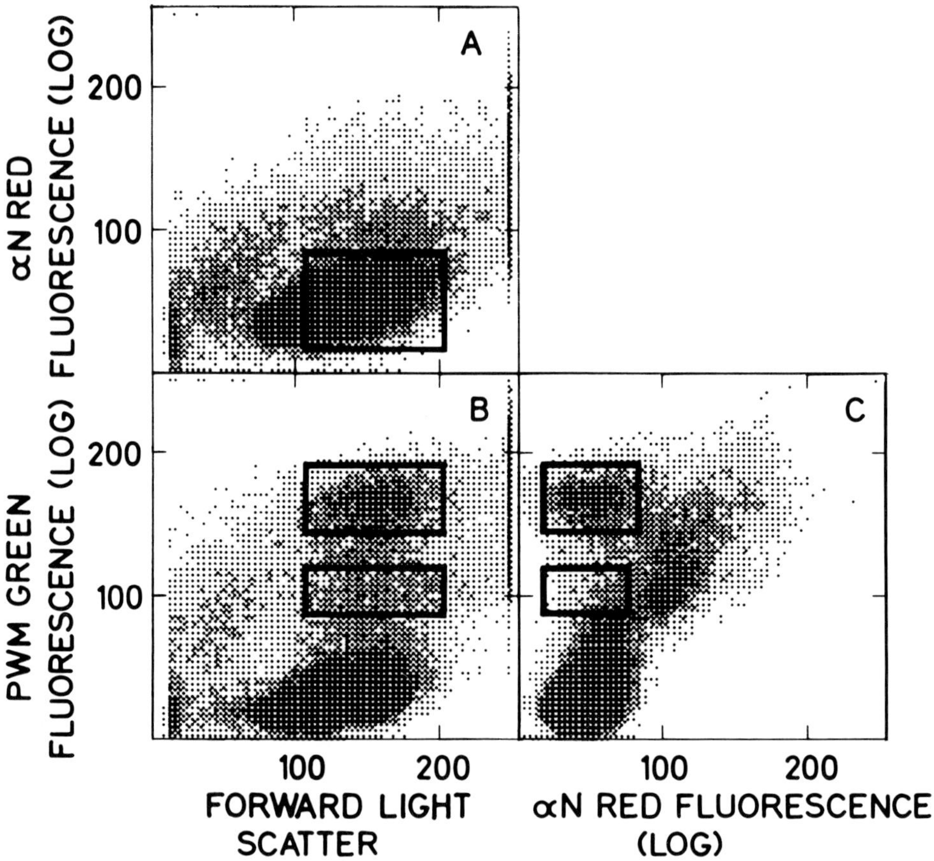


Fig. 1. Correlated intensity plots generated by the FACS for light density fetal liver cells labeled with FITC-PWM and a rhodamine-containing antineutrophil (αN) antibody sandwich. (A) forward light scatter intensity versus red fluorescence (αN-binding), (B) forward light scatter intensity versus green fluorescence intensity (PWM-binding), and (C) red fluorescence intensity versus green fluorescence intensity. The density of spots indicates relative cell number using 3 arbitrary levels. Boxes indicate selection windows used for sorting. For both fractions the same windows for light scatter and red fluorescence intensity were used. The 1L fraction enriched for CFC was contained in the uppermost box with respect to green fluorescence and the 1R fraction enriched for CFU-S was contained in the lower box. Both fluorescence scales are 3 decades log and the light scatter scale is linear (arbitrary units). Figures indicate Channel number.

TABLE 4. Ability of colonies produced by fractionated cells to produce secondary colonies[1]

| Fraction | Number of primary colonies transferred | Number of primary colonies producing secondary colonies | Total number of secondary colonies produced | | |
|---|---|---|---|---|---|
| | | | Pure-erythroid | Mixed-erythroid | Granulocyte-macrophage |
| 1 R | 38 | 11 | 218 | 17 | 62 |
| 1 L | 43 | 1 | 0 | 0 | 9 |

[1]Individual, sequential primary colonies, dispersed and recultured after 7 days in methylcellulose cultures. Secondary colony numbers determined from duplicate agar cultures of cells from individual primary colonies cells after 7 days' incubation. All cultures (1 ml) contained 20% fetal calf serum and were stimulated by 0.1 ml SCM.

rophage colonies being obtained from 4 of the 7 primary colonies.

## DISCUSSION

Using the in vivo spleen colony assay for hemopoietic multipotential cells (CFU-S), a number of reports have documented marked heterogeneity in the ability of primary CFU-S to produce secondary CFU-S capable of spleen colony formation (Siminovitch et al., 1963; Metcalf and Moore, 1971; Vogel et al., 1968).

Although it has been proposed that the observed heterogeneity in CFU-S renewal ability may be the result of a random-event component (Till et al., 1964), alternative explanations involving variations in extrinsic factors acting upon CFU-S have also been proposed (reviewed by Metcalf and Moore, 1971). That CFU-S do not represent a homogeneous population has been shown in a number of different systems, all of which indicate that CFU-S exist as a heterogeneous population with differing intrinsic abilities to produce secondary CFU-S (Worton et al., 1969; Rosendaal et al., 1976; Mauch et al., 1980). All of these data have been taken to indicate that CFU-S exist as a continuum of cells varying in self-renewal capacity. Within this continuum, early CFU-S would be less committed to differentiate and have a high self-renewal capacity whereas CFU-S further along the continuum would be more committed to differentiation and have a lower self-renewal potential.

It has been proposed that in vitro mixed-erythroid colony formation represents proliferation and differentiation from cells equivalent to CFU-S (Metcalf et al., 1978; Humphries et al., 1979; Johnson, 1980). As with CFU-S, secondary colony formation by mixed-erythroid colony cells has been shown to vary markedly between individual primary colonies (Metcalf et al., 1978; Humphries et al., 1981). Recently it has been proposed that this heterogeneity reflects a general model of stem cell differentiation in which the choice to self-replicate or not is ultimately determined at the single-cell level by a mechanism involving a random-event component that is intrinsic to the stem cell (Humphries et al., 1981).

The present experiments show that secondary colony formation by primary mixed-erythroid colonies varies with the geographical site from which the colony-forming cells were obtained. Spleen-derived mixed-erythroid colonies are superior at producing secondary mixed-erythroid colonies than bone marrow-derived mixed-erythroid colonies. These data suggest that external factors (e.g., the spleen versus bone marrow environment) predetermine the ability of mixed-erythroid colonies to produce secondary colonies within the uniform environment presented by the in vitro cultures, this view being at variance with that proposed by Humphries et al. (1981).

Similarly, the ability to fractionate fetal liver mixed-erythroid colonies with differing capacities for secondary colony formation supports the concept that mixed-erythroid colony-forming cells have pre-existing differences in self-renewal capacity.

The observations with long-term marrow culture supernatants provide a means to determine how these pre-existing self-renewal capacities may be generated. Thus mixed-erythroid colony-forming cells that may have been exposed to a similar factor in vivo might be expected to produce larger numbers of secondary colonies when subsequently cultured in vitro.

Further experiments are required to determine the nature of the active component in the long-term culture medium and its possible in vivo relevance. In addition, further in vitro experiments are required to determine if in the continued presence of these conditioned media primary, secondary, and tertiary mixed-erythroid colony-forming cells can be induced to undergo continual self-renewal. Whatever the outcome of such experiments, the present data further confirm the similarity in the behavior of in vitro mixed-erythroid colony-forming cells and CFU-S.

## ACKNOWLEDGMENTS

The authors wish to thank Cathy Quilici, Judy Tarrant, Luba Panczak, and Elizabeth Viney for expert technical assistance. A portion of this work was performed while G. R. Johnson was the recipient of the University of Toronto W. P. Caven Fellowship at the Ontario Cancer Institute. This work was supported by the Carden Fellowship Fund of the Anti-Cancer Council of Victoria, the National Health and Medical Research Council, Canberra, and the National Cancer Institute, Bethesda, Maryland (grant CA-25972). G. M. Keller was supported by the Medical Research Council (Canada) and the National Cancer Institute of Canada.

## LITERATURE CITED

Dexter, T. M., T. D. Allen, and L. G. Lajtha (1977) Conditions controlling the proliferation of haemopoietic stem cells in vitro. J.Cell Physiol., *91*:335–344.

Hara, H., and M. Ogawa (1978) Murine hemopoietic colonies in culture containing normoblasts, macrophages, and megakaryocytes. Am.J.Hematol., *4*:23–34.

Humphries, R. K., A. C. Eaves, and C. J. Eaves (1981) Self-renewal of hemopoietic stem cells during mixed colony formation in vitro. Proc.Natl.Acad.Sci. U.S.A., *78*:3629–3633.

Humphries, R. K., P. B. Jacky, F. J. Dill, A. C. Eaves, and G. J. Eaves (1979) CFU-S in individual erythroid colonies derived in vitro from adult mouse marrow. Nature, *279*:718–720.

Johnson, G. R. (1980) Colony formation in agar by adult bone marrow multipotential hemopoietic cells. J.Cell Physiol., *103*:371–383.

Johnson, G. R. (1981) Is erythropoiesis an obligatory step in the commitment of multipotential hematopoietic stem cells? In Baum, S. J., Ledney, G. D., Khan, A. eds.: Experimental Hematology Today 1981. Basel: S. Karger, pp. 13–20.

Johnson, G. R., and D. Metcalf (1977) Pure- and mixed-erythroid colony formation in vitro stimulated by spleen conditioned medium with no detectable erythropoietin. Proc.Natl.Acad.Sci. U.S.A., *74*:3879–3882.

Mauch, P., J. S. Greenberger, L. Botnick, E. Hannon, and S. Hellman (1980) Evidence for structured variation in self-renewal capacity within long-term bone marrow cultures. Proc.Natl.Acad.Sci. U.S.A., *77*:2927–2930.

Metcalf, D., and G. R. Johnson (1978) Production by spleen and lymph node cells of conditioned medium with erythroid and other hemopoietic colony-stimulating activity. J.Cell Physiol., *96*:31–42.

Metcalf, D., G. R. Johnson, and T. Mandel (1978) Colony formation in agar by multipotential hemopoietic cells. J.Cell Physiol., *98*:401–420.

Metcalf, D., and M. A. S. Moore (1971) Haemopoietic Cells. North-Holland Research Monographs: Frontiers of Biology. Amsterdam: North-Holland, vol. 24, pp. 70–108.

Nicola, N. A., D. Metcalf, H. von Melchner, and A. W. Burgess (1981) Isolation of murine fetal hemopoietic progenitor cells and selective fractionation of various erythroid precursors. Blood (in press).

Rosendaal, M., G. S. Hodgson, and T. R. Bradley (1976) Haemopoietic stem cells are organized for use on the basis of the generation age. Nature, *264*:68–69.

Siminovitch, L., E. A. McCulloch, and J. E. Till (1963) The distribution of colony-forming cells among spleen colonies. J.Cell Comp.Physiol., *62*:327–336.

Till, J. E., and E. A. McCulloch (1961) A direct measurement of the radiation sensitivity of normal mouse bone marrow cells. Radiat. Res., *14*:213–222.

Till, J. E., E. A. McCulloch, and L. Siminovitch (1964) A stochastic model of stem cell proliferation, based on the growth of spleen colony-forming cells. Proc.Natl.Acad.Sci. U.S.A., *51*:29–36.

Trentin, J. J. (1970) Influence of hematopoietic organ stroma (hematopoietic inductive microenvironments) on stem cell differentiation. In Gordon, A. S. ed.: Regulation of Hematopoiesis. New York: Appleton-Century-Crofts, pp. 161–186.

Vogel, H., H. Niewisch, and G. Matioli (1968) The self-renewal probability of hemopoietic stem cells. J.Cell Physiol., *72*:221–228.

Worton, R. G., E. A. McCulloch, and J. E. Till (1969) Physical separation of hemopoietic stem cells from cells forming colonies in culture. J.Cell Physiol., *74*:171–182.

Wu, A. M., J. E. Till, L. Siminovitch, and E. A. McCulloch (1967) A cytological study of the capacity for differentiation of normal hemopoietic colony-forming cells. J.Cell Physiol., *69*:177–184.

JOURNAL OF CELLULAR PHYSIOLOGY SUPPLEMENT 1:31–36 (1982)

# Detection In Vitro of a Unique, Multipotent Hemopoietic Progenitor

G.M. KELLER AND R.A. PHILLIPS*
*The Ontario Cancer Institute and Department of Medical Biophysics, University of Toronto, Toronto, Ontario M4X 1K9, Canada*

The maintenance of mature cells in any cell renewal system such as skin, testis, and blood requires ultimately the presence of pluripotent stem cells whose unique characteristics are their abilities both to self-renew and to make differentiated progeny. In the blood-forming system, numerous studies have identified stem cells in the various lineages in this complex pathway (for review, see Till and McCulloch, 1980). The proposed relationships between the various known stem cells are shown in Figure 1. As indicated in this figure, the major differences between the various progenitors are in their potential for differentiation and in their capacity for self-renewal. Although it is clear that this system is precisely regulated, little is known about the specific factors that regulate differentiation and self-renewal. One of the major controversies in hemopoiesis has revolved around whether differentiation can be directed extrinsically either by factors or exposure to a hemopoietic microenvironment (Curry and Trentin, 1967) or whether the pattern of differentiation is programmed into a cell and is determined by solely stochastic processes (Till et al., 1964).

## FACTORS IMPLICATED IN THE REGULATION OF STEM CELLS

Several investigators have identified factors that can either stimulate resting stem cells to enter DNA synthesis (Byron, 1978) or inhibit proliferating cells, perhaps causing them to enter a quiescent state (Lord, 1978). Although there is indirect evidence that these factors that are detected in vitro are involved in the natural in vivo regulation of stem cells, their precise role in the regulation of self-renewal and/or differentiation remains uncertain. Analysis of the cellular basis of the anemia in mice of genotype $Sl/Sl^d$ clearly indicates that the hemopoietic microenvironment, i.e., a short-range cellular interaction, is important in regulating hemopoiesis (McCulloch et al., 1965; Bernstein, 1970). It is not clear whether the microenvironment simply provides a milieu permissive for differentiation or whether it can direct cells to differentiate along a specific pathway.

In addition to the factors mentioned above, many investigators have obtained evidence that thymus-derived cells play some role in regulating hemopoiesis. In early transplantation studies, Goodman and her colleagues demonstrated that addition of thymocytes to bone marrow cells can promote the growth of hemopoietic stem cells in particular strains of mice (Salinas and Goodman, 1972). Frindel et al. (1976) have shown that the stimulation of T lymphocytes by various antigens and mitogens can induce the proliferation of hemopoietic stem cells. More recently, Sharkis et al. (1978) showed that elimination of Thy1 positive cells from bone marrow by treatment with antibody and complement severely reduced the ability of the bone marrow to cure congenitally anemic mice having a stem cell defect. They could reconstitute the curing ability of depleted bone marrow by the addition of thymocytes from normal, but not anemic, donors (Wiktor-Jedrzejczak et al., 1977).

All of the experiments described above involve the use of indirect assays for the study of hemopoietic stem cells. Even the most direct assay for hemopoietic stem cells, the formation of colonies in the spleens of irradiated mice, the CFU-s assay (Till and McCulloch, 1961), requires stem cells to proliferate many times to produce colonies containing approximately one million cells. Thus, investigators must attempt, by looking at the progeny in the spleen colonies, to infer what happened during the early stages of stem cell differentiation. It is obvious that significant advances in understanding the regulation of self-renewal and differentiation could occur if there were in vitro assays that allowed the study of stem cells at early stages of differentiation.

*To whom correspondence/reprint requests should be addressed.

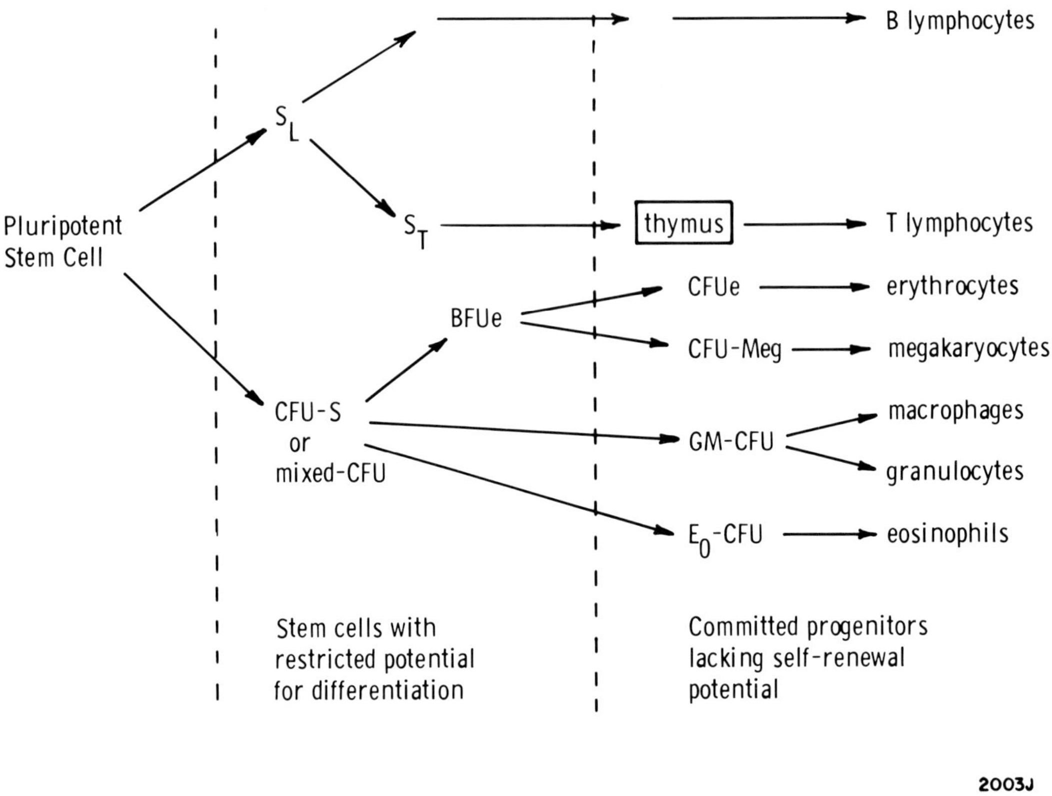


Fig. 1. Proposed lineage diagram for the differentiation of cells in the hemopoietic system. $S_L$, lymphoid committed stem cell; $S_T$, stem cell restricted to T cell differentiation; CFU-s, spleen colony-forming unit; BFUe, burst-forming unit dependent on erythropoietin; CFUe, progenitor for erythrocytes; CFU-Meg, progenitor committed to megakaryocyte production; GM-CFU, progenitor for granulocytes and macrophages; $E_O$-CFU, progenitor for eosinophils.

In recent years, several investigators have described in vitro assays detecting multipotent hemopoietic progenitors. Axelrad and his colleagues first described a colony assay that detected progenitors with the capacity to produce both megakaryocytes and erythrocytes; he termed this progenitor a burst-forming unit (BFU-e) (McLeod et al., 1976). Later, Johnson and Metcalf (1978) developed conditions that generated colonies containing erythrocytes, granulocytes, macrophages, and megakaryocytes; because the colonies always contained erythroid cells and some, but not all, of the other differentiated cells, they termed the progenitor a mixed/erythroid-CFU (Mix/E-CFU). That the assay for large erythroid colonies also detects multipotent stem cells with some potential for self-renewal is best demonstrated by the studies of Humphries et al. (1979), who showed that these in vitro colonies contain CFU-s. However, one of the puzzling features of these in vitro assays was their relatively small potential for self-renewal. If the Mix/E-CFU assay detected stem cells of the hemopoietic system, one would expect the colonies to contain large numbers of pluripotent progenitors. Most experiments on BFU-e and Mix/E-CFU have demonstrated relatively little ability for self-renewal (for an exception see the paper by Johnson et al. in this volume).

## EXPERIMENTAL RESULTS

To analyze the poor self-renewal potential of the multipotent cells detected in the CFU-s and Mix/E-CFU assays, we have established an in vitro system in which to study the early events in the differentiation of stem cells and the factors regulating their self-renewal. The experiments described below involve the study of bone marrow from mice pretreated with hydroxyurea. The rationale for using this initial cell source was the report of Rosendaal et al. (1979) that stem cells surviving the treatment with hydroxyurea have extensive self-renewal potential. Cells were grown in methylcellulose as described by Iscove et al. (1980) using medium containing 30% fetal calf serum. In addition the cultures contained conditioned medium from mitogen-stimulated spleen cells as described by Johnson and Metcalf (1978) and $5 \times 10^6$ syngeneic thymocytes (see below).

To measure the number of colony-forming cells in the colonies from the primary culture, single colonies from the primary culture were picked with a Pasteur pipette, diluted in medium, made into a single cell suspension, and replated in the same conditions described above except that $10^5$ irradiated bone marrow cells were used instead of thymocytes. The reason for using irradiated bone marrow cells was the report by Humphries et al. (1981) that such an addition markedly enhances colony-forming potential especially when cells are plated at low numbers. The number of colonies in the secondary dishes at day 7 was used as a measure of the total colony-forming potential of the primary colonies.

All the colonies from the primary dish were picked and individually replated; the results, with one exception, were identical to those reported by others (Johnson and Metcalf, 1978; Humphries et al., 1981). Both the granulocyte/macrophage colony forming units (GM-CFU) and the Mix/E-CFU gave small numbers of secondary colonies. However, a small colony similar in morphology to a GM colony had much greater replating potential. Morphologically, the primary difference between the colonies with high replating potential and G/M colonies was that the former lacked a halo of cells and were smaller and more compact than the colonies containing granulocytes and macrophages (Fig. 2). Compact colonies containing only 1,000–5,000 cells/colony can give rise to

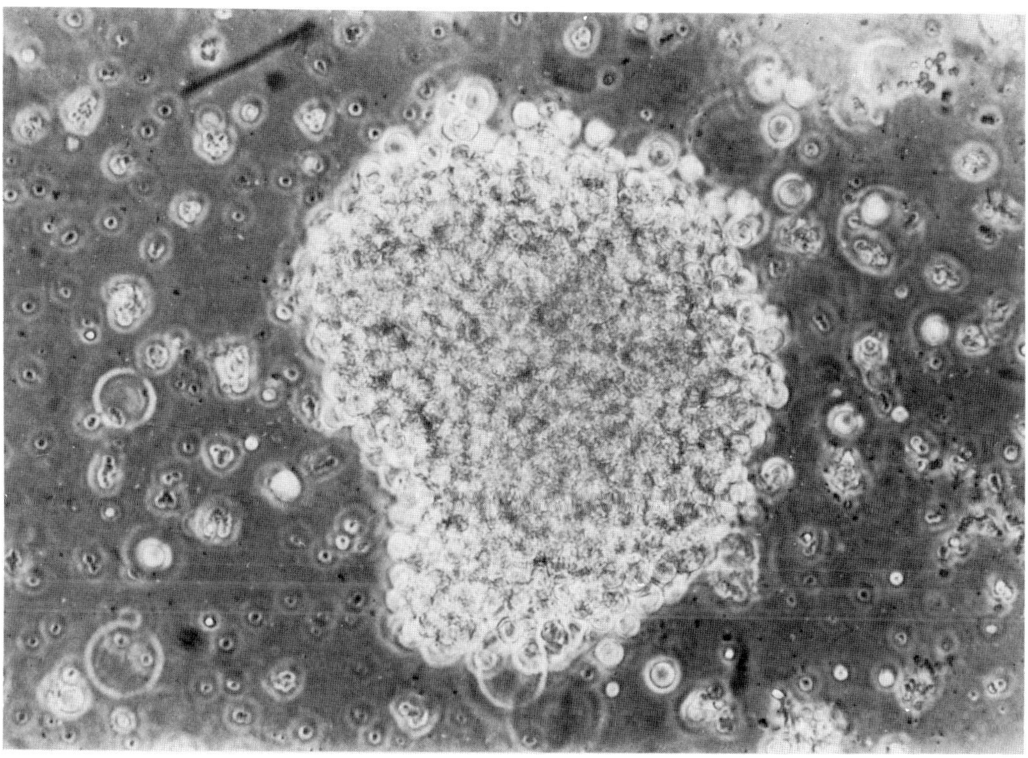

Fig. 2. Photomicrograph of a day eight compact colony (× 250). Note the absence of a halo characteristic of a granulocyte-macrophage colony.

TABLE 1. Replating potential of compact and mixed/erythroid colonies

| Age of 1°[1] colony (days) | Morphology of 1° colony | Number of 1° colonies giving 2° colonies | 2° cols/1° col | | | |
|---|---|---|---|---|---|---|
| | | | Total | | Mix/E | |
| | | | Mean[2] | Range | Mean[2] | Range |
| 6 | Compact | 32/88 | 63 | 6–750 | 0 | — |
| 8 | Compact | 53/94 | 94 | 4–750 | 9 | 2–32 |
| | Mix/E | 11/17 | 15 | 2–104 | 0 | — |
| 10 | Compact | 19/49 | 203 | 22–800 | 11 | 2–117 |
| | Mix/E | 10/33 | 14 | 3–120 | 0 | |

[1]1°, primary culture; 2°, secondary culture.
[2]Geometric mean of the 2° colonies from the positive 1° colonies.

large numbers of secondary colonies as shown in Table 1. In this experiment, compact colonies picked on days 6, 8, and 10 produced a mean number of secondary colonies of 63, 94, and 203, respectively. In contrast, Mix/E colonies picked on days 8 and 10 produced only about 15 secondary colonies. Although most of the secondary colonies are granulocyte/macrophage in nature, there are a few Mix/E colonies (confirmed by benzidine staining of hemoglobin) and colonies with the same morphology as the original compact colony. The secondary compact colonies have not yet been tested for their replating potential. Nevertheless, the heterogeneity of colony types in the secondary dish clearly indicates that the compact colony-forming unit is multipotent. When compact colonies are picked and the cells smeared on a slide for morphological studies, only immature blasts are observed; less than 5% of the cells show any features characteristics of myeloid differentiation.

Because of the proposed importance of a thymus-derived cell in the regulation of hemopoiesis, we investigated the need for thymocytes in the primary culture. The data in Table 2 show that the addition of thymocytes has little effect on the total number of primary colonies obtained from plating different numbers of bone marrow cells. Similarly, addition of thymocytes does not enhance the production of Mix/E colonies. However, when colonies are initiated from a small number of bone marrow cells, the addition of thymocytes increases the number of compact colonies approximately two-fold. When higher numbers of bone marrow cells are plated, greater than $1 \times 10^5$ per dish, the addition of thymocytes has a smaller effect. The data in Table 2 also show that the conditions for measuring Mix/E and compact colonies are suboptimal in that the number of colonies obtained is not a linear function of the number of bone marrow cells plated. In fact, approximately the same number of both types of colonies were obtained on all plates containing between $1 \times 10^5$ and $5 \times 10^5$ bone marrow cells.

TABLE 2. Effect of thymocytes on colony formation

| Number of BM cells plated | Thymocytes | Number of colonies[1] | | |
|---|---|---|---|---|
| | | Total | Mix/E | Compact |
| $5 \times 10^4$ | – | 13 | 1.0 | <0.1 |
| $5 \times 10^4$ | + | 24 | 1.4 | 1.9 |
| $1 \times 10^5$ | – | 25 | 2.0 | 0.9 |
| $1 \times 10^5$ | + | 30 | 2.6 | 2.1 |
| $2.5 \times 10^5$ | – | 52 | 1.8 | 2.2 |
| $2.5 \times 10^5$ | + | 55 | 2.2 | 3.2 |
| $5 \times 10^5$ | – | 91 | 3.5 | 2.3 |
| $5 \times 10^5$ | + | 87 | 1.3 | 3.3 |

[1]Average number observed on 10 replicate plates.

In addition to increasing the number of compact colonies, the addition of thymocytes results in an increase in the replating potential of primary compact colonies. Table 3 summarizes an experiment in which individual colonies were grown in the presence and absence of thymocytes, picked and replated as described above. It is clear from these data that the addition of thymocytes results in a 3- to 4-fold increase in the number of secondary colonies per primary compact colony, but this increase with added thymocytes is restricted to colonies obtained when small numbers of bone marrow cells are plated to generate primary colonies. It is also important to note that even in the absence of thymocytes, compact colonies replate better than any other type of colony either with or without thymocytes.

Another interesting feature of compact colonies is that they contain multipotent spleen colony-forming stem cells. The presence of CFU-s in compact colonies was detected in the

TABLE 3. Effect of thymocytes on the replating potential of compact colonies

| Number of BM cells in 1° culture | Number of 2° colonies/1° colony[1] | |
|---|---|---|
| | − Thymocytes | + Thymocytes |
| $1 \times 10^5$ | 54 | 245 |
| $2.5 \times 10^5$ | 67 | 195 |
| $5 \times 10^5$ | 54 | 58 |

[1]Compact colonies were picked from plates grown under the conditions shown. Fifteen colonies were pooled and replated. The data are expressed as the number of secondary colonies per primary colony.

following experiment. In vitro colonies were grown from bone marrow cells derived from CBA/T6 donors. At 8 days, a large number of compact colonies were picked and pooled. The equivalent of 15 compact colonies was injected into individual, irradiated (900 rads) CBA recipients. At 10 days, spleen colonies were counted and karyotyped for the presence of the T6 chromosome marker. The recipients had an average of 7 colonies per spleen, and all of 5 colonies examined had the T6 marker, identifying them as being derived from the injected compact colonies. This experiment clearly shows the occurrence in compact colonies of multipotent stem cells.

## DISCUSSION

Because we do not yet know the full differentiative potential of the compact colony-forming units, we cannot place this progenitor in a lineage diagram. On the basis of the in vitro data, it is likely that this colony-forming unit is closely related to a subset of Mix/E colonies described by others (Johnson and Metcalf, 1978). Gregory (1976) has described erythroid colonies that mature at different times in culture. It is possible that if compact colonies could be observed for very long times, in excess of 12 days, they may develop into large, Mix/E colonies. At present, the requirement for thymocytes in the primary culture creates suboptimal conditions such that the colonies deteriorate after 10 days in culture.

Two observations suggest that cell interactions are required to initiate the formation of compact colonies. First, the nonlinearity of the assay for compact-CFU supports this suggestion. Second, the enhancement of compact colonies following the addition of thymocytes to small numbers of bone marrow cells is further support for cell interactions during the initiation and/or proliferation of compact colonies.

The possibility that cell interactions may be involved in the growth of compact colonies, together with the analysis of cells within compact colonies, suggests some interesting possibilities regarding the differentiation and self-renewal of cells within compact colonies. By morphological analysis, the majority of cells within a compact colony are undifferentiated blasts. In some compact colonies, over half of the cells have colony-forming potential on replating. However, because the majority of the secondary colonies are unipotent (Keller, in preparation), one must conclude that the majority of the blasts in the original compact colony are also committed to a specific differentiation pathway. This observation suggests that within the microenvironment of the compact colony, either differentiation signals are not sufficient to induce the differentiation of the committed progenitors or that other stimuli are encouraging self-renewal over differentiation. In other words, it is possible that a pluripotent progenitor can differentiate into a committed granulocyte precursor and that the committed precursor can still extensively self-renew as long as it is in the presence of appropriate self-renewal factors.

## ACKNOWLEDGMENTS

This research was supported by the National Cancer Institute of Canada and the Medical Research Council. G.M.K. is the recipient of an MRC fellowship.

## LITERATURE CITED

Bernstein, S.E. (1970) Tissue transplantation as an analytic and therapeutic tool in the treatment of hereditary anemias. Am. J. Surg., *119:*448–451.

Byron, J. (1978) Histamine $H_2$-receptor and the hematopoietic stem cell (CFU-S). In: Differentiation of Normal and Neoplastic Hematopoietic Cells. B. Clarkson, P.A. Marks, and J.E. Till, eds. Cold Spring Harbor Laboratory, New York, pp. 121–130.

Curry, J.L., and J.J. Trentin (1967) Hemopoietic spleen colony studies. I. Growth and differentiation. Dev. Biol., *15:*395–413.

Frindel, E., E. Leuchars, and A.J.S. Davies (1976) Thymus dependency of bone marrow stem cell proliferation in response to certain antigens. Exp. Hematol., *4:*275–284.

Gregory, C.J. (1976) Erythropoietic sensitivity as a differentiation marker in the hemopoietic system: Studies of three erythropoietic colony responses in culture. J. Cell Physiol., *89:*289–302.

Humphries, R.K., A.C. Eaves, and C.J. Eaves (1981) Self-renewal of hemopoietic stem cells during mixed colony formation in vitro. Proc. Natl. Acad. Sci. U.S.A., *78:*3629–3633.

Humphries, R.K., P.B. Jacky, F.J. Dill, A.C. Eaves, and C.J. Eaves (1979) CFU-S in individual erythroid colonies derived in vitro from adult mouse bone marrow. Nature, *279:*718–720.

Iscove, N.N., L.J. Guilbert, and C. Weyman (1980) Complete replacement of serum in primary cultures of erythropoietin-dependent red cell precursors (CFU-E) by albumin, transferrin, iron, unsaturated fatty acid, lecithin, and cholesterol. Exp. Cell Res., *126:*121–126.

Johnson, G.R., and D. Metcalf (1978) In vitro clonal analysis of fetal hepatic hemopoiesis. In: Differentiation of Normal and Neoplastic Hematopoietic Cells. B. Clarkson, P.A. Marks, and J.E. Till, eds. Cold Spring Harbor Laboratory, New York, pp. 49–62.

Lord, B.I. (1978) Cellular and architectural factors influencing the proliferation of hematopoietic stem cells. In: Differentiation of Normal and Neoplastic Hematopoietic Cells. B. Clarkson, P.A. Marks, and J.E. Till, eds. Cold Spring Harbor Laboratory, New York, pp. 775–788.

McCulloch, E.A., L. Siminovitch, J.E. Till, E.S. Russell, and S.E. Bernstein (1965) The cellular basis of the genetically determined hemopoietic defect in anemic mice of genotype $Sl/Sl^d$. Blood, *26:*399–410.

McLeod, D.L., M.M. Shreeve, and A.A. Axelrad (1976) Induction of megakaryocyte colonies with platelet formation in vitro. Nature, *261:*492–494.

Rosendaal, M., G.S. Hodgson, and T.R. Bradley (1979) Organization of hemopoietic stem cells: The generation-age hypothesis. Cell Tissue Kinet., *12:*17–29.

Salinas, F.A. and J.W. Goodman (1972) Relative effect of thymocytes from irradiated donors on hemopoiesis in P → $F_1$ chimeras. Proc. Soc. Exp. Biol. Med., *140:*439–443.

Sharkis, S.J., W. Wiktor-Jedrzejczak, A. Ahmed, G.W. Santos, A. McKee, and K.W. Sell (1978) Anti-theta sensitive regulatory cell (TSRC) and hematopoiesis: Regulation of differentiation of transplanted stem cells in $W/W^v$ anemic and normal mice. Blood, *42:*802–817.

Till, J.E., and E.A. McCulloch (1961) A direct measurement of the radiation sensitivity of normal mouse bone marrow. Radiat. Res., *14:*213–222.

Till, J.E., and E.A. McCulloch (1980) Hemopoietic stem cell differentiation. Biochim. Biophys. Acta, *605:*431–459.

Till, J.E., E.A. McCulloch, and L. Siminovitch (1964) A stochastic model of stem cell proliferation, based on the growth of colony-forming cells. Proc. Natl. Acad. Sci. U.S.A., *51:*29–36.

Wiktor-Jedrzejczak, W., S. Sharkis, A. Ahmed, K.W. Sell, and G.W. Santos (1977) Theta-sensitive cell and erythropoiesis: Identification of a defect in $W/W^v$ anemic mice. Science, *196:*313–315.

JOURNAL OF CELLULAR PHYSIOLOGY SUPPLEMENT 1:37–43 (1982)

# Cell Lineages in Hematopoietic Neoplasia Studied With Glucose-6-Phosphate Dehydrogenase Cell Markers

PHILIP J. FIALKOW*
*Department of Medicine, University of Washington, Seattle, Washington 98195*

Hematologic stem cells pluripotent for lymphoid and myelogenous cells have not been identified directly in humans. However, several indirect approaches have been used to study this question. For example, the existence of a progenitor pluripotent for lymphoid as well as myeloid cells is suggested by the finding of both cell types in individual, presumably clonal, colonies grown in vitro from human marrow (Messner et al., 1981). The indirect method used in the work reviewed here is based upon study of hematologic neoplasms arising in females whose X-chromosome inactivation mosaicism is delineated with glucose-6-phosphate dehydrogenase (G6PD) markers. Since the G6PD locus is on the X chromosome and undergoes inactivation, only one of the two G6PD genes is active in somatic cells of females. Thus, women with the usual G6PD gene ($Gd^B$) on one X chromosome and the common variant $Gd^A$ on the other X have two populations of cells, one producing A-type G6PD and the other B-type G6PD. The two enzymes can easily be distinguished from one another by starch-gel or cellulose-acetate electrophoresis.

The presence of G6PD-marked cellular mosaicism allows studies of such questions as whether a tumor develops from one or from many cells. For example, if a neoplasm develops from a single cell in a $Gd^B/Gd^A$ heterozygote, it will display either type B or type A G6PD (a single-enzyme phenotype); whereas one arising from multiple cells may show both B and A types (a double-enzyme phenotype). Hierarchal relationships of differentiated cells also may be elucidated with this system. For example, if a tumor has arisen in a pluripotent stem cell of type A, all descendants of that progenitor will be of type A. This paper briefly discusses studies done with G6PD of blood-cell neoplasms with emphasis on work done in the author's laboratory. A more extensive review appears elsewhere (Fialkow, 1980).

## CHRONIC MYELOGENOUS LEUKEMIA

Whereas both B and A types of G6PD were found in nonhematopoietic tissues from 23 women who were heterozygous for G6PD and had Philadelphia chromosome (Ph)-positive chronic myelogenous leukemia (CML), only one type of G6PD was observed in their granulocytes (13 patients had type B and 10 had type A enzyme) (Fialkow et al., 1967, 1977, 1978b, unpublished). Conversely, in $Gd^B/Gd^A$ women without leukemia the granulocytes almost always show both B and A types of G6PD (Fialkow, 1973). The postulate based on these findings that CML is a clonal disease is strongly supported by studies with other isoenzyme and chromosome markers (references in Fialkow, 1980). It is important to emphasize, however, that the statement that CML (or any other disease) is clonal applies only to the stage at which it is studied. Conceivably, many cells were altered initially, but by the time leukemia is evident only one clone has evolved.

The same single-enzyme G6PD phenotypes found in the granulocytes have also been found in erythrocytes, platelets, blood monocytes/macrophages, and eosinophils from patients with CML (Fialkow et al., 1977, 1980; Koeffler et al., 1980). Similarly, Ph has been detected in erythrocyte precursors and in macrophages from such patients (Clein and Flemans, 1966; Rastrick et al., 1968; Golde et al., 1977). Thus, CML involves a multipotent hematologic stem cell. However, marrow fibroblasts do not appear to arise from this CML stem cell (references in Fialkow, 1980).

## LYMPHOCYTES IN CML

Based primarily on the reports that phytohemagglutinin (PHA)-stimulated blood lymphocytes from patients with CML were Ph-

*To whom correspondence/reprint requests should be addressed.

0021-9541/82/SUPPL. 1-0037$02.50 

negative, many people concluded that lymphocytes do not arise from the CML stem cell. G6PD studies indicate that this supposition is correct for at least the majority of circulating T lymphocytes when the disease is in clinical remission (Fialkow et al., 1978a). However, E-rosette-positive lymphocytes studied when the leukemia is under poor control had the same single-enzyme G6PD phenotype as the leukemic granulocytes. Either these cells were T lymphocytes that arose from the CML stem cell or they were other cells of the CML clone that had acquired T-lymphocyte attributes during evolution of the leukemia.

Several types of investigations indicate that cells of the CML clone differentiate into B-lymphocytes. For example, non-E-rosetting lymphocytes from two G6PD heterozygous patients with CML displayed the same single-enzyme phenotype found in the CML granulocytes, indicating that some lymphoid cells arose from the leukemic clone (Fialkow, et al., 1978a). Surface or cytoplasmic immunoglobulin (Ig) was detected in up to 46% of the cells in these non-T-lymphocyte populations, suggesting that cells from the CML clone differentiated into B lymphocytes. To investigate this possibility further, Epstein-Barr virus (EBV)-transformed B-lymphoblastoid cell lines were established from a $Gd^B/Gd^A$ patient with Ph-positive CML in the chronic phase. Chromosomes and G6PD were then studied to determine if progenitor B-lymphoid cells for any of the cell lines originated from the CML clone. In nine of the lines, at least 93% of the cells were Ph-positive (Martin et al., 1980). Each of these lines displayed only B-type G6PD, the same enzyme that was found in the leukemic clone of this patient. In the four lines that were studied, between 33 and 90% of the cells were positive for cytoplasmic Ig. These observations of lymphoblastoid cell lines which synthesize Ig and are Ph-positive and display the same G6PD phenotype as the leukemic clone confirm the supposition reached from studies of this patient's non-E-rosetting lymphocytes that leukemic precursors in patients with CML in the chronic phase are capable of B-lymphoid differentiation.

Studies in mice indicate that the lymphoid and myeloid systems have in common a pluripotent stem cell (Wu et al., 1968; Edwards et al., 1970; Abramson et al., 1977). The G6PD findings in lymphocytes in patients with CML indicate that there are similar stem cells to those in rodents and that CML involves a stem cell pluripotent for lymphocytes as well as for myelogenous cells. It is not known yet whether progeny of the CML stem cell can differentiate into mature Ig-secreting plasma cells.

Of the 74 B-lymphoblastoid cell lines established from the $Gd^B/Gd^A$ patient discussed above, 18 (24%) displayed G6PD type A. Since the CML clone in this patient showed B-type enzyme, the progenitors for at least these 18 lymphoid cell lines originated from nonclonal, presumably normal B-lymphoid cells. As in the case of the nonclonal T-lymphocytes, either these cells were long lived and antedated the development of CML or they arose during the course of the disease from pluripotent stem cells not involved by the leukemia.

In its chronic phase, CML is relatively "benign." However, in patients surviving several years after the disease is first detected, it usually transforms to an acute leukemia-like state, "blast crisis." The occurrence of lymphocytes that arise from the CML stem cell may explain why in approximately one third of cases of blast crisis the leukemic cells have morphologic, enzymatic, immunologic, and other characteristics which resemble those found in the common types of acute lymphoblastic leukemia (Boggs, 1974; McCaffrey et al., 1975; Janossy et al., 1976; Sarin et al., 1976). Less frequently, the cells express cytoplasmic Ig but not surface Ig (i.e., they express the phenotype of "pre-B-lymphoid" cells) (Greaves et al., 1979; Lebien et al., 1979; Vogler et al., 1979). In the three cases evaluated for Ig light chains, only a single type of chain was found in the leukemic blasts. These findings contrast with the Ig diversity found in the Ph-positive lymphoblastoid cell lines established from the patient with CML in the chronic phase (Martin et al., 1980) (i.e., classes of Ig synthesized by the cell lines differed from line to line). This observation suggests that Ig restriction is not fixed in stem cells in the CML clone, and that these cells may be responsive to the induction factors which normally regulate B-lymphoid differentiation. Furthermore, the data suggest that Ig restriction occurs at some stage in differentiation between the pluripotent stem cell involved in the chronic phase of CML and the more differentiated blast cell present in the acute phase of the disease. The results also support the suggestion based on chromosome findings that a single CML subclone evolves to the blastic transformation in many patients.

## SIGNIFICANCE OF Ph

Studies of G6PD and chromosomes in B-lymphoblastoid cell lines established from the patient with Ph-positive leukemia who was heterozygous for G6PD provide evidence for the existence of cells derived from the abnormal clone that are Ph-negative (Fialkow et al., 1981b). Thus, acquisition of Ph may not be a sufficient cause for the disease. It is possible that at least two steps are involved in the pathogenesis of Ph-positive CML. In an early step an abnormality occurs which leads to proliferation of a clone of genetically unstable pluripotent hematopoietic stem cells. The genetic instability predisposes to the development of Ph that occurs in a later step. Acquisition of Ph in a cell(s) leads to a further growth advantage and to the manifestations we recognize as being CML (Fialkow et al., 1981b).

## AGNOGENIC MYELOID METAPLASIA WITH MYELOFIBROSIS

This disorder is characterized by fibrosis in the marrow with consequent marrow failure and blood formation in the spleen and other organs without an obvious underlying disease such as CML or polycythemia vera. The basic abnormality is unknown; the disorder is often classified with CML and polycythemia vera as a myeloproliferative neoplasm, although the putative neoplastic cell has not been identified. However, some investigators feel that agnogenic myeloid metaplasia is a generalized hyperplasia in response to an unknown stimulant. Results of studies with G6PD in four patients indicate that at least in these patients the disorder is clonal and involves a hematologic stem cell multipotent for granulocytes, erythrocytes, and platelets (Kahn et al., 1975; Jacobson et al., 1978; Fialkow et al., unpublished). The clonal development of agnogenic myeloid metaplasia cannot be explained simply if the disease results from a generalized hyperplasia.

The cause of marrow fibrosis, the predominant clinical feature of agnogenic myeloid metaplasia is not known; most investigators previously considered it to be a part of the same process that affected the blood cells. G6PD findings suggest that this is not a correct interpretation. Marrow fibroblasts have not been tested directly, but the observations in two patients that such cells grown in culture had a normal double-enzyme G6PD phenotype strongly suggest that the myelofibrosis is a secondary abnormality (Jacobson et al., 1978; Fialkow et al., unpublished). Similar conclusions were reached in marrow fibrosis in CML (Fialkow et al., 1977) and in a patient with acute myelofibrosis (Van Slyck et al., 1970). The marrow fibrosis may represent a secondary response to marrow injury or to stimulatory factors produced by the abnormal hematologic cells.

## OTHER MYELOPROLIFERATIVE DISORDERS

G6PD studies in two patients with polycythemia vera (Adamson et al., 1976), in one patient with Ph-negative CML (Fialkow et al., 1980), and in three patients with essential thrombocythemia (Fialkow et al., 1981a) indicate that the disease in these patients involved stem cells that are multipotent for granulocytes, erythrocytes, and platelets and the results also suggest that they are clonal disorders. Detailed studies of lymphocytes have not been reported in patients with these myeloproliferative disorders.

## ACUTE NONLYMPHOCYTIC LEUKEMIA

Eight G6PD heterozygotes with acute nonlymphocytic leukemia (ANLL) have been studied (Wiggans et al., 1978; Fialkow et al., 1979, 1981c, unpublished). In each case, only a single-enzyme G6PD phenotype was found in the leukemic blast cells, suggesting that the disorders in these patients were clonal. In the patients who were studied when their disease entered remission, there was a return to a normal double-enzyme phenotype in granulocytic cells. These observations support the suggestion based on chromosomal studies that during remission of ANLL the marrow is repopulated by normal stem cells, and they are in contrast with the findings in CML of persistence of single-enzyme G6PD phenotypes throughout clinical remission induced by conventional chemotherapy (Fialkow et al., 1977, 1978b). However, a return to a double-enzyme phenotype was observed in granulocytic cells of a patient with CML treated with intensive chemotherapy (Singer et al., 1980a). The observations of double-enzyme phenotypes in remission of ANLL or CML indicate that normal granulopoietic potential and, therefore, normal stem cells were present but not expressed during the acute phases of the diseases.

The question of whether ANLL involves a multipotent stem cell or a progenitor restricted to differentiate along the granulocytic pathway was evaluated in six patients. Two of them were elderly women and a single-enzyme phenotype was found not only in the blast cells, but also in the erythrocytes and platelets. Thus, in these two older patients, the abnormal clone arose in a stem cell multipotent for red cells, granulocytes, and platelets. In contrast, findings in erythroid cells in the four younger patients with ANLL indicate that the leukemia is expressed in cells with restricted differentiative ability. Nucleated erythroid cells from marrow displayed a double-enzyme phenotype and both variant enzyme and B-type erythroid bursts were detected in cultures of peripheral blood mononuclear cells (Fialkow et al., 1979, 1981c, unpublished). Thus, these erythroid cells arose from presumably normal stem cells. The observations indicate that in these younger people, the ANLL is expressed in cells with restricted differentiative ability. Two mechanisms could be responsible for this restriction: The progenitors for the leukemia were multipotent stem cells in which erythroid differentiation was suppressed, or they were cells analogous to granulocyte colony-forming units (i.e., progenitors restricted to granulocytic/monocytic differentiation).

The difference between the young and elderly patients in pathogenesis of ANLL may be reflected in the age-related variations in clinical features. The G6PD test results confirm the suggestion based on chromosomal studies that ANLL can arise in a multipotent stem cell in some patients (Jensen and Killman, 1971; Blackstock and Garson, 1974). In both instances, the patients on whom this suggestion was based were elderly and may well have had preleukemic syndromes. In fact, in the cytogenetic studies, the patients were selected for chromosome evaluation of erythroid cells because their marrows showed more than the usual degree of erythroid proliferation. In addition, at least one of these patients showed sideroblastic changes, suggesting that the disease may have evolved from a preleukemic stage.

It can be speculated that the heterogeneity in stem cell involvement in patients with ANLL reflects etiologic differences and can also have clinical implications. Patients with marked erythroid hyperplasia, a preleukemic syndrome, or exposure in adult life to an oncogen may develop ANLL different from that seen in many children and younger adults. More G6PD heterozygotes with ANLL must be studied to determine the relative frequency of origin and expression in progenitors with restricted, as compared to multipotent, differentiation as well as the relationship of pattern of cell involvement to other findings such as cell type and cytogenetics. The results could be important for the development of classifications of ANLL that provide a greater degree of guidance for therapy and prognosis than is given by currently described schemes.

## CHRONIC LYMPHOCYTIC LEUKEMIA

The proliferating cells in the majority of patients with this leukemia are B-lymphocytes that have monoclonal surface Ig restricted to one light and one heavy chain class and to one idiotype (references in Fialkow, 1980). The fact that in each case of chronic lymphocytic leukemia the billions of neoplastic B-lymphocytes generally all synthesize a single molecular species of Ig suggests that at the time of diagnosis all the CLL cells are clonal. This supposition has been confirmed with G6PD (Fialkow et al., 1978c). These studies have also shown that granulocytes, erythrocytes, platelets, and at least some T lymphocytes do not arise from the leukemic progenitor cells. Thus, CLL is expressed in progenitors with differentiative expression restricted to the B-lymphocyte pathway and the disease stands in contrast to chronic myeloproliferative syndromes such as CML or polycythemia vera, all of which involve multipotent stem cells.

## APPLICATIONS OF THE G6PD SYSTEM TO THE STUDY OF HEMATOLOGIC CELL REGULATION IN NEOPLASIA

The G6PD system has been used to demonstrate the persistence of residual but suppressed normal stem cells in patients with polycythemia vera or CML (Prchal et al., 1978; Singer et al., 1980a). The system has also been used to delineate abnormalities in hematopoietic regulation in the two diseases (Prchal et al., 1978; Singer et al., 1979; Adamson et al., 1980; Singer et al., 1980b, 1981). The results of these G6PD studies suggest that polycythemia vera is a slowly progressive neoplasm in which the growth of normal committed stem cells is suppressed. Currently detectable chromosomal abnormalities are not required as the immediate cause of the clonal proliferation of the disease. In vitro studies suggest that maturation of normal erythroid elements is inhibited at a stage between BFU-E and CFU-E progenitors. The studies have also delineated a mechanism(s) to explain the suppression of normal granulopoiesis in patients with polycythemia vera: CFU-C from the polycythemia vera clone are less responsive to physiologic feedback control than are their normal counterparts and members of the abnormal clone may elaborate a substance which inhibits the growth of normal CFU-C. The clinical phenotype appears to depend upon differences in the regulatory control of cells that appear later in the maturation process than CFU-C and BFU-E. These differences therefore probably lie in processes that lead to amplification of the numbers of the mature progeny. And finally, CFU-C from the polycythemia vera clone can be physically separated from their normal counterparts. Some of these findings have implications not only for the pathogenesis of the disease but also for its therapy. Thus, determination of enzyme phenotypes of hematopoietic-cell colonies derived from G6PD heterozygotes with myeloproliferative disorders provides information about the balance between abnormal and normal cell lines in these diseases. Perhaps this system can be used for preliminary evaluation of therapeutic modalities designed to selectively inhibit abnormal stem cells or to selectively stimulate normal progenitors.

The studies have revealed notable differences between polycythemia vera and CML. It is more difficult to demonstrate residual normal stem cells in vitro in CML. Thus far, it has not been shown that CFU-C from the abnormal clone are less responsive to physiologic feedback control than their normal counterparts. Furthermore, CFU-C from the leukemic clone have not been physically separated from normal CFU-C.

## POTENTIAL APPLICATIONS OF THE G6PD SYSTEM TO STUDY OF LONG-TERM HEMATOLOGIC CELL CULTURES

The technique for long-term culture of marrow cells in vitro described by Dexter (1977) has been applied to human marrow (Gartner and Kaplan, 1980). A feature necessary to the maintenance of long-term growth of hematologic cells in this system is the development of an adherent stromal layer. It is hoped that G6PD assays of hematologic cells grown in this system will provide the opportunity to define better the level at which residual normal stem cells are suppressed in patients with clonal disorders like CML or polycythemia vera, the mechanisms of suppression, the nature of the growth advantage of neoplastic clones, and the changes which occur with therapy and progression of disease (e.g., loss of normal committed stem cells, of pluripotent progenitors, or both).

Some preliminary experiments have been done. For example, cells that grew in suspension in long-term marrow cell cultures from a G6PD heterozygote with ANLL and granulocytic colonies grown from these cells manifested the same single-enzyme G6PD phenotype found on direct assay of the leukemic blast cells. These findings indicate that progeny of the patient's malignant cells were maintained in the long-term marrow culture system. In contrast to the single-enzyme phenotype of the cells grown in suspension, a double-enzyme phenotype was found in the stromal layer cells, indicating the latter's origin from a progenitor not involved by the leukemia.

Other studies of the origin of the stromal layer in Dexter-type cultures of human marrow cells suggest a common stem cell for hematopoietic and microenvironmental cells (Keating et al., 1982). Long-term marrow-cell cultures were grown from 12 patients who received marrow grafts from opposite-sex siblings. The presence or absence of a Y chromosome in the stromal cells was used to determine the donor or host-cell origin of the stromal layer in the long-term cultures. By three months after transplantation, virtually all of the stromal cells were donor in origin. This is in contrast to marrow fibroblasts which other workers have shown to be of host origin (Wilson et al., 1978; Golde et al., 1980).

Three lines of evidence suggested that the stromal cells in the long-term marrow cultures were composed in part of endothelial-like cells: 1) electromicroscopic studies revealed the presence of junctional complexes between stromal cells, a feature more in keeping with endothelial cells than with fibroblasts; 2) approximately 25% of stromal cells manifested factor VIII associated antigen, a feature thought to be specific for endothelial cells; 3) analysis of the culture medium demonstrated interstitial and basal lamina collagen types, some of which are known to be synthesized by endothelial cells (Striker et al., 1980). Thus, the results indicate that the stromal microenvironment in long-term, Dexter-like cultures of human marrow cells is transplantable and consists in part of endothelial-like cells. The data are consistent with the hypothesis that marrow hematologic cells and microenvironmental cells have a common progenitor stem cell. This hypothesis can be tested by examining stromal-cell layers from patients with clonal disorders of pluripotent stem cells such as CML.

## ACKNOWLEDGMENTS

Work done in the author's laboratory was supported by grants CA 16448 and GM 15253 from the National Cancer Institute and the National Institute of General Medical Sciences, respectively, National Institutes of Health, and by the Veterans Administration.

## LITERATURE CITED

Abramson, S., R.G. Miller and R.A. Phillips (1977) The identification in adult bone marrow of pluripotent and restricted stem cells of the myeloid and lymphoid system. J. Exp. Med., *145:*1567–1579.

Adamson, J.W., P.J. Fialkow, S. Murphy, J.F. Prchal, and L. Steinmann (1976) Polycythemia vera: Stem-cell and probable clonal origin of the disease. N. Engl. J. Med., *295:*913–916.

Adamson, J.W., J.W. Singer, P. Catalana, S. Murphy, N. Lin, L. Steinmann, C. Ernst, and P.J. Fialkow (1980) Polycythemia vera: Further in vitro studies of hematopoietic regulation. J. Clin. Invest., *66:*1363–1368.

Blackstock, A.M., and C.M. Garson (1974) Direct evidence for involvement of erythroid cells in acute myeloblastic leukaemia. Lancet, *2:*1178.

Boggs, D.R. (1974) Hematopoietic stem cell theory in relation to possible lymphoblastic conversion of chronic myeloid leukemia. Blood, *44:*449–453.

Clein, G.P., and R.J. Flemans (1966) Involvement of the erythroid series in blastic crisis of chronic myeloid leukemia: Further evidence for the presence of Philadelphia chromosome in erythroblasts. Br. J. Haematol., *12:*754–758.

Dexter, T.M., T.D. Allen, and L.G. Lajtha (1977) Conditions controlling the proliferation of haemopoietic stem cells in vitro. J. Cell Physiol. *91:*335–344.

Edwards, G.E., R.G. Miller, and R.A. Phillips (1970) Differentiation of rosette-forming cells from myeloid stem cells. J. Immunol. *105:*719–729.

Fialkow, P.J. (1973) Primordial cell pool size and lineage relationships of five human cell types. Ann. Hum. Genet., *37:*39–48.

Fialkow, P.J. (1980) Clonal and stem cell origin of blood cell neoplasms. In Lobue, J., Gordon, A.S., Silber, R., and Muggia, F.M., eds.: Contemporary Hematology and Oncology. New York: Plenum Press, Vol. 1, pp. 1–46.

Fialkow, P.J., A.M. Denman, R.J. Jacobson, and M.N. Lowenthal (1978a) Chronic myelocytic leukemia: origin of some lymphocytes from leukemic stem cells. J. Clin. Invest., *62:*815–823.

Fialkow, P.J., A.M. Denman, J.W. Singer, R.J. Jacobson, and M.N. Lowenthal (1978b) Human myeloproliferative disorders: Clonal origin in pluripotent stem cells. In Clarkson, B., Marks, P.A., and Till, J.E., eds.: Differentiation of Normal and Neoplastic Hematopoietic Cells. Cold Spring Harbor, New York: Cold Spring Harbor Laboratory, Vol. 5, pp. 131–144.

Fialkow, P.J., G.B. Faguet, R.J. Jacobson, K. Vaidya, and S. Murphy (1981a) Evidence that essential thrombocythemia is a clonal disorder with origin in a multipotent stem cell. Blood *58:*916–919.

Fialkow, P.J., S.M. Gartler, and A. Yoshida (1967) Clonal origin of chronic myelocytic leukemia in man. Proc. Natl. Acad. Sci. U.S.A., *58:*1468–1471.

Fialkow, P.J., R.J. Jacobson, and T. Papayannopoulou (1977) Chronic myelocytic leukemia: Clonal origin in a stem cell common to the granulocyte, erythrocyte, platelet and monocyte/macrophage. Am. J. Med., *63:*125–131.

Fialkow, P.J., R.J. Jacobson, J.W. Singer, R.A. Sacher, R.W. McGuffin, and J.R. Neefe (1980) Philadelphia chromosome ($Ph^1$)-negative chronic myelogenous leukemia (CML): A clonal disease with origin in a multipotent stem cell. Blood, *56:*70–73.

Fialkow, P.J., P.J. Martin, V. Najfeld, G.K. Penfold, R.J. Jacobson, and J.A. Hansen (1981b) Evidence for a multistep pathogenesis of chronic myelogenous leukemia. Blood, *58:*158–163.

Fialkow, P.J., V. Najfeld, A.L. Reddy, J. Singer, and L. Steinmann (1978c) Chronic lymphocytic leukaemia: Clonal origin in a committed B-lymphocyte progenitor. Lancet, *2:*444–446.

Fialkow, P.J., J.W. Singer, J.W. Adamson, R.L. Berkow, J.M. Friedman, R.J. Jacobson, and J.W. Moohr (1979) Acute nonlymphocytic leukemia: Expression in cells restricted to granulocytic and monocytic differentiation. N. Engl. J. Med., *301:*1–5.

Fialkow, P.J., J.W. Singer, J.W. Adamson, K. Vaidya, L.W. Dow, J. Ochs, and J.W. Moohr (1981c) Acute nonlymphocytic leukemia: Heterogeneity of stem cell origin. Blood, *57:*1068–1073.

Gartner, S., and H.S. Kaplan (1980) Long-term culture of human bone marrow cells. Proc. Natl. Acad. Sci. U.S.A., *77:*4756–4759.

Golde, D.W., C. Burgaleta, R.S. Sparkes, and M.J. Cline (1977) The Philadelphia chromosome in human macrophages. Blood, *49:*367–370.

Golde, D.W., W.G. Hocking, S.G. Quan, R.S. Sparkes, and R.P. Gale (1980) Origin of human bone marrow fibroblasts. Br. J. Haematol. *44:*183–187.

Greaves, M.F., W. Verbi, B.R. Reeves, A.V. Hoffbrand, H.C. Drysdale, L. Jones, L.S. Sacker, I. Somaratunga (1979) "Pre-B" phenotypes in blast crisis of $Ph^1$ positive (CML): Evidence for a pluripotential stem cell "target." Leuk. Res. *3:*181–191.

Jacobson, R.J., A. Salo, and P.J. Fialkow (1978) Agnogenic myeloid metaplasia: A clonal proliferation of hematopoietic stem cells with secondary myelofibrosis. Blood, *51:*189–194.

Janossy, G., M. Roberts, and M.F. Greaves (1976) Target cell in chronic myeloid leukemia and its relationship to acute lymphoid leukemia. Lancet, *2:*1058–1060.

Jensen, M.K., and S-A. Killmann (1971) Additional evidence for chromosome abnormalities in the erythroid precursors in acute leukaemia. Acta. Med. Scand., *189:*97–100.

Kahn, A., J.F. Bernard, D. Cottriau, J. Marie, and P. Boivin (1975) Gd(–) Abrami: A deficient G-6-PD variant with hemizygous expression in blood cells of a woman with primary myelofibrosis. Humangenetik, *30:*41–46.

Keating, A., J.W. Singer, G. Striker, P. Killen, W.H. Raskind, A.C. Salo, J. Sanders, E.D. Thomas, D. Thorning, and P.J. Fialkow (1982) In vitro evidence for donor origin of the hematopoietic microenvironment following marrow transplantation in man. (submitted for publication.)

Koeffler, H.P., A.M. Levine, M. Sparkes, R.S. Sparkes (1980) Chronic myelocytic leukemia: Eosinophils involved in the malignant clone. Blood, *55:*1063–1065.

LeBien, T.W., J. Hozier, J. Minowada, and J.H. Kersey (1979) Origin of chronic myelocytic leukemia in a precursor for pre-B lymphocytes. N. Engl. J. Med., *301:*144–147.

Martin, P.J., V. Najfeld, J.A. Hansen, G.K. Penfold, R.J. Jacobson, and P.J. Fialkow (1980) Involvement of the B-lymphoid system in chronic myelogenous leukaemia. Nature, *287:*49–50.

McCaffrey, R., T.A. Harrison, R. Parkman, and D. Baltimore (1975) Terminal deoxynucleotidyl transferase activity in human leukemic cells and in normal human thymocytes. N. Engl. J. Med., *292:*775–780.

Messner, H.A., C.A. Izaguirre, and N. Jamal (1981) Identification of T lymphocytes in human mixed hemopoietic colonies. Blood, *2:*402–405.

Prchal, J.F., J.W. Adamson, S. Murphy, L. Steinmann, and P.J. Fialkow (1978) Polycythemia vera: The in vitro response of normal and abnormal stem cell lines to erythropoietin. J. Clin. Invest. *61:*1044–1047.

Rastrick, J.M., P.H. Fitzgerald, and F.W. Gunz (1968) Direct evidence for presence of Ph[1] chromosome in erythroid cells. Br. Med. J., *1:*96–98.

Sarin, P.S., P.N. Anderson, and R.C. Gallo (1976) Terminal deoxynucleotidyl transferase activities in human blood leukocytes and lymphoblast cell lines: High levels in lymphoblast cell lines and in blast cells of some patients with chronic myelogenous leukemia in acute phase. Blood, *47:*11–20.

Singer, J.W., J.W. Adamson, Z.A. Arlin, S.J. Kempin, B.D. Clarkson, and P.J. Fialkow (1981) Chronic myelogenous leukemia: In vitro studies of hematopoietic regulation on a patient undergoing intensive chemotherapy. J. Clin. Invest., *67:*1593–1598.

Singer, J.W., J.W. Adamson, C. Ernst, N. Lin, L. Steinmann, S. Murphy, and P.J. Fialkow (1980b) Polycythemia vera: Physical separation of normal and neoplastic committed granulocyte-macrophage progenitors. J. Clin. Invest., *66:*730–735.

Singer, J.W., Z.A. Arlin, V. Najfeld, J.W. Adamson, S.J. Kempin, B.D. Clarkson, and P.J. Fialkow (1980a) Restoration of nonclonal hematopoiesis in chronic mylogenous leukemia (CML) following a chemotherapy-induced loss of the Ph[1] chromosome. Blood, *56:*356–360.

Singer, J.W., P.J. Fialkow, J.W. Adamson, L. Steinmann, C. Ernst, A. Murphy, and K.J. Kopecky (1979) Polycythemia vera: Increased expression of normal committed granulocytic stem cells in vitro after exposure of marrow to tritiated thymidine. J. Clin. Invest., *64:*1320–1324.

Striker, G.E., J.M. Harlan, and S.M. Schwartz (1980) Human endothelial cells in vitro. Methods Cell Biol. *21A:*135–151.

Van Slyck, E.J., L. Weiss, and M. Dully (1970) Chromosomal evidence for the secondary role of fibroblastic proliferation in acute myelofibrosis. Blood, *36:*729–735.

Vogler, L.B., W.M. Christ, A.S. Vinson, M.G. Brattain, and M.S. Coleman (1979) Philadelphia-chromosome-positive pre-B-cell leukemia presenting as blast crisis of chronic myelogenous leukemia. Blood, *54:*1164–1170.

Wiggans, R.G., R.J. Jacobson, P.J. Fialkow, P.V. Woolley, J.S. Macdonald, and P.S. Schein (1978) Probable clonal origin of acute myeloblastic leukemia following radiation and chemotherapy of colon cancer. Blood, *52:*659–663.

Wilson, F.D., B.R. Greenberg, N.P. Konrad, A.K. Klein, and P.A. Walling (1978) Cytogenetic studies on bone marrow fibroblasts from a male-female hematopoietic chimera: Evidence that stromal elements in human transplantation recipients are of host type. Transplantation, *25:*87–88.

Wu, A.M., J.E. Till, L. Siminovitch, and E.A. McCulloch (1968) Cytological evidence for a relationship between normal and hematopoietic colony-forming cells and cells of the lymphoid system. J. Exp. Med., *127:*455–464.

JOURNAL OF CELLULAR PHYSIOLOGY SUPPLEMENT 1:45–51 (1982)

# The Growth of Large Megakaryocyte Colonies From Human Bone Marrow

H. A. MESSNER,* N. JAMAL, AND C. IZAGUIRRE
*The Ontario Cancer Institute and Institute of Medical Science, University of Toronto, Toronto, Ontario M4X 1K9, Canada*

The growth of large, compact megakaryocyte colonies in cultures of human bone marrow is promoted by fresh human plasma and medium conditioned by phyto-hemagglutinin stimulated leukocytes (PHA-LCM). These colonies are typically composed of large cells with translucent cytoplasma, surrounded by a highly refractile border. In addition, they may also contain smaller cells of similar morphology. Independent of their size, all cells react positively with antibodies directed against human factor VIII antigen. The frequency of megakaryocyte colonies may vary for different individuals from 1–35 colonies per $10^5$ mononuclear bone marrow cells. The observed linear relationships between the number of cultured cells and the frequency of colonies suggests a single cell origin. The described culture conditions also support the development of a larger megakaryocyte component within multilineage mixed colonies, so that it will now be feasible to investigate the mechanisms involved in directing pluripotent cells towards megakaryocytopoiesis.

Advances in cell culture technology have provided tools to study control mechanisms that govern human pluripotent hemopoietic progenitors (1–4) as well as precursors that are restricted to one lineage of hemopoietic differentiation (5–7). The most recent addition to these colony assays was reported for human megakaryocyte progenitors (CFU-M) present in peripheral blood and bone marrow (8–12). Megakaryocyte colonies grown in plasma clot are usually small and disperse containing up to 30 cells. Their recognition was difficult and improved only recently since antibodies directed against platelet glycoproteins were used for their detection by immunofluorescence microscopy (10–13).

We are now reporting improved culture conditions that promote growth of large, compact megakaryocyte colonies and enhance the frequency of megakaryocytes within multilineage colonies. Megakaryocytes grown under these conditions reacted positively with antibodies raised in rabbits against human factor VIII antigen as previously described for direct preparations for human megakaryocytes (13,14) and cultured megakaryocytes of other species (13,15).

## MATERIALS AND METHODS

### Preparation of cells and culture conditions

Bone marrow samples from 26 consenting normal volunteers were aspirated into heparinized syringes. The specimens were centrifuged with 60% Percoll to obtain a mononuclear cell suspension of density less than 1.077 gm/ml. These cells were routinely suspended in a mixture of Iscove's modified Dulbecco's MEM (IDMEM, Gibco, Grand Island, NY), 5 $\times$ $10^{-5}$ M 2-mercaptoethanol, 30% fresh human plasma, 5% of a medium conditioned by peripheral leukocytes (PHA-LCM) (16) of a patient with hemochromatosis in the presence of phytohemagglutinin (PHA-HA 15 Wellcome), and methylcellulose at a final concentration of 0.9%.

Human plasma was collected under two different conditions. Whenever possible, fresh human plasma was obtained using heparin as anticoagulant at a concentration of 2,000 units per 50 cc of peripheral blood. Plasma samples originally anticoagulated with ACD were sup-

*To whom correspondence/reprint requests should be addressed.

plemented with 4,000 units of heparin for each 100 ml.

The influence of fetal calf serum (FCS), combinations of FCS and human plasma, and human serum on colony growth was assessed in parallel control experiments. In some studies, sheep erythropoietin (EPO, Connaught Laboratories, Willowdale, Ontario) was added in increasing concentrations up to 4 units per culture plate.

One milliliter aliquots of this culture mixture were placed into 35 mm Petri dishes (Lux) and incubated for 12 to 24 days at 37°C in humidified atmosphere supplemented with 5% $CO_2$. The plates were examined by inverted microscopy for megakaryocyte colonies and colonies of other cell lineages.

### Preparation of E-rosette depleted bone marrow populations

Bone marrow cell suspensions were incubated with AET-treated sheep red blood cells (Woodlyn Laboratories, Kitchener, Ontario) at 4°C for 1 hour (17). E-rosette forming cells were subsequently removed by centrifugation in 60% Percoll.

### Preparation for cytological assessment and examination by immunofluorescence

Colonies were individually removed from the plates using a micropipette. They were placed on glass slides and gently blown to spread cells adequately for examination. Slides were then fixed in cold acetone for 10 minutes, washed in phosphate buffered saline (PBS), and either stained with Wright stain or further prepared for immunofluorescence. The cells were exposed for 30 minutes at room temperature to antibodies (dilution 1:30) raised in rabbits against human factor VIII antigen (DAKO-immunoglobulins, Copenhagen, Denmark). The slides were washed extensively in PBS, stained for 30 minutes with affinity purified goat antirabbit IgG (dilution 1:30) conjugated to FITC (TAGO, Burlingame, CA), and carefully washed in PBS prior to mounting with a cover glass using PBS-glycerol (1:9) and sealing with nail polish. The slides were analyzed with a suitably equipped Zeiss Universal fluorescence microscope.

## RESULTS

The addition of fresh human plasma to bone marrow cultures increased the viscous support provided by methylcellulose to a semisolid state and reproducibly supported the growth of megakaryocyte colonies. Using an inverted microscope of high optical quality, they can readily be distinguished from multilineage colonies, erythroid bursts, and granulocytic colonies present in the same culture plates. Megakaryocyte colonies are compact and may contain from 20–500 cells (Fig. 2). These cells are usually large with translucent, hyaline cytoplasma and a distinct cell border of high refractility. In addition, one can identify smaller cells of similar morphology among the large cells.

Colonies of this phenotype can be recognized after approximately 10 days of culture. They usually increase in size as well as in proportion of large cells for further 6 to 8 days and may persist subsequently up to day 24 of culture.

The size and typical morphology of individual cells facilitates discrimination from granulocytic colonies and erythroid bursts. They are also distinguishable from macrophages that may have similar size but display in contrast a typical yellowish-brownish and granular cytoplasma.

### Frequency of megakaryocyte colonies

Twenty-six normal bone marrow samples were examined for the presence of these colonies. They were identified in 23 specimens at a frequency that varied from 1–35 per $1 \times 10^5$ mononuclear cells.

The assay was found to be quantitative with extrapolation of the regression line through zero (Fig. 3). The frequency of megakaryocyte colonies was dependent upon addition of fresh human plasma, PHA-LCM, and EPO.

### Influence of human plasma

Human serum or FCS supported the development of a small number of megakaryocyte colonies. The plating efficiency was significantly enhanced with increasing concentrations of fresh human plasma. Dose response curves comparing human plasma and serum obtained from a single donor with FCS are depicted in Figure 4. Maximum plating efficiency was obtained at a concentration of 30% human plasma. Further addition of plasma or FCS in various concentrations did not enhance colony formation. The stimulatory activity present in human plasma was maintained after heat inactivation but declined with prolonged storage at 4°C, thus requiring storage at −20°C. The stimulatory activity observed in individual plasma samples was also identified in pools of AB and O plasma.

In addition to growth support of megakaryocyte colonies, human plasma promoted the development of granulocytic colonies, erythroid bursts, and multilineage mixed hemopoietic colonies.

### Influence of PHA-LCM

Megakaryocyte colonies were occasionally observed without added PHA-LCM. Addition of PHA-LCM regularly increased the plating efficiency and size of megakaryocytic colonies.

The optimal concentration of PHA-LCM was dependent upon the composition of the target cell suspensions (Fig. 5). Mononuclear bone marrow cells grew optimally at 5% PHA-LCM; the plating efficiency was rapidly reduced with

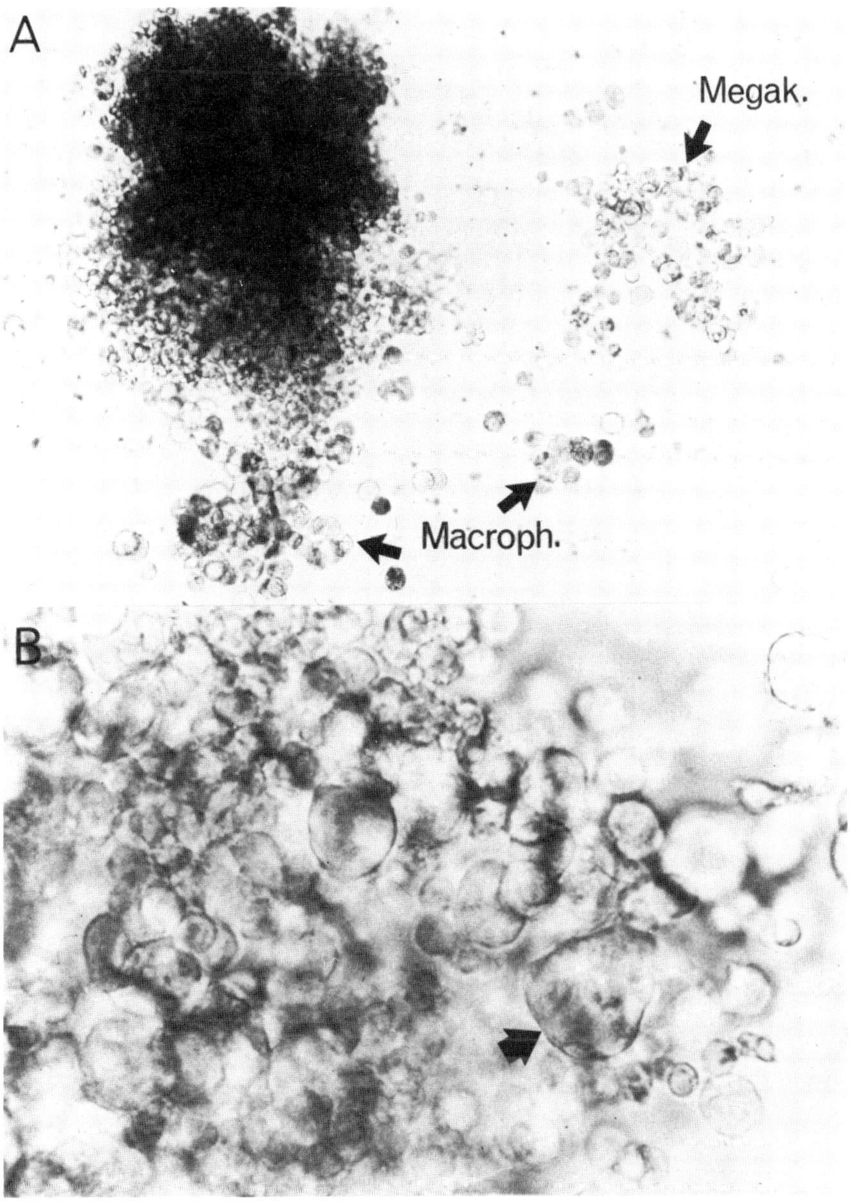


Fig. 1. (A) Picture of part of a culture plate containing a multilineage colony, a megakaryocyte colony, and clusters of macrophages. (B) Enlargement of the right margin of the multilineage colony. The arrow ind cates an example of a large, translucent megakaryocyte ai.iong cells of small size.

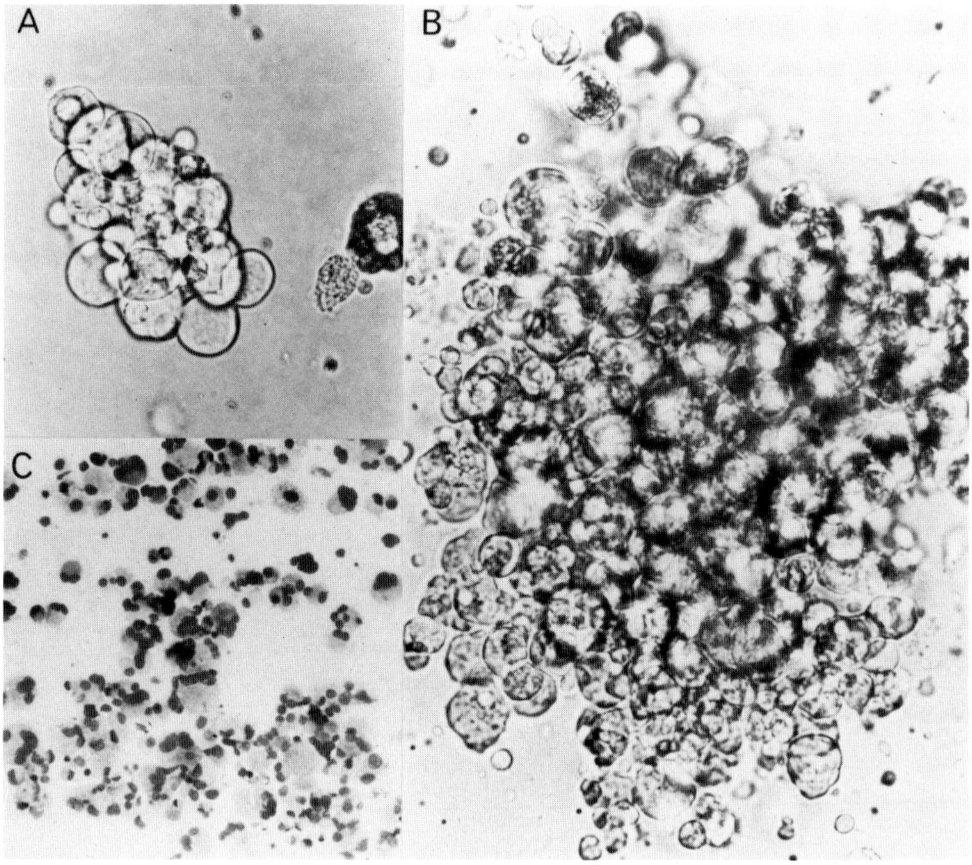


Fig. 2. (A) Small megakaryocyte colony containing approximately 20 cells of heterogeneous size with typical morphology. (B) Large megakaryocyte colonies containing in excess of 500 cells. (C) Wright-stain preparation of a large megakaryocyte colony. Cells with varying degrees of nuclear segmentation can be observed.

increasing concentrations of PHA-LCM. Mononuclear bone marrow cells that were also depleted of E-rosette forming cells usually maintained optimal growth at higher PHA-LCM concentrations.

## Influence of erythropoietin

Twelve of the twenty-six samples were plated without and with 1 unit of EPO. Pure megakaryocyte colonies developed consistantly at lower frequency in cultures that contained EPO (Fig. 6). Increasing concentrations of EPO appeared to further reduce the frequency of megakaryocyte colonies (Table 1). However, the total number of mixed and megakaryocyte colonies observed in individual plates that contained increasing concentrations of erythorpoietin remained very similar.

## Megakaryocytes in multilineage colonies

Previously described culture conditions promoted growth of multilineage colonies (2,4) that usually contained a small number of megakaryocytes. The use of human plasma increased the megakaryocytic component within mixed colonies, and colonies containing several hundred megakaryocytes can now readily be observed. Detection of megakaryocytes in mixed colonies is now usually feasible by direct observation with an inverted microscope (Fig. 1).

Two types of mixed colonies can be recognized. In the presence of EPO, colonies may contain erythroid and granulocytic cells as well as megakaryocytes. Some mixed colonies are only composed of erythroblasts and megakaryocytes or granulocytes and megakary-

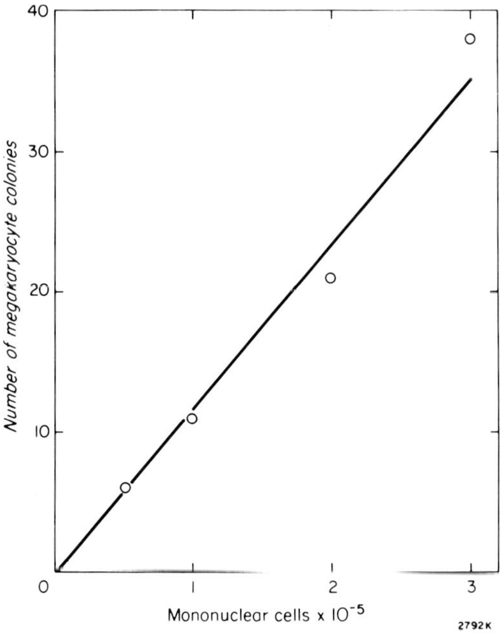


Fig. 3. Plating efficiency of CFU-M present in increasing numbers of plated mononuclear bone marrow cells. A linear relationship is observed.

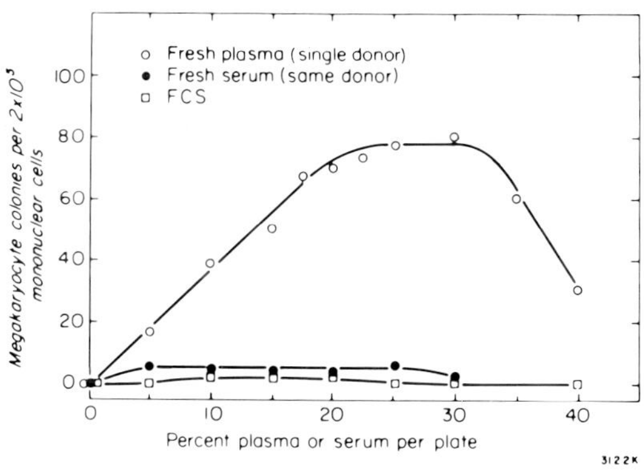


Fig. 4. Formation of megakaryocyte colonies in the presence of increasing concentrations of fresh human plasma obtained from a single donor, fresh serum from the same individual, and FCS.

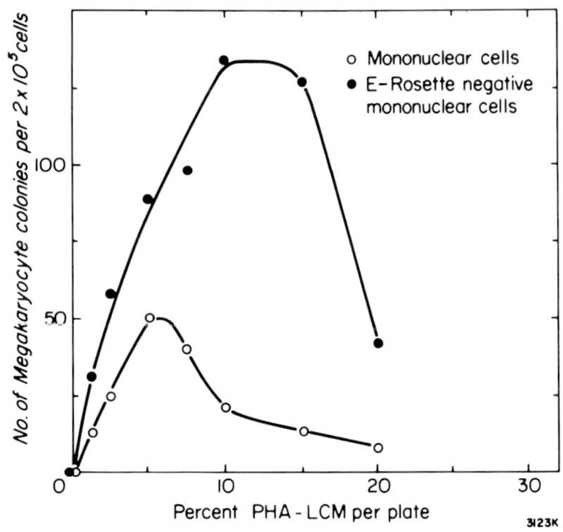


Fig. 5. Influence of increasing concentrations of PHA-LCM on the formation of megakaryocyte colonies grown from mononuclear bone marrow cells and from the same cell suspension after depletion of E-rosette forming cells.

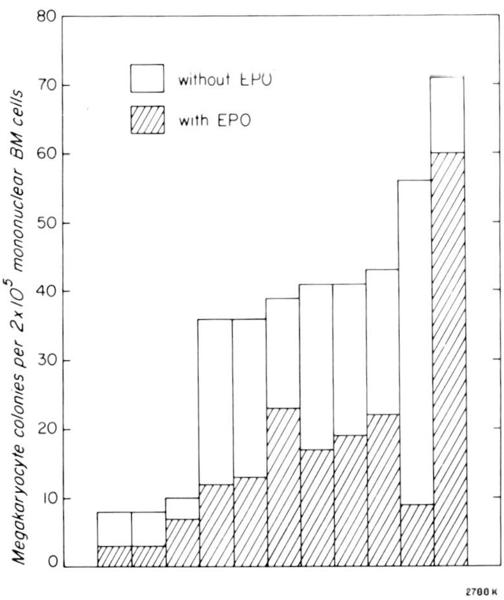


Fig. 6. Frequency of megakaryocyte colonies with and without EPO (represented by the total column).

ocytes (Fig. 1). The latter association is most commonly observed in cultures without EPO.

### Identification of cells by immunofluorescence

Light microscopic examination of megakaryocyte colonies yielded a heterogeneity of cells with differing size, cytoplasmic granulation, and nuclear segmentation. In order to identify these cells as megakaryocytes, slides were incubated with antibodies raised in rabbits against human factor VIII antigens. Independent of size and nuclear segmentation, a bright cytoplasmic fluorescence was observed. When mixed colonies were examined under similar conditions, only megakaryocytes were visualized, thus providing a useful marker for their positive identification (Fig. 7).

## DISCUSSION

Megakaryocytes can now be grown in culture within multilineage colonies derived from pluripotent progenitors and as large, tightly aggregated pure megakaryocyte colonies. Optimal growth for both types of colonies was observed in the presence of fresh human plasma and PHA-LCM. Erythropoietin was not required for the development of megakaryocyte colonies or mixed colonies that contained megakaryocytes in association with granulocytes. The addition of EPO consistently reduced the frequency of megakaryocytic colonies while promoting the maturation of erythroid components within multilineage colonies and thus facilitating their recognition. The reduced plating efficiency for megakaryocytic colonies in the presence of EPO may be related to a phenotypical change to a mixed colony, rather than a direct inhibitory effect.

Further detailed assessment of this phenomenon is essential to determine whether or not a bipotent progenitor, restricted to erythroid and megakaryocytic lineages, exists in human, as it was previously postulated for murine, hemopoiesis (18). Our observation of mixed colonies containing granulocytes and megakaryocytes would not support this restriction.

The frequency, size, and cohesiveness of the described megakaryocytic colonies varies somewhat from assays previously reported. The differences may reflect altered culture conditions, such as the use of human plasma rather than serum, the addition of PHA-LCM, and the viscous support provided by methylcellulose rather than plasma clot.

TABLE 1. Plating efficiency of megakaryocyte colonies and multilineage colonies under the influence of increasing concentrations of EPO

| | Frequency per $10^5$ mononuclear cells | | |
|---|---|---|---|
| EPO (units per plate) | Multilineage colonies | Megakaryocyte colonies | Total |
| 0.0 | 0 | 36 | 36 |
| 0.25 | 26 | 20 | 46 |
| 0.5 | 32 | 15 | 47 |
| 1.0 | 32 | 12 | 44 |
| 2.0 | 28 | 6 | 34 |
| 4.0 | 32 | 6 | 38 |

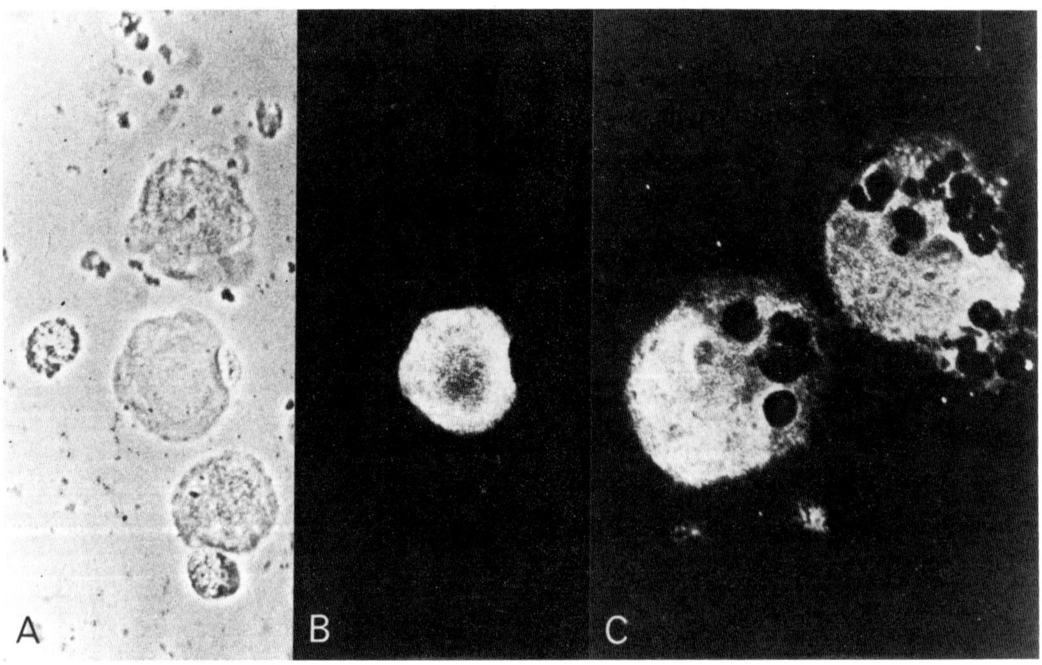


Fig. 7. Immunofluorescence analysis of megakaryocytes in colonies with antibodies directed against factor VIII antigen. (A) Phase contrast picture of cells within a multilineage colony. (B) Positive cytoplasmic reaction of one small mononuclear cell within the same frame. (C) Positive reaction of two multisegmented megakaryocytes of large size.

Alternatively it is conceivable that the colonies grown in various systems represent progeny of different subpopulations of megakaryocytic progenitors, a phenomenon that is well documented for subsets of precursors within the erythroid lineage (19). These questions will require further characterization of the colony forming cells by determining their physical and biological properties.

For instance, one such property may be associated with differing sensitivity of progenitors to erythropoietin. The observation of such a difference may resolve the current differences between data presented by Mazur et al. (10–12) supported by our study and the report by Vainchenker (8,9). The requirement for stimulators, such as molecules present in PHA-LCM, may disclose further differences in clonogenic subpopulations and may also provide means to test physiological growth regulators as suggested by stimulatory activities for megakaryocyte colonies present in serum of some patients with aplastic anemia and thrombocytopenia (11,12).

## ACKNOWLEDGMENTS

The authors wish to thank Miss J. Luker for assisting in the preparation of the manuscript. Grant support was provided by the Medical Research Council of Canada and the National Cancer Institute.

## LITERATURE CITED

1. Fauser AA, Messner HA: Granulo-erythropoietic colonies in human bone marrow, peripheral blood, and cord blood. Blood, *52:*1243, 1978.
2. Fauser AA, Messner HA: Identification of megakaryocytes, macrophages, and eosinophils in colonies of human bone marrow containing neutrophilis granulocytes and erythroblasts. Blood, *53:*1023, 1979.
3. Messner HA, Fauser AA: Culture studies of human pluripotent hemopoietic progenitors. Blut, *41:*327, 1980.
4. Messner HA, Fauser AA: Megakaryocytes in human mixed hemopoietic colonies. In: Megakaryocytes In Vitro. Levine R, Williams N, Evatt BL, eds. Elsevier-North-Holland Biomedical Press, New York pp. 161–169, 1982.
5. Senn JS, McCulloch EA, Till JE: Comparison of colony forming ability of normal and leukemic human marrow in cell culture. Lancet, *2:*597, 1967.
6. Tepperman AD, Curtis JE, McCulloch EA: Erythropoietic colonies in cultures of human marrow. Blood, *44:*659, 1974.
7. Iscove NN, Sieber F and Winterhalter KH: Erythroid colony formation in cultures of mouse and human bone marrow: Analysis of the requirement for erythropoietin by gel filtration and affinity chromatography on agarose-concanavalin A. J. Cell Physiol., *83:*309, 1974.
8. Vainchenker W, Guichard J, Breton-Gorius J: Growth of human megakaryocyte colonies in culture from fetal, neonatal and adult peripheral blood cells: Ultrastructural analysis. Blood Cells, *5:*25, 1979.
9. Vainchenker W, Bouguet J, Guichard J, Breton-Gorius J: Megakaryocyte colony formation from human bone marrow precursors. Blood, *54:*940, 1979.
10. Mazur EM, Hoffman R, Chasis J, Marchesi S, Bruno E: Immunofluorescent identification of human megakaryocyte colonies using an antiplatelet glycoprotein antiserum. Blood, *57:*277, 1981.
11. Hoffman R, Mazur EM, Bruno E, Floyd V: Assay of an activity in the serum of patients with disorders of thrombopiesis that stimulates formation of megakaryocytic colonies. N. Engl. J. Med., *305:*533, 1981.
12. Mazur EM, Hoffman R, Bruno E. Regulation of human megakaryocytopoiesis: An In Vitro analysis. J. Clin. Invest., *68:*733, 1981.
13. Rabellino EM, Nachman RL, Williams N, Winchester RJ and Ross GD: Human megakaryocytes. I. Characterization of the membrane and cytoplasmic components of isolated marrow megakaryocytes. J. Exp. Med., *149:*1273, 1979.
14. Hoyer LW, de los Santos RP, Hoyer JR: Antihemophilic factor antigen: Localization in endothelial cells by immunofluorescent microscopy. J. Clin. Invest., *52:*2737, 1973.
15. Nachman R, Levine R, Jaffe EA: Synthesis of factor VIII antigen by cultured guinea pig megakaryocytes. J. Clin. Invest., *60:*914, 1977.
16. Aye MT, Niho Y, Till JE, McCulloch EA: Studies of leukemic cell populations in culture. Blood, *44:*205, 1974.
17. Minden MD, Buick RN, McCulloch EA: Separation of blast cell and T-lymphocyte progenitors in the blood of patients with acute myeloblastic leukemia. Blood, *54:*186, 1979.
18. McLeod DL, Shreeve ML, Axelrad AA: Chromosome marker evidence for the bipotentiality of BFU-E. Blood, *56:*318, 1980.
19. Gregory CJ, Eaves AC: Human marrow cells capable of erythropoietic differentiation in vitro: Definition of three erythroid colony responses. Blood, *49:*855, 1977.

JOURNAL OF CELLULAR PHYSIOLOGY SUPPLEMENT 1:53–64 (1982)

# G-CSF: Its Relationship to Leukemia Differentiation-Inducing Activity and Other Hemopoietic Regulators

MALCOLM A.S. MOORE*
*Department of Developmental Hematopoiesis, Sloan-Kettering Institute for Cancer Research, New York, New York 10021*

The murine myelomonocytic leukemia WEHI-3B exists as differentiation-inducible ($D^+$) and noninducible ($D^-$) cell lines. Both lines produce a CSF species that stimulates exclusively the formation of neutrophil granulocyte colonies. This G-CSF copurifies with a mast cell growth factor but can be separated from M- and GM-CSF. NZB bone marrow is unresponsive to G-CSF stimulation. WEHI-3B $D^+$ cells can be induced to terminal granulocyte differentiation by a factor present in murine and human postendotoxin serum that is different from G-CSF present in WEHI-3B $D^+$ or $D^-$ CM since the latter has little or no leukemia differentiation-inducing activity. Endotoxin treatment of *C. parvum* primed mice leads to simultaneous induction of serum activities with selective action on myeloid leukemic cells, a serum differentiation inducing activity and a leukemic colony inhibitory activity. These factors act synergistically to block leukemic stem cell self-renewal. The results suggest that a variety of inducible factors may have potent and selective antileukemic activity.

In normal myelopoiesis, colony-stimulating factors (CSF) are required at each step in the proliferation and differentiation sequence from the CFU-c to mature neutrophil or macrophage; thus it is not possible to distinguish between a proliferation-inducing versus a differentiation-promoting action. However in the majority of human acute myeloid leukemias, CSF promotes proliferation but not differentiation (Moore et al., 1974). This observation does not necessarily imply that the maturation block in leukemia is irreversible and much attention has been paid to various myeloid leukemic cell line models that indeed can be induced to terminal differentiation. The most extensively studied has been the mouse myeloid leukemic cell line M-1 (Ichikawa, 1969), which can be induced to granulocyte and/or macrophage differentiation by compounds as diverse as DMSO, actinomycin-D, endotoxin, phorbol esters, dexamethasone, and various protein inducers (Weiss and Sachs, 1978; Honma et al., 1977, 1978; Sachs, 1978). The human promyelocytic leukemic cell line HL-60 (Collins et al., 1977) can likewise be induced to terminal differentiation to granulocytes by exposure to polar compounds (Collins et al., 1979; Fontana et al., 1980) or retinoids (Breitman et al., 1980; Honma et al., 1980a), and to macrophages, following treatment with phorbol esters (Territo and Koeffler, 1981; Rovera et al., 1979). The existence of physiological inducers of myeloid leukemic cell differentiation has been documented in various systems with the most consistent being the observation that serum collected 3–6 hours following endotoxin administration could induce terminal differentiation of the M-1 cell line (Lotem et al., 1980) and also a murine myelomonocytic leukemic cell line WEHI-3 (Metcalf, 1980). Since endotoxin induces a dramatic increase in serum CSF levels and high concentrations of pure GM-CSF were shown to induce leukemic cell differentiation (Metcalf, 1979), it was assumed that the differentiation-inducing protein was indeed CSF. More recently biochemical characterization of postendotoxin serum has shown that the differentiation factor could be separated from GM-CSF (MGI-1) (Lotem et al., 1980) and was termed MGI-2, or could be separated from the bulk of serum CSF but coeluted with a minor species of CSF that stimulated only granulocyte colony formation (Bur-

*To whom correspondence/reprint requests should be addressed.

gess and Metcalf, 1980). The cellular origin of endotoxin-induced differentiation protein is unclear, but macrophages are likely candidates and in this context evidence is pointing to the capacity of myeloid leukemic cells to produce their own differentiation-inducing protein. Indeed differentiation-inducing agents such as endotoxin, phorbol esters, and glucocorticoids appear to act indirectly, promoting differentiation by inducing leukemic cells to produce their own differentiation protein (Falk and Sachs, 1980; Honma et al., 1978; Weiss and Sachs, 1978).

The capacity of leukemic cells to produce hematopoietic growth regulating factors, either constitutively or following appropriate induction is well documented and may be central to the selective advantage of the leukemic population. In this context cell lines developed from the WEHI-3 myelomonocytic leukemia have proved of particular value in analysis of growth regulator production by neoplastic hematopoietic cells as well as regulator response of leukemic cells. The myelomonocytic leukemia WEHI-3 originated in a BALB/c mouse that had undergone mineral oil injections intended to induce plasma cell tumor development (Metcalf et al., 1969, Warner et al., 1969). The tumor was composed of a mixed population of monocytic and granulocytic cells. On transplantation of the tumor, four distinct sublines developed, two of which retained the original chloroma appearance and were distinguishable by karyotype (one diploid and one tetraploid). Two WEHI-3 continuous cell lines have been developed. One by Ralph et al. (1976) from the 125th passage of the WEHI-3 subline B (the hypodiploid line). This line is not inducible to mature granulocytes or macrophages although it is Fc- and C′-receptor positive, phagocytic, Thy-1-antigen positive and secreting plasminogen activator, and lysozyme (Ralph et al., 1976). While incapable of extensive differentiation the most significant feature of the cell line is its capacity to produce a wide spectrum of biologically relevant molecules that influence hemopoiesis and immune responses. The cell line produces GM-CSF (Ralph et al., 1978), macrophage CSF, and neutrophil (G) CSF (Williams et al., 1978), eosinophil CSF (Metcalf et al., 1974), megakaryocyte CSF (Williams et al., 1981), erythroid burst-promoting activity (Iscove, 1982), erythropoietin (Kubanek et al., 1981), mast-cell growth factor (Yung et al., 1981), Interleukin I (Lachman et al., 1977), endogenous pyrogen (Bodel, 1978), prostaglandin E (Kurland et al., 1979), lysozyme and plasminogen activator (Ralph et al., 1976) and has receptors for and responds to lactoferrin inhibition (Broxmeyer and Ralph, 1977).

While it may be argued that production of these various regulatory macromolecules reflects oncogenic transformation, it should be noted that all the features of the cell line are features displayed by subpopulations of macrophages under appropriate stimulation. Indeed the most neoplastic feature of the cell line is that most of the factors are produced continuously and constitutively rather than as a result of lymphokine or adjuvant induction.

A second cell line was independently developed from WEHI-3B at an early stage of in vivo passage and in contrast to the preceding cell line it retains its hypodiploid karyotype and can be induced to terminal granulocyte and/or macrophage differentiation (Metcalf, 1979,1980). To distinguish this line from that of Ralph et al. (1976), we are adopting the nomenclature WEHI-3B-D$^+$ for the former and WEHI-3B-D$^-$ for the latter.

The present study was designed to investigate the interrelationships of the various factors produced by WEHI-3 leukemic cells with particular emphasis on the G-CSF activity and its relationship to leukemia differentiation-inducing activities induced by endotoxin in mouse and human sera.

## METHODS

### Preparation of conditioned medium from WEHI-3 and L cells

W3-D$^+$,W3-D$^-$ or L cells were propagated in 75 cm$^2$ plastic flasks in McCoy's 5a medium containing 2.0% fetal calf serum. Medium was harvested at 3–4 day intervals and concentrated 5-fold by Amicon (Lexington, MA) UM 10 ultrafiltration. Semipurified CSA preparations were prepared by concentrating serum-free conditioned media, equilibration with 0.02 M Na phosphate buffer, pH 8.0 and then passed through a DEAE Sephadex A25 column. Non-bound protein was washed through the column with 5-fold bed volume of buffer followed by a single step elution of bound protein with 1 M NaCl in equilibrating buffer. These conditions have been shown to produce partial separation of G-CSF from GM- and M-CSF (Williams et al., 1978).

### Assay for mast cell growth factor (MCGF)

Pure cell lines of mast cells were derived from marrow cells obtained from long term mouse bone marrow cultures as previously described (Tertian et al., 1981). Mast cell lines were passaged in Con A stimulated spleen CM at $10^5$ cells/ml every 4–8 days. Samples of 0.1 ml volumes containing $10^4$ mast cells were distributed to wells of 96 well flatbottom microtiter plates (Costar, Cambridge, MA). Conditioned media to be tested for MCGF activities were added to triplicate or quadruplicate wells at appropriate dilutions. Culture plates were incubated for 72 hours at 37°C in a humidified incubator and at the end of this incubation period, 5 μCi tritiated thymidine were added to each well. The cultures were incubated for an additional 4 hours and the amount of $H^3$Tdr incorporated was quantitated in a liquid scintillation counter. The relative amounts of $H^3$TdR incorporated were expressed as a stimulation index (SI) defined as SI = Average cpm of test samples/Average cpm of medium controls.

### Mouse CFU-c assay

The 2.5 × $10^4$–$10^5$ mouse bone marrow cells were cultured in 1 ml of 0.3% agar in McCoy's 5A modified medium containing 10% fetal calf serum. Semipurified WEHI-3D⁻ cell CM or LCCM were used as standard sources of CSF. Cultures were scored after 7 days incubation and, where indicated, individual colonies were removed by micropipette, placed on glass slides, and stained with aceto-orcein prior to morphological typing.

### Assay for induction of WEHI-3D⁺ differentiation

Cells of the WEHI-3B-D⁺ line (and in some experiments the nondifferentiating WEHI-3B-D⁻ cells) were plated at 300 cells per ml in McCoy's 5A medium plus 0.3% agar and 10% FCS in the presence of titrated activities under test for both colony inhibition and colony differentiation. After 7 days, the total colonies and total diffuse colonies are scored. Metcalf (1979,1980) has reported that under these conditions, differentiation of WEHI-3D⁺ cells to mature neutrophils and macrophages is associated with conversion of the compact, undifferentiated leukemic blast colonies to diffuse, dispersed, differentiated clones. Where necessary morphology was directly determined on individually isolated colonies.

### Human post-endotoxin serum

Highly purified lipopolysaccharide extracted from *S. abortus equi* is currently under study in Phase I clinical trials to determine the maximum tolerated dose and toxicity of endotoxin given by the intravenous route to patients with advanced cancer. Serum samples were obtained immediately preceding the iv injection of 1.0 μg/M² of the endotoxin and at 0.5, 1, 2, 6, 8, 12, and 24 hours thereafter. The serum samples were assayed at 1% and 10% v/v against 300 WEHI-3D⁺ cells in 1 ml agar cultures and colony inhibition and colony differentiation measured at 7 days of culture.

### Mouse post-endotoxin serum

B6D2F$_1$ mice weighing 25–30 g were injected intravenously with 5 μg of endotoxin (lipopolysaccharide W E. coli 0127:B8, Difco Labs). Blood was collected 3 hours later and the serum separated.

### Induction of serum activities following *C. parvum*-endotoxin treatment

The CD-1 female mice weighing 25–30 g were injected with *C. parvum* (Wellcome Research Labs) 1 mg/0.5 ml ip diluted in saline. On day 9 or 10, endotoxin (lipopolysaccharide W *E. coli* 0127:B8, Difco Labs) 10 μg was injected iv. Blood was collected 1–5 hours later and the serum separated. Batches of serum were pooled, centrifuged at 39,000g for 1 hour, filtered (Nalgen 0.45 μ), and tested for in vitro activity against WEHI-3D⁺ leukemic cells or normal bone marrow. This protocol was developed for the induction of a serum factor-tumor necrosis factor (TNF), which induces necrosis in experimental tumors (Carswell et al., 1975). *C. parvum*-endotoxin serum and partially purified TNF have proved inhibitory to the cloning of normal mouse CFU-C as well as a variety of murine and human neoplastic cell lines, including WEHI-3D⁻ cells (Shah et al., 1978). Further characterization of TNF serum was undertaken using batch chromatography on DEAE A-50 equilibrated in Tris buffer 0.05 M containing 0.2 M NaCl at pH 8.0. Tris buffer (the same as the starting buffer) was then added in an amount equal to twice the volume of the original swollen gel. The fraction collected (fraction 0.2 M, referring to the NaCl content of the buffer) contained 80–90% of the total recovered protein, very little TNF activity, and the bulk of leukemia differentiation-inducing activity. Additional buffer containing 1.0 M NaCl was then added and the fraction eluted, which contained only 10–20% of the total protein, had the bulk of TNF activity. Further purification was obtained on a Sephadex G-100 column.

### Reagents

Prostaglandins were purchased from Sigma Chemical Co., St. Louis, MO. Stock solutions were prepared in absolute ethanol and stored at −70°C. 13-Cis retinoic acid (Hoffman-La Roche Inc.) was also dissolved in ethanol and used only when freshly prepared.

### Mice

Balb/c, NZB and B6D2F, mice of either sex were obtained from the Jackson Laboratories, Bar Harbor, ME and used at 8–12 weeks of age.

## RESULTS

### Neutrophil CSF (G-CSF) and mast cell growth factor (MCGF) production by WEHI-3 cell conditioned medium

After stepwise column chromatography of WEHI-3D⁻ CM, the breakthrough fraction stimulated predominantly neutrophil-granulocyte colony formation, whereas the fraction eluted with 1 M NaCl was found to induce primarily macrophage colony growth (Table 1). Incomplete separation of the various CSFs was evident, but with titration the overwhelming majority of colonies stimulated by eluted material were granulocytic whereas a comparable dilution of the breakthrough material stimulated almost exclusively macrophage colony formation. The DEAE fractions of WEHI-3D⁻ CM enriched respectively for G-CSF or M-CSF were then assayed for MCGF activity. The unseparated but concentrated CM contained significant MCGF as indicated by the stimulation index (Table 2) and was a more active source than Con A stimulated mouse spleen CM as reported previously (Yung et al., 1981). MCGF copurified with G-CSF and appeared in the DEAE breakthrough fraction with little activity coeluting with M-CSF (Table 2).

### Genetic restriction in response to WEHI-3 CM

We previously have reported that the numbers of myeloid progenitors in NZB bone marrow that were responsive to CSF in L cell-conditioned medium or endotoxin serum were at least normal, but at all ages granulocyte-macrophage precursors were poor responders in cultures stimulated by WEHI-3D⁻ CM (Kincade et al., 1979). The basis of this defective response remains obscure but could imply that the CSF produced by the nondifferentiating WEHI-3D⁻ cell line were defective relative to CSF produced by normal cells. Since the WEHI-3D⁺ cell line retains the capacity to differentiate to mature granulocytes, we compared the CSFs produced by the D⁻ and D⁺ cells with respect to colony stimulation of Balb/c and NZB marrow. Table 3 shows that the CM from both cell lines had comparable capacity to stimulate Balb/c marrow colony formation (note that the 1:8 dilution of the W3D⁻ DM corresponded to the 1:1 dilution of the unconcentrated W3D⁺ CM. Concentrated W3D⁺ CM was found to be inhibitory at higher concentrations). Both sources of CM were defective in stimulating NZB marrow colony formation with maximum colony incidence being respectively 12% and 33% of Balb/c marrow values with D⁻ and D⁺CM. Earlier studies have shown that prostaglandin E inhibits CFU-C proliferation (Kurland et al., 1979) and that this inhibition is selectively directed against

TABLE 1. Morphological analysis of colonies stimulated by fractions of WEHI-3D⁻ cell conditioned medium

| Sources of CSF | Concentration[1] | CFU-c/$10^5$ cells | Percent colony type at day 7 G | G-M | M |
|---|---|---|---|---|---|
| Unfractionated | 1.00 | 283 ± 26 | 29 | 42 | 26 |
| W-3D⁻ CM | 0.25 | 207 ± 18 | 19 | 29 | 48 |
| " | 0.067 | 105 ± 12 | 7 | 36 | 57 |
| " | 0.016 | 24 ± 8 | 0 | 3 | 97 |
| DEAE breakthrough | 20.0 | 127 ± 12 | 51 | 30 | 16 |
| fraction | 5.0 | 100 ± 10 | 50 | 33 | 14 |
| " | 1.25 | 25 ± 5 | 68 | 27 | 3 |
| " | 0.0625 | 17 ± 4 | 80 | 18 | 0 |
| " | | | | | |
| DEAE eluate | 5.0 | 212 ± 21 | 11 | 40 | 49 |
| (1 M NaCl) | 1.25 | 173 ± 14 | 12 | 24 | 64 |
| " | 0.31 | 50 ± 5 | 3 | 16 | 81 |
| " | | | | | |

[1]Concentration is quoted in relation to the original volume (10% WEHI-3 CM). The WEHI-3 CM was separated stepwise on DEAE Sephadex A-25, pH 8.0. For morphological analysis 30–40 sequential colonies were typed as G, granulocyte; G-M, granulocyte-macrophage; or M, macrophage.

TABLE 2. Presence of MCGF in G-CSF enriched fractions of DEAE-Sephadex separated WEHI-3 CM

| Source of MCGF | Concentration (percent) | MCGF activity (stimulation index) |
|---|---|---|
| Unfractionated WEHI-3D⁻ CM | 10 | 531.1 |
| DEAE breakthrough fraction | 50 | 686.3 |
| (20 μg protein/ml; G-CSF) | | |
| " | 25 | 734.5 |
| " | 12 | 679.3 |
| " | 6 | 372.9 |
| DEAE eluate, 1 M NaCl | 50 | 6.5 |
| (20 μg protein/ml; M-CSF) | 25 | 1.5 |
| " | 12 | 1.1 |
| " | | |

TABLE 3. Influence of conditioned media from WEHI-3 leukemic cells on colony formation by Balb/c and NZB marrow cells

| | Colonies per 5 × $10^4$ BALB/C marrow | | | | Colonies per 5 × $10^4$ NZB marrow | | | |
|---|---|---|---|---|---|---|---|---|
| Concentration of CM[1] | W3D⁻ CM | W3D⁺ CM | W3D⁻ CM + PGE[2] | W3D⁺ CM + PGE | W3D⁻ CM | W3D⁺ CM | W3D⁻ CM + PGE | W3D⁺ CM + PGE |
| 1 : 1 | 76 ± 11 | 49 ± 5 | 46 ± 8 | 17 ± 7 | 9 ± 2 | 16 ± 2 | 2 ± 1 | 0 |
| 1 : 2 | 74 ± 12 | 29 ± 3 | 44 ± 6 | 5 ± 2 | 5 ± 2 | 18 ± 8 | 5 ± 2 | 0 |
| 1 : 4 | 57 ± 6 | 20 ± 4 | 36 ± 2 | 2 ± 0 | 5 ± 1 | 10 ± 3 | 0 | 0 |
| 1 : 8 | 54 ± 3 | 13 ± 2 | 38 ± 14 | 1 ± 1 | 6 ± 1 | 7 ± 6 | 0 | 0 |
| 1 : 16 | 43 ± 2 | 9 ± 5 | 19 ± 2 | 0 | 6 ± 2 | 9 ± 3 | 0 | 0 |
| 1 : 32 | 20 ± 1 | 5 ± 3 | 11 ± 2 | 0 | 2 ± 1 | 0 | 0 | 0 |

[1]W3D⁻ CM is a × 10 concentrate of serum-free CM from the nondifferentiating cell line. W3D⁺ CM is unconcentrated CM from the differentiating line.
[2]Cultures were exposed to $10^{-6}$ M prostaglandin $E_2$ or $10^{-7}$ M prostaglandin $E_1$.

colonies stimulated by M-CSF and to a lesser extent by GM-CSF but G-CSF-stimulated colonies are refractory to PGE inhibition (Pelus et al., 1979). The addition of PGE at a concentration of $10^{-6}$ M to Balb/c marrow cultures stimulated either by W-3D⁻ or D⁺ CM led to 42 ± 6% and 81 ± 12% reduction, respectively, in colony formation indicating that the PGE insensitive colony formation was due to the G-CSF activity in the two CM (Table 3), a conclusion substantiated by morphological analysis of the colonies. The low levels of colony formation seen with W3D⁺ and D⁻ CM and NZB marrow may be attributed to contaminating M-CSF, a possibility suggested by the observation that PGE totally eliminated NZB colony formation induced by W3D⁺ CM at all concentrations and by W3D⁻ CM at 1:4 and lower concentrations.

## Inducers of differentiation of WEHI-3D⁺ leukemic cells

Previous reports have documented that mouse serum obtained 1.5–3 hours postendotoxin treatment was a potent source of a factor capable of inducing terminal differentiation of WEHI-3 (Metcalf, 1980) and M-1 (Lotem et al., 1980) leukemic cells. Data presented in Figure 1 confirms these reports and shows that serum obtained 3 hours postendotoxin treatment was a potent inducer of WEHI-3D⁺ cells with 100% differentiation of colonies seen with a 1:2 dilution of the serum. Administration of bacterial products to mice can lead to induction of a variety of bioregulatory macromolecules including colony-stimulating factors and activities inhibitory to normal or leukemic hemopoietic cell proliferation. In the latter category, the sequential administration of *C. parvum* then endotoxin to mice induces an acute elevation in the serum of a glycoprotein termed tumor necrosis factor (TNF). TNF induces necrosis of transplanted murine tumors and is toxic in vitro for neoplastic murine and human cell lines (Green et al., 1976).

Serum obtained 1–5 hours after endotoxin treatment of *C. parvum*-primed mice (TNF serum) was markedly inhibitory to cloning of WEHI-3D⁺ cells when used at high concentrations, an observation consistent with our earlier observation on the antiproliferative effect of TNF serum on WEHI-3D⁻ cells (Shah et al., 1978). Semipurified TNF activity obtained by sequential DEAE batch chromatography and Sephadex G-100 separation retained the capacity to inhibit WEHI-3D⁺ cloning but had

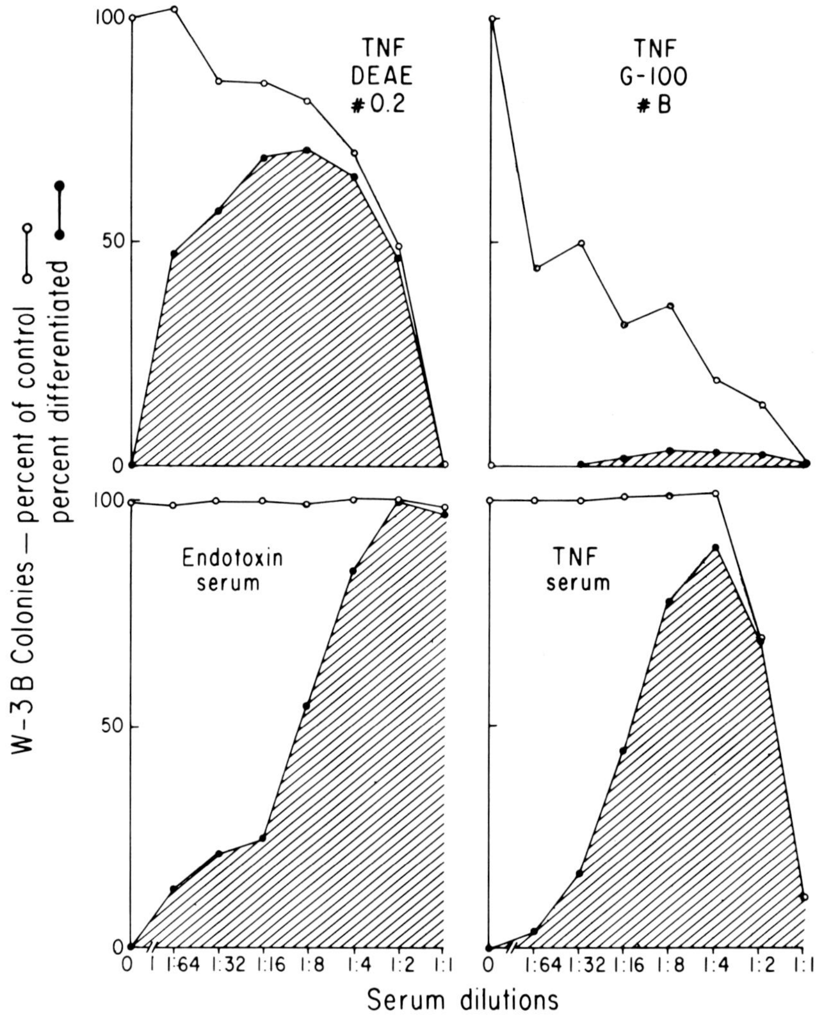


Fig. 1. Influence of 3 hour postendotoxin mouse serum and of 3 hour postendotoxin-*C. parvum* sensitized mouse serum (TNF serum) on cloning and differentiation of WEHI-3B ($D^+$) myelomonocytic leukemic cells. Cells plated at 300 per 1 ml semisolid agar culture and scored after 7 days for the percent of colonies converting from tight to diffuse (differentiated). Note colony inhibition with high concentrations of TNF but not of endotoxin serum. TNF serum fractions subject to DEAE batch elution with 0.2 M NaCl (top left panel) contained the majority of differentiation-inducing protein (DF) but still had colony-inhibitory activity of a TNF type. Fractions of TNF serum subject to G-100 separation with maximum TNF activity had very little DF but profoundly inhibited WEHI-3 cloning (top right panel).

minimal differentiation-inducing capacity (Fig. 1). The bulk of differentiation-inducing activity in *C. parvum*-endotoxin serum eluted from DEAE with 0.2 M NaCl but was contaminated by a leukemia colony-inhibitory activity, possibly of TNF type.

Burgess and Metcalf (1980) have reported that differentiation factor (GM-DF) in postendotoxin serum could not be separated from a CSF species exclusively stimulating neutrophil colonies. Since WEHI-3D$^+$ and D$^-$ CM are potent sources of G-CSF (Tables 1–3), we investigated the capacity of these conditioned media, concentrates thereof, and partially purified G-CSF to induce differentiation of WEHI-3D$^+$ cells. Ten-fold concentrated W-3D$^-$ CM or LCCM were added to cultures of WEHI-3D$^+$ cells over a range of concentrations to a maximum of 992 and 1470 units of CSF activity, respectively (Fig. 2). Some induction of differentiation was seen with the higher concentrations of WEHI-3D$^-$ CM, which served as a po-

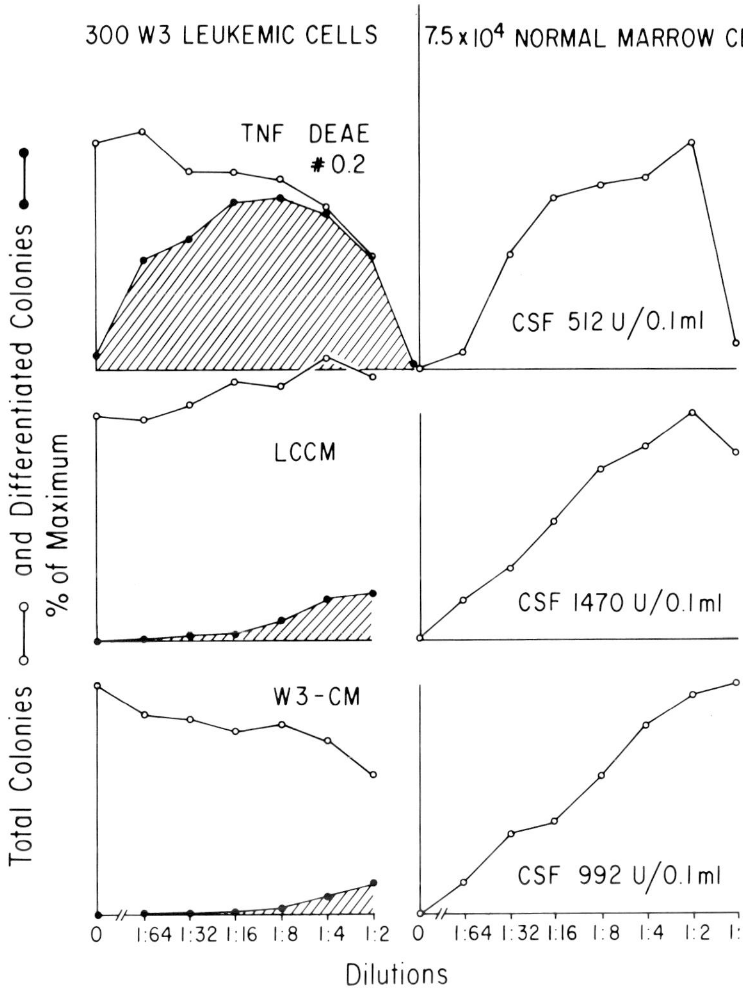


Fig. 2. Comparison of CSF activity versus leukemic cell differentiation factor in TNF serum, mouse L cell-conditioned medium (LCCM) and in WEHI-3B ($D^-$) myelomonocytic leukemic cell conditioned medium. CSF activity, in units, was measured on the linear portion of the dose response curve in cultures of $7.5 \times 10^4$ B6D2F$_1$ mouse bone marrow cells. Note that colonies stimulated by W-3-CM were mostly granulocytic, those by LCCM were mostly macrophage, and those stimulated by the TNF serum fraction subject to DEAE batch elution with 0.2 M NaCl were granulocyte, macrophage, and mixed. The ability of these three sources of CSF to induce differentiation of WEHI-3B ($D^+$) cells was measured in agar cultures containing 300 leukemic cells. TNF serum fractions stimulated extensive differentiation of leukemic colonies whereas the sources of G-CSF (W3-CM) and of macrophage-CSF (LCCM) had very little differentiation-inducing activity.

tent source of G-CSF; however an equivalent concentration of LCCM also induced a similar degree of differentiation of the leukemic cells yet was devoid of G-CSF activity (Fig. 2 and Table 4). The potent differentiation-inducing capacity of the DEAE 0.2 M NaCl eluate of *C. parvum*-endotoxin (TNF) serum could not be correlated with its G- or M-CSF activity, which was one third to one half the activities in the W-3 and LCCM (Fig. 2). Furthermore, this serum fraction was as effective in stimulating NZB marrow colony formation as that of G-CSF responsive mouse strains (data not shown). The data shown in Figure 2 also support an earlier observation (Shah et al., 1978) that normal marrow CFU-c are inhibited by TNF, since at the highest concentration used, the serum has little detectable colony stimulating activity. Comparison with colony inhibition of the leukemic cells by the same serum

TABLE 4. Influence of various activities on cloning and Differentiation of WEHI-3D$^+$ leukemic cells[1]

| Activity | Colonies percent control | Colonies percent differentiated |
|---|---|---|
| Media control | 100 | 0.3 |
| W-3D$^-$ CM 1:2 | 78 | 9.5 |
| W-3D$^+$ CM 1:2 | 107 | 2.0 |
| LCCM 1:2 | 92 | 7.0 |
| Prostaglandin E $10^{-7}$ M | 85 | 0.4 |
| W-3D$^-$ CM + PGE $10^{-7}$ M | 74 | 13.5 |
| LCCM + PGE $10^{-7}$ M | 88 | 14.2 |
| Mouse Postendotoxin serum 1:1 | 108 | 100.0 |
| Human Postendotoxin serum 1:1 | 89 | 59.0 |
| Mouse C.parvum-endotoxin serum 1:2 | 69 | 100.0 |
| Mouse normal serum | 142 | 0.0 |
| C. parvum-endotoxin serum, DEAE 0.2 M, 1 mg protein | 72 | 97.0 |
| " " " " " 0.1 mg protein | 72 | 24.0 |
| " " " " " " +PGE $10^{-7}$ M | 51 | 12.0 |
| " " " " " " +AC-D 5 ng | 86 | 28.0 |
| Actinomycin-D 5 ng | 96 | 5.0 |
| Endotoxin 1 μg | 90 | 1.0 |
| Retinoic acid $10^{-7}$ M | 36 | 67.0 |

[1]Three hundred WEHI-3D$^+$ cells plated in the presence of varying activities.

fraction confirms that leukemic cells are 10–100 fold more sensitive to TNF growth inhibition. A reproducible inhibition of WEHI-3D$^+$ cloning was also seen when cells were exposed to high concentrations of WEHI-3D$^-$ CM (Fig. 2 and Table 4). This was not detectable with normal marrow cloning nor when an equivalent concentration of LCCM was used. Conditioned medium obtained from mass culture of WEHI-3D$^+$ cells did not inhibit W-3D$^+$ cloning nor did it induce differentiation (Table 4).

Studies with the M-1 murine myeloid leukemic cell line and the human promyelocytic leukemic cell line Hl-60 have demonstrated a plethora of agents capable of directly inducing differentiation or acting synergistically with other inducing agents (Sachs, 1978, Breitman et al., 1980, Honma et al., 1980b). The 13-cis retinoic acid at low concentrations proved inhibitory to WEHI-3D$^+$ cloning and induced differentiation (Table 4). Actinomycin-D was inhibitory at a concentration of 10 ng but did not induce differentiation at that or lower concentrations. Endotoxin was lacking in any significant direct action on cell cloning or differentiation. Prostaglandin E at concentrations of $10^{-7}$–$10^{-6}$ M was only slightly inhibitory to WEHI-3 cloning and did not induce differentiation. No significant enhancement of leukemic cell differentiation was seen when either Actinomycin-D or prostaglandin E were added to leukemic cell cultures in the presence of suboptimal concentrations of serum differentiation-inducing activity (Table 4) and only slight enhancement of the differentiation-inducing capacity of high concentrations of WEHI-3D$^-$ CM and LC-CM was seen in the presence of $10^{-7}$ M PGE.

### Induction of leukemia differentiating activity in human postendotoxin sera

In conjunction with a clinical trial to determine the maximum tolerated dose of highly purified endotoxin in patients with advanced cancer, serum samples were obtained at various intervals after intravenous administration of 1 μg of endotoxin/M$^2$ and the samples were tested for ability to inhibit or differentiate WEHI-3D$^+$ cells. Preendotoxin human serum, like normal mouse serum (Table 4,5), did not inhibit colony formation nor induce differentiation. Significant serum colony inhibitory activity was seen within 30 minutes of endotoxin treatment but thereafter the human sera were only slightly suppressive for colony formation. Induction of a differentiation factor in the sera was clearly evident 2 hours post endotoxin, peaking by 4 hours and returning to baseline values by 24 hours (Table 5).

## DISCUSSION

Burgess and Metcalf (1980) have suggested that a differentiation-inducing protein (GM-DF) for myeloid leukemic cells could be a subspecies of CSF restricted to inducing normal neutrophil differentiation and distinct from

TABLE 5. Induction of leukemia differentiating activity in human post-endotoxin sera[1]

| Time post endotoxin (hours) | W-3D$^+$ percent control | Colonies percent differentiated | WBC $\times 10^3$ |
|---|---|---|---|
| 0 | 96.4 ± 13 | 3 ± 4 | 8.2 ± 2.0 |
| 0.5 | 64.0 ± 22 | 9 ± 10 | 8.4 ± 1.0 |
| 2 | 82.6 ± 14 | 42 ± 36 | 2.7 ± 1.6 |
| 4 | 89.0 ± 16 | 59 ± 28 | 5.6 ± 4.3 |
| 6 | 87.6 ± 14 | 54 ± 32 | 7.7 ± 3.1 |
| 8 | 73.4 ± 18 | 32 ± 27 | 16.8 ± 6.5 |
| 12 | 76.8 ± 14 | 24 ± 33 | 16.8 ± 9.6 |
| 24 | 81.0 ± 12 | 3 ± 2 | 17.1 ± 7.7 |

[1]Data based on study of five patients receiving 1 μg/m² of endotoxin. Sera (10% v/v) assayed against a target population of 300 WEHI-3D$^+$ leukemic cells.

the bulk of GM-CSF, which possesses low (Metcalf, 1979) or no (Lotem et al., 1980) GM-DF activity. These conclusions have been based on extensive purification of biological activities from mouse postendotoxin serum. We elected to approach this problem in a different way, namely, to demonstrate that conditioned medium from the WEHI-3 leukemic cell line contained a CSF species that stimulated neutrophil granulopoiesis but did not contain significant M- or GM-CSF activity. G-CSF activity could not be separated from a mast cell growth factor under conditions used in this study, but additional approaches have been used to separate G-CSF and MCGF (Yung, Y.-P. and Moore, M.A.S., unpublished observations). G-CSF from WEHI-3B-D$^-$ cells did not induce differentiation of D$^+$ WEHI-3B cells except when used at high concentrations. When compared to the GM-DF activity in postendotoxin, post *C. parvum*-endotoxin and DEAE fractions of mouse serum, differentiation activity of W-3 CM was very low relative to G-CSF levels. Indeed equivalent differentiation inducing activity was found in L-cell CM that contained no detectable G-CSF activity and was predominantly a source of macrophage CSF. We cannot exclude the possibility that high concentrations of G- and M-CSF, like GM-CSF, may have a low capacity to induce WEHI-3B-D$^+$ differentiation although it is more likely that our semipurified sources of CM are contaminated by a GM-DF distinct from conventional CSF species.

The use of a nondifferentiating leukemic cell line as a source of a G-CSF may be questioned since it is possible that the progression to a differentiation noninducible state resulted from a qualitative change in an autocrine growth factor such that it retained a capacity to induce neutrophil differentiation in normal progenitors (i.e., G-CSF) but failed to cause leukemic cell differentiation (possibly acting only as an autoproliferation factor). For this reason, we compared the factors produced by the differentiation-inducible D$^+$ WEHI-3 cell line with that of the D$^-$ line and found them to be similar with respect to production of a G-CSF active against normal CFU-c but incapable of inducing significant leukemic cell differentiation. Furthermore, the G-CSF produced by both lines failed to stimulate colony formation by NZB bone marrow CFU-c, a genetic defect in responsiveness that is intrinsic to the NZB target cells rather than to the cell-line source of G-CSF (Kincade et al., 1979). It should be noted that NZB bone marrow colony formation is normal to hypersensitive to stimulation by GM-CSF and macrophage CSF.

Bacteria such as C. parvum or BCG or bacterial lipopolysaccharides are potent inducers of a variety of biological response modifiers including various CSF species, interferons, prostaglandins, tumor necrosis factor (TNF), and leukemia differentiation factor. We have shown that TNF can be separated from CSF, interferon and prostaglandin (Shah et al., 1978), and in the present study we have shown it to be biochemically and functionally distinct from GM-DF. The apparent lack of species specificity of TNF and its selective (quantitative if not qualitative) ability to inhibit proliferation of transformed cells, including WEHI-3B, suggests that acquisition of in vivo sensitivity to TNF may prove a most important marker of leukemic transformation and eventually a therapeutic agent for erradicating emerging leukemic clones. The ability of *C. parvum* plus endotoxin to induce high serum levels of TNF and of GM-DF with similar kinetics points to the importance of synergism studies with various biological response modifiers since TNF inhibits leukemic cell clonal expansion by a direct antiproliferative (clonal inhibition) action, whereas GM-DF inhibits leukemic cell self-renewal by promoting terminal differentiation. Extrapolation of this approach to human leukemia therapy is feasible since TNF inhibition of human leukemic has been demonstrated (Shah et al., 1978) and our present study shows that a serum GM-DF active against WEHI-3B-$D^+$ cells is induced in human serum between 2–6 hours following adminstration of highly purified endotoxin. Further studies are required to better define the spectrum of leukemic cell types that can be induced to differentiation with human sources of GM-DF, but it is of interest that the kinetics of production and biological activity of the human and mouse GM-DF are so similar.

Myeloid cell lines clearly differ in their responsiveness to direct or indirect inducers of cell differentiation. In this context, the dexamethasone induction of M-1 cell differentiation is blocked by indomethacin—a prostaglandin synthesis inhibitor—and this block is counteracted by addition of exogenous PGE, or $E_2$, but not $F_{2\alpha}$ (Honma et al., 1979). Since M-1 cells produce PGE upon exposure to dexamethasone, LPS and protein inducers (Honma et al., 1980b) differentiation induction may be mediated by PGE. Indeed in the M-1 system, PGE induces increased lysozyme synthesis and synergises with low concentrations of dexamethasone in induction of phagocytosis, locomotive activity, and morphological maturation of M-1 cells (Honma et al., 1980b). Our present study has shown that the differentiating line of WEHI-3B cells is insensitive to PGE-mediated growth inhibition and differentiation, in contrast to the extreme sensitivity of the nondifferentiating WEHI-3B-$D^-$ cells that show 50% inhibition of agar cloning with $10^{-9}$ M $PGE_1$ (Kurland and Moore, 1977). The latter cell line is also capable of constitutive production of prostaglandin E and can be induced to produce high levels following exposure to LPS or an exogenous source of CSF (Kurland et al., 1979). It is possible that the differential sensitivities of the WEHI-3B-$D^+$ line versus the WEHI-3B-$D^-$ and M-1 lines to prostaglandin E reflects the predominant monoblastic-macrophage lineage of the latter in contrast to the predominant granulocytic differentiation of the former.

Prostaglandin E inhibition of normal CFU-c proliferation is predominantly directed at progenitors committed to monocyte-macrophage differentiation whereas granulocyte-committed progenitors are insensitive even to high concentrations of PGE (Pelus et al., 1979). Human acute and chronic myeloid leukemic CFU-c are generally insensitive to growth inhibition and differentiation induction when exposed to high concentrations of prostaglandin E (Pelus et al., 1980) and thus appear more related to WEHI-3B-$D^+$ clonogenic cells. Sensitivity of normal and leukemic human CFU-c to prostaglandin E correlates well with expression of Ia-like antigenic determinants on the myeloid progenitors and cell cycle status (Pelus et al., 1982), so the possibility of synergism between PGE and other induces of leukemic cell differentiation that may modulate cell proliferation or Ia antigen expression remains possible. We observed some enhancement of differentiation and decrease in proliferation when WEHI-3B-$D^+$ cells were exposed to WEHI-3$D^-$ or L-cell CM in the presence of $10^{-7}$ M PGE, but no synergism was seen with the differentiation protein in *C. parvum*-endotoxin serum. Differentiation of M-1 cells has been reported following their exposure to low concentrations of Actinomycin-D (Honma et al., 1978) and this drug can sensitize $D^-$ clones of M-1 cells to other differentiation inducing agents either in vivo or in vitro. We failed to observe direct or synergistic differentiation induction of WEHI-3B-$D^+$ cells following actinomycin-D treatment, possibly since the cell line was relatively insensitive to actinomycin-D growth inhibition.

It is unlikely that the results obtained with differentiation-inducible murine or human leukemic cell lines can be extrapolated to the majority of primary myeloid leukemias, at least in so far as induction of terminal differentiation is concerned; nevertheless biological response modification in vivo or in vitro may modify the neoplastic phenotype sufficiently so as to eliminate the competitive advantage of the leukemic clone relative to the normal hematopoietic stem cell population.

## ACKNOWLEDGMENTS

The author wishes to express his appreciation for the excellent technical assistance of Mrs. A.P. Sheridan and Mr. R. Hernandez. I wish to thank Richard Eger and Dr. Y.-P. Yung for their assistance in the characterization of G-CSF and MCGF, Dr. L. Old for his generous collaboration in the TNF study, and Drs. Oettgen and Gabrilov for provision of the human postendotoxin sera. This work was supported by Institute Core grant CA-08748 from the National Institutes of Health, NCI grants 19052, 17085, and the Gar Reichman Foundation.

## LITERATURE CITED


Bodel, P. (1978) Spontaneous pyrogen production by mouse histiocytic and myelomonocytic tumor cell lines in vitro. J. Exp. Med., *147:*1503–1516.

Breitman, T., S. Selonick, and S. Collins (1980) Induction of differentiation of the human promyelocytic leukemia cell line (HL-60) by retinoic acid. Proc. Natl. Acad. Sci. U.S.A., *77:*2936–2940.

Broxmeyer, H.E., and P. Ralph (1977) Regulation of a mouse myelomonocytic leukemia cell line in culture. Cancer Res., *37:*3578–3584.

Burgess, A.W. and D. Metcalf (1980) Characterization of a serum factor stimulating the differentiation of myelomonocytic leukemic cells. Int. J. Cancer, *26:*647–654.

Carswell, E.A., L.J. Old, R.L. Kassel, S. Green, N. Fiore, and B. Williamson (1975) An endotoxin-induced serum factor that causes necrosis of tumors. Proc. Natl. Acad. Sci. U.S.A. *72:*3660–3670.

Collins, S.J., R.C. Gallo, R.E. Gallagher (1977) Continuous growth and differentiation of human myeloid leukemic cells in suspension culture. Nature, *270:*347–349.

Collins, S.J., F.W. Ruscetti, R.E. Gallagher, and R.C. Gallo (1979) Normal functional characteristics of cultured human promyelocytic leukemia cells (HL-60) after induction of differentiation by dimethyl sulfoxide. J. Exp. Med., *149:*969–974.

Falk, A., and L. Sachs (1980) Clonal regulation of the induction of macrophage- and granulocyte-inducing proteins for normal and leukemic myeloid cells. Int. J. Cancer, *26:*595–601.

Fontana, J.A., D.G. Wright, E. Shiffman, B.A. Corcoran and A.B. Deisseroth (1980) Development of chemotactic responsiveness in myeloid precursor cells: Studies with a human leukemia cell line. Proc. Natl. Acad. Sci. U.S.A., *77:*3664–3668.

Green, S., A.E. Dobrjansky, E.A. Carswell, R.L. Kassel, L.J. Old, N. Fiore, and M.K. Schwartz (1976) Partial purification of a serum factor that causes necrosis of tumors. Proc. Natl. Acad. Sci. U.S.A., *73:*381–389.

Honma, Y., T. Kasukabe, and M. Hozumi (1978) Relationship between leukemogenicity and in vivo inducibility of normal differentiation in mouse myeloid leukemia cells. J. Natl. Cancer Inst., *61:*837–841.

Honma, Y., T. Kasukabe, and M. Hozumi (1979) Inhibition of differentiation of cultured mouse myeloid leukemia cells by nonsteroidal anti-inflammatory agents and counteraction of the inhibition by prostaglandin E. Cancer Res., *39:*2190–2194.

Honma, Y., T. Kasukabe, J. Okabe, and M. Hozumi (1977) Glucocorticoid-induced differentiation of cultured mouse myeloid leukemia cells. Gann, *68:*241–246.

Honma, Y., T. Kasukabe, M. Hozumi and Y. Koshihara (1980b) Regulation of prostaglandin synthesis during differentiation of cultured mouse myeloid leukemia cells. J. Cell Physiol., *104:*349–357.

Honma, Y., K. Takenaga, T. Kasukabe and M. Hozumi (1981a) Induction of differentiation of cultured human promyelocytic leukemia cells by retinoids. Biochem. Biophys. Res. Comm., *95:*507–512.

Ichikawa, Y. (1969) Differentiation of a cell line of myeloid leukemia. J. Cell Physiol., *74:*223–224.

Iscove, N. (1982) J. Cell Physiol. (in this supplement).

Kincade, P.W., G. Lee, G. Fernandes, M.A.S. Moore, N. Williams, and R.A. Good (1979) Abnormalities in clonable B lymphocytes and myeloid progenitors in autoimmune NZB mice. Proc. Natl. Cancer Inst. U.S.A., *76:*3464–3468.

Kubanek, B., W. Heit, and I.N. Rich (1981) Continuous erythropoietin production by macrophages. Exp. Hematol., (suppl. 9) *9:*59 (Abstr.).

Kurland, J., and M.A.S. Moore (1977) Modulation of hemopoiesis by prostaglandins. Exp. Hematol., *5:*357–373.

Kurland, J.I.,, L.M. Pelus, P. Ralph, R.S. Bockman, and M.A.S. Moore (1979) Induction of prostaglandin E synthesis in normal and neoplastic macrophages: Role for colony-stimulating factor(s) distinct from effects on myeloid progenitor cell proliferation. Proc. Natl. Acad. Sci. U.S.A., *76:*2326–2330.

Lachman, L.B., M.P. Hocker, G.T. Blyden, and T. Handschumacher (1977) Preparation of lymphocyte-activating factor from continuous murine macrophage cell lines. Cell Immunol., *34:*416–419.

Lotem, J., J.H. Lipton, and L. Sachs (1980) Separation of different molecular forms of macrophage and granulocyte-inducing proteins for normal and leukemic myeloid cells. Int. J. Cancer, *25:*763–771.

Metcalf, D. (1979) Clonal analysis of the action of GM-CSF on the proliferation and differentiation of myelomonocytic leukemic cells. Int. J. Cancer, *24:*616–623.

Metcalf, D. (1980) Clonal extinction of myelomonocytic leukemic cells by serum from mice injected with endotoxin. Int. J. Cancer, *25:*225–233.

Metcalf, D., M.A.S. Moore and N. Warner (1969) Colony formation in vitro by myelomonocytic leukemic cells. J. Natl. Cancer Inst., *43:*983–1001.

Metcalf, D., J. Parker, H.M. Chester, and P.W. Kincade (1974) Formation of eosinophilic-like granulocytic colonies by mouse bone marrow cells in vitro. J. Cell Physiol., *84:*275–289.

Moore, M.A.S., G. Spitzer, N. Williams, D. Metcalf, and J. Buckley (1974) Agar culture studies in 127 cases of untreated acute leukemia: The prognostic value of reclassification of leukemia according to in vitro growth characteristics. Blood, *44:*1–12.

Pelus, L.M., H.E. Broxmeyer, B.D. Clarkson, and M.A.S. Moore (1980) Abnormal responsiveness of granulocyte-macrophage committed colony-forming cells from patients with chronic myeloid leukemia to inhibition by prostaglandin E. Cancer Res., *40:*2512–2515.

Pelus, L.M., H.E. Broxmeyer, J.I. Kurland, and M.A.S. Moore (1979) Regulation of macrophage and granulocyte proliferation: Specificities of prostaglandin E and lactoferrin. J. Exp. Med., *150:*277–292.

Pelus, L.M., S. Saletan, R. Silver, and M.A.S. Moore (1982) Expression of Ia-antigens on normal and chronic myeloid leukemic human granulocyte-macrophage colony-forming cells (CFU-GM) is associated with the regulation of cell proliferation by prostaglandin E. Blood (in press).

Ralph, P., M.A.S. Moore, and K. Nilsson (1976) Lysozyme synthesis by established human and murine histiocytic lymphoma cell lines. J. Exp. Med., *143:*1528–1533.

Ralph, P., H.E. Broxmeyer, M.A.S. Moore, and I. Nakoinz, (1978) Induction of myeloid colony-stimulating activity in murine monocyte tumor cell lines by macrophage activators and in a T-cell line by concanavalin A. Cancer Res., *38:*1414–1419.

Rovera, G., T.G. O'Brien, and L. Diamond (1979) Induction of differentiation in human promyelocytic leukemia cells by tumor promoters. Science, *204:*868–870.

Sachs, L. (1978) Control of normal cell differentiation and the phenotypic reversion of malignancy in myeloid leukemia. Nature, *274:*535–539.

Shah, R.G., S. Green, and M.A.S. Moore (1978) Colony-stimulating and inhibiting activities in mouse serum after C. parvum-endotoxin treatment. J. Reticuloendothel. Soc., *23:*29–41.

Territo, M.C., and H.P. Koeffler (1981) Induction by phorbol esters of macrophage differentiation in human leukaemia cell lines does not require cell division. Br. J. Haematol., *47:*479–483.

Tertian, G., T-P Yung, D. Guy-Grand, and M.A.S. Moore (1981) Long-term in vitro culture of murine mast cells. I Description of a growth factor-dependent culture technique. J. Immunol., *127:*788–794.

Warner, N., M.A.S. Moore, and D. Metcalf (1969) A transplantable myelomonocytic leukemia in Balb/c mice: Cytology, karyotype, and muramidase content. J. Natl. Cancer Inst., *43:*963–968.

Weiss, B. and L. Sachs (1978) Indirect induction of differentiation in myeloid leukemic cells by lipid A. Proc. Natl. Acad. Sci., U.S.A., *75:*1374–1378.

Williams, N., R.R. Eger, M.A.S. Moore, and N. Mendelsohn (1978) Differentiation of mouse bone marrow precursor cells into neutrophil granulocytes by an activity separated from WEHI-3 conditioned medium. Differentiation, *11:*59–63.

Williams, N., H. Jackson, P. Ralph, and I. Nakoinz (1981) Cell interactions influencing murine marrow megakaryopoiesis: Nature of the potentiating cell in bone marrow. Blood, *57:*157–163.

Yung, Y.P., R. Eger, G. Tertian, and M.A.S. Moore (1981) Long-term in vitro culture of murine mast cells: Purification of a mast cell growth factor and its dissociation from TCGF. J. Immunol., *127:*794–799.

JOURNAL OF CELLULAR PHYSIOLOGY SUPPLEMENT 1:65–78 (1982)

# Molecules Stimulating Early Red Cell, Granulocyte, Macrophage, and Megakaryocyte Precursors in Culture: Similarity in Size, Hydrophobicity, and Charge

N. N. ISCOVE,* C. A. ROITSCH, N. WILLIAMS, AND L. J. GUILBERT

*Basel Institute for Immunology, CH-4005, Basel, Switzerland (N.N.I., C.A.R.), Department of Physiology, University of Melbourne, Parkville, 3052 Victoria, Australia (N.W.), and Department of Microbiology and Immunology, Albert Einstein College of Medicine, Bronx, New York 10461 (L.J.G.)*

Molecules in conditioned medium from stimulated lymphocyte populations or from certain cell lines are known to stimulate cells committed to various hemopoietic lineages as well as pluripotential cells to form colonies in culture. In this study, the relationship between molecules active on pluripotential cells and early cells committed to granulocyte, macrophage, megakaryocyte, or red cell production was explored using techniques of chemical separation. After separation on the basis of charge, or after sequential purification using methods of high resolving power based on hydrophobicity and size, these activities remained associated with one another. The observations provide support for a model which proposes that pluripotential hemopoietic precursor cells as well as their early committed progeny may all be responsive to a single lineage-indifferent factor. Responsiveness to "lineage-specific" factors such as erythropoietin is proposed to be a feature only of later cells after they have made the appropriate receptors as part of their differentiation program.

Conditioned media from lectin- or antigen-stimulated populations of lymphocytes have a variety of actions on hemopoietic cells. In semisolid cultures such media stimulate colony ("burst") formation by pluripotential myeloid/erythroid precursors (BFU-M/E) (Johnson and Metcalf, 1977; Hara and Ogawa, 1978; Iscove, 1978; Fauser and Messner, 1978; Schreier and Iscove, 1980) and by more restricted erythroid precursors (BFU-E) (Aye, 1977; Axelrad et al., 1978; Iscove, 1978; Nathan et al., 1978; Schreier and Iscove, 1980). These activities will be referred to here as myeloid/erythroid burst-promoting activity (M/E-BPA), and erythroid burst-promoting activity (E-BPA), or more generally as BPA. In addition, these conditioned media also stimulate growth of granulocyte/macrophage colony-forming cells (GM CFC) (Parker and Metcalf, 1974; Schreier and Iscove, 1980) and of megakaryocyte precursors (CFU-Mk) (Metcalf et al., 1975). These colony-stimulating activities will be called GM-CSA and Mk-CSA.

Conditioned media may contain a variety of different growth factors each acting on different kinds of hemopoietic precursor cells. Alternatively, it has been suggested that a single molecular class (having "BPA") could have activity on pluripotential cells as well as on their immediate but more restricted progeny (Iscove and Guilbert, 1978; Iscove, 1978). In this model, growth and differentiation of early cells would depend on a factor that is "lineage indifferent." Responsiveness to "lineage-specific" factors, on the other hand, would develop only with subsequent maturation down the different lineages. Erythropoietin is one example of a lineage-specific factor acting exclusively on red cell precursors. Another example is CSF-1, which is specific for later cells in the mononuclear phagocyte pathway (Guilbert and Stanley, 1980).

It is implicit in this model that growth and maturation down any particular pathway will require the participation of two factors, one operating early in the pathway, the other late.

*To whom correspondence/reprint requests should be addressed.

A dual requirement of this nature has already been demonstrated for development down the erythroid pathway (E-BPA early, erythropoietin late) (Axelrad et al., 1978; Wagemaker, 1978; Iscove, 1978; Wagemaker, 1980) and the megakaryocyte pathway (Mk-CSA early, Mk-Potentiator late) (Williams et al., 1982). In this paper, results are presented of additional tests of this model. In addition to their activity-stimulating granulocyte and macrophage colony formation, preparations of BPA are shown to stimulate net increases in the number of granulocyte/macrophage precursors as well as of pluripotential and committed erythroid precursors during the course of 8 days of liquid culture of mouse bone marrow cells. The WEHI-3 line of continuously growing mouse macrophages (Ralph et al., 1976) is known to secrete BPA (Iscove and Schreier, 1979b; Wagemaker, 1980) along with GM-CSA (Ralph et al., 1976; Williams et al., 1978) and Mk-CSA (Williams et al., 1982). Preparations of BPA partially purified from WEHI-3 conditioned medium augmented macrophage colony growth in the presence of pure CSF-1, mimicking the action of highly purified GM-CSF from lung cells. Finally, the results are presented of purification of BPA on the basis of charge, and sequentially at high resolution on the basis of hydrophobicity and size. Despite very extensive purification away from other contaminating proteins, the resulting preparation continued to display activity jointly on erythroid, granulocyte/macrophage, and megakaryocyte precursors.

## MATERIALS AND METHODS

### Production of BPA

Spleen cells from C57BL mice ($10^6$/ml) were cultured with either concanavalin A (1 μg/ml) or pokeweed mitogen (Gibco, Grand Island, NY, dissolved as directed, and used at a dilution of 1/1000). The medium (IMDM, Iscove et al., 1980) contained 800 μg/ml BSA, 30 μg/ml transferrin·2Fe, 100 μg/ml soybean lipid, and 25 μg/ml cholesterol. Conditioned medium was harvested at 6 days.

The WEHI-3 cells were adapted to culture in the laboratory of Dr. P. Ralph (Ralph et al., 1976) and reached Basel after 2 years of passaging in the laboratory of Drs. R. Bradley and S. Hasthorpe. They were cloned for use in the present study (clone G-4), but each of 8 clones had identical properties to those of the parent line. They were passaged continuously in IMDM containing transferrin·2Fe (15 μg/ml) and FCS (0.2%). The cells grow without the serum but secretion of BPA is greatly reduced. Ten-liter batches of conditioned medium were produced by expanding the cells in medium containing transferrin·2Fe (10 μg/ml) and FCS (0.1%). Medium was harvested when growth had ceased and cells were beginning to die. The conditioned medium was concentrated 200- to 1000-fold over Amicon UM-10 or YM-10 membranes, centrifuged to remove sediment, adjusted to pH 8.5 for stability (Burgess et al., 1980), filter-sterilized, and stored at 4°C with apparent long-term stability.

### Purification of BPA

All columns were run 4°C. Measurement of pH was always at room temperature.

Lectin-stimulated spleen cell-conditioned medium was purified on concanavalin A-Sepharose and Sephadex G-150 as detailed earlier (Iscove, 1978). The resultant material was concentrated by ultrafiltration over an Amicon Diaflo UM-10 membrane to a volume of 2 ml, equilibrated by dialysis against running buffer, and applied to a 28 cm × 1.6 cm column of DEAE-Sepharose (Pharmacia, Uppsala, Sweden). The column was equilibrated and developed with tetraethylethylenediamine·HCl 25 mM, pH 6.0, NaCl 50 mM + polyethylene glycol (PEG, MW 6,000) 50 mg/L, followed by 1 column volume of NaCl 1 M in the same buffer. The BPA and GM-CSA emerged unretarded at 1 column volume, whereas the material eluted with 1 M NaCl had macrophage-CSA and most of the O.D.$_{280}$, but no BPA.

Conditioned medium from WEHI-3 cells was heterogeneous in behavior on concanavalin A-Sepharose: 50% of BPA and GM-CSA was retarded, and 50% was unretarded. This step was therefore not used for activity from this source. For ion exchange chromatography, 3 ml of concentrated conditioned medium were dialyzed for 3 hr at room temperature with agitation against imidazole HCl buffer 25 mM pH 7.0, NaCl 50 mM, and PEG 10 mg/L. The sample was then adjusted if necessary to match this buffer in conductivity and pH using $H_2O$ and HCl. It was then applied to a 28 cm × 1.6 cm column of DEAE-Sepharose (Pharmacia, Uppsala, Sweden) equilibrated in the same buffer, followed by 150 ml of starting buffer; 100 ml of 1 M NaCl in starting buffer; and finally 200 ml of equilibrating buffer. Fractions were titrated to pH 8–8.5 with a concentrated solution of Tris base and filter-sterilized.

As the initial step in a purification sequence,

50 ml of 350-fold concentrated conditioned medium (equivalent to 18 L of original conditioned medium) were applied to an 86 cm × 2.5 cm column of Sephacryl S-200 (Pharmacia, Uppsala, Sweden) and eluted with HEPES. NaOH buffer 4 mM, pH 7.4, NaCl·16 M, and PEG 10 mg/ml. Smaller scale separations on an 88 cm × 1.6 cm column of Sephadex G-75 Superfine (Pharmacia) were carried out with the same buffer. Fractions were filter-sterilized.

Active fractions from Sephacryl S-200 were pooled and concentrated (UM-10) to 30 ml. Ammonium sulfate (A.S.) powder was added to a final concentration of 1.6 M, PEG to 10 mg/L and concentrated Tris base to a pH of 7.4. A small amount of precipitate was devoid of significant activity and was removed. The sample was applied to a 40 cm × 2.5 cm column of phenyl Sepharose (Pharmacia) equilibrated in A.S. 1.6 M, Tris base to pH 7.4, and PEG 10 mg/L. The sample was followed by 200 ml equilibrating buffer (eluting peak I in Fig. 4); 200 ml A.S. 0.8 M, Tris-HCl 25 mM pH 7.4, and PEG (eluting peak II); 200 ml Tris base 0.1 M pH 10.6, ethylene glycol 25% v/v, and PEG (eluting peak III); 200 ml ethanol 50% v/v in water (eluting peak IV); and 300 ml of equilibrating buffer. Alkaline fractions were brought to pH 8–8.5 with HCl. The fractions were pooled according to the 4 peaks in optical density (see Fig. 4), and each pool was concentrated (UM-10) to 2–5 ml. The pools were dialyzed thoroughly against NaCl·16 M, HEPES NaOH 4 mM pH 7.4, EDTA 2 mM, and PEG 10 mg/L for 24 hr, filter-sterilized, and stored at 4°C until assayed. Peak II (2 ml) contained a total BPA activity corresponding to about 4 L of the starting conditioned medium.

For reverse-phase high pressure liquid chromatography, 1.8 ml of the dialyzed peak II material was applied to a 0.46 cm × 10 cm Brownlee MPLC RP-18 column (Brownlee Laboratories, Santa Clara, CA) packed with 10 μm Lichrosorb ($C_{18}$-silane on silica support). The column was eluted with a nonlinear 10–40% gradient of n-propanol in 0.012 M HCl at 0.8 ml/min (Alvarez et al., 1981). Fractions (1.2 ml) were collected with aseptic precautions into sterile tubes containing an amount of 20 mM Tris-HCl pH 9.0 with NaCl and PEG to give finally a pH of 8–8.5, NaCl·2 M and PEG 100 mg/L. Fractions were stored at 4°C with apparent stability. They were assayed at very high dilution, where final concentrations of propanol in the cultures were innocuous.

Active fractions were pooled, concentrated to 1 ml over a UM-10 membrane and then to 0.35 ml in a dialysis sac surrounded by PEG (MW 20,000). This material (0.2 ml) was applied to a 0.75 cm × 60 cm high performance size exclusion column (TSK G3000SW, Toyo Soda, Amsterdam) and eluted at 0.1 ml/min with Tris-HCl 10 mM pH 8.0, NaCl 0.15 M, and PEG 10 mg/L. Fractions (0.16 ml) were collected with aseptic precautions into sterile tubes and stored at 4°C.

Active fractions were pooled and concentrated in a dialysis sac against PEG to 0.15 ml (with considerable loss of activity and UV-absorbing material). This volume was then applied again to the reverse-phase RP-18 column and eluted with a nonlinear 10–40% gradient of n-propanol in 0.012 M HCl at 0.8 ml/min. Fractions were collected and assayed as before.

### Liquid cultures

Bone marrow was flushed from the femurs of C57BL mice into IMDM and then diluted without washing into complete culture medium. This contained BSA 1%, transferrin·2Fe 30 μg/ml, and soybean lipid 100 μg/ml with cholesterol 60 μg/ml. Cells were cultured for 8 days in a volume of 1 ml in upright, loosely capped 15 ml Falcon plastic tubes (2051) in a humid 5% $CO_2$ atmosphere.

Increase in the numbers of primitive precursors (GM-CFC, BFU-M/E, and BFU-E) did not require serum and was optimum at an initial seeding of 8 × $10^4$ nucleated cells. Increase in CFU-E required serum (FCS) (even in the presence of BSA, transferrin, and lipid) and was better at 10% than at 5%. Optimum initial density for CFU-E increase was 4 × $10^4$ cells.

### Colony assays

BPA and GM-CSA were assayed in 1 ml methyl cellulose cultures (Iscove and Schreier, 1980) containing 8–10 × $10^4$ unwashed C3H·HeJ bone marrow cells, IMDM, BSA, transferrin·2Fe, oleic acid, dipalmitoyl phosphatidyl choline, cholesterol, 4% FCS, and erythropoietin 1 U/ml as detailed elsewhere (Iscove et al., 1980), along with the material to be tested. CFU-E were counted at 2 days of culture and other colonies at day 10 or 11. Lipids were sonicated at a final pH of 6.8, where dispersion is much easier than at the pH of 5.1 specified earlier (Iscove et al., 1980). Human urinary erythropoietin was purified on DEAE cellulose, concanavalin A-Sepharose, and Sephadex G-150 as detailed earlier (Iscove et al., 1980) and was free of BPA and GM-CSA.

The concanavalin A-Sepharose column was routinely purged with 50 mM boric acid·NaOH pH 7.0, not sodium borate·HCl as given earlier (Iscove et al., 1980) and which does not work. Erythropoietin recovery was best when buffers for Con A-Sepharose and Sephadex chromatography contained a relatively high concentration (500 mg/L) of PEG 6000.

Cultures for tests of synergy with CSF-1 were performed as above but without serum and with soybean lipid 100 μg/ml (Astec FLPV, Associated Concentrates, Woodside, NY) plus cholesterol 25 μg/ml (optimum concentrations) as the lipid source. Macrophage colonies were counted after 8 days of culture.

Megakaryocyte colonies were grown in 1.3 ml agar cultures of $1 \times 10^5$ C3H·HeJ bone marrow cells in IMDM containing BSA 0.1%, transferrin·2Fe 30 μg/ml, soybean lipid 100 μg/ml, and cholesterol 25 μg/ml, with or without FCS (Williams et al., 1982). As a source of MK-Potentiator, cultures also contained 50 μl of crude serum-free (0.03% BSA) conditioned medium from the P388D1 macrophage cell line stimulated with 0.1 μg/ml LPS (Williams et al., 1981). Acetyl cholinesterase-positive colonies were scored after 7 days of culture as described (Williams et al., 1981).

### Transformation of colony counts into units of activity

As evident in Figure 1, colony counts are not linearly related to the amounts of activity added to the culture. Rather, the relationship is a sigmoidal one between counts and the logarithm of the activity concentration.

When colony-stimulating activity is chemically separated and the results of assays on fractions are plotted as raw colony counts, the resulting activity profiles give a distorted impression of the activity distribution. In effect, the relative amount of activity tends to be exaggerated in the lower range of colony counts and underestimated in the higher range. It is therefore helpful to convert colony counts into an index that is directly proportional to the concentration of factor present. This can be done if the mathematical relationship between colony counts and activity concentration can be specified.

In the case of myeloid/erythroid, macrophage, and granulocyte colonies, the relationship is well approximated by the equation

$$p = \frac{100}{(k/a) + 1}, \tag{1}$$

where p is the percent of maximum colony count, a is the amount of activity added (e.g., volume), and k is a constant. The equation can be rewritten as

$$p = \frac{100}{ke^{-z} + 1}, \tag{2}$$

where $z = \ln(a)$. This form now expresses the relationship between p and ln(a). The curve of p versus z given by (2) is a symmetrical sigmoid with a nearly linear portion in the range $20 < p < 80$, and a slope of 25 at the midpoint of the linear portion (where $a = k$ and $p = 50$). The empirically measured slopes were 24 for granulocyte + macrophage (GM), 27 for M/E, and 15 for total M/E + E from the data given in Figure 1, and 23 for megakaryocytes (Mk). These slopes were determined by least squares fits of the equation

$$p = m + n \ln(a) \tag{3}$$

to the data on the linear portions of the curves, where a is the relative amount of activity (e.g., volume) and n is the slope.

Where there was good fit to equation (1) (GM, Mk, M/E), it was used to give the desired transformation. Defining 1 activity unit/culture as the amount required to give 50% of maximum colony count, and substituting $p = 50$ and $a = 1$ in (1), k is specified as 1. Rearranging,

$$a = \frac{1}{(100/p) - 1}, \tag{4}$$

where a is the number of units and p is the percent of maximum colony counts.

In the case of total M/E + E bursts where fit to equation (1) is poor, it is possible to make use of the linear portion of the relation between percent maximum colony count and the logarithm of factor concentration. Defining a unit of activity as above and substituting $p = 50$ and $a = 1$ in equation (3), m is specified as 50. Rearranging,

$$a = e^{(p-50)/n} \tag{5}$$

This equation is best restricted to colony counts between 20 and 80% of maximum.

These transformations have the advantage of allowing profiles to be plotted that give a more realistic quantitative impression of the distribution of activity. However, it must be cautioned that the transformations also ex-

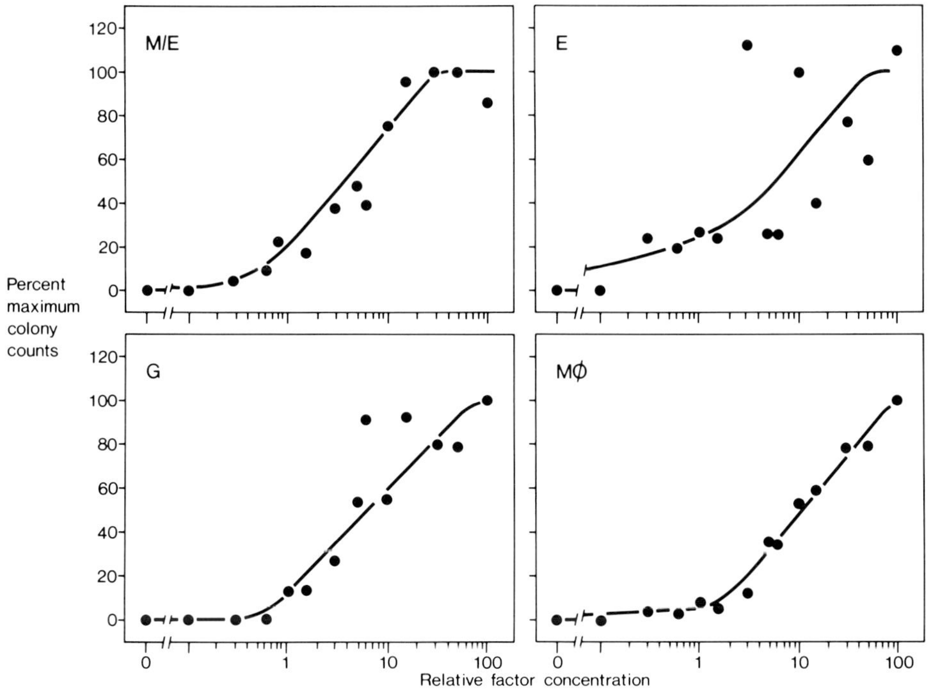


Fig. 1. Effect of WEHI-3 conditioned medium on numbers of mixed myeloid/erythroid bursts (M/E), pure or mixed megakaryocyte/erythrocyte bursts (E), giant disperse granulocytic colonies (G), and macrophage or mixed granulocyte/macrophage colonies (Mϕ). Conditioned medium was partially purified on Sephadex G-75. Active fractions were pooled, concentrated by ultrafiltration, and titrated in methyl cellulose cultures of $10^5$ bone marrow cells. Data from 4 experiments on titration of 2 individual preparations were normalized and pooled. Mean maximum counts were M/E 19, E 19, Mϕ 84, and G 9. Relative factor concentration is an arbitrary scale expressing relative volumes of the concentrate included in the cultures.

aggerate the magnitude of possibly insignificant differences in higher colony counts.

## RESULTS

If BPA-containing preparations are active on precursors of BFU-M/E, BFU-E, and GM-CFC, then they should influence the numbers of such progenitors in liquid cultures of bone marrow cells. To test for such an effect, adult mouse bone marrow cells were set up in liquid culture. Parallel aliquots of cells were also assayed for their initial progenitor cell content by culturing them in methyl cellulose. After 8 days of liquid culture, cells were harvested and assayed for colony-forming cells again in methyl cellulose. The pooled results from several experiments are given in Table 1. There was a net increase in the number of both granulocyte/macrophage precursors (about 2-fold) and early erythroid precursors (about 10-fold), which was dependent on the presence of the BPA preparation. Mixed BFU-M/E comprised about 30% of the total input erythroid BFU, but this proportion fell to about 8% of the output BFU, signifying a 2.7-fold increase of the BFU-M/E population in the presence of BPA. Table 1 also shows that whereas the BPA preparation was sufficient to support increase of primitive precursors, the more mature CFU-E required the additional presence of erythropoietin to achieve net increase of about 4-fold. A 14-fold increase in total CFU-E was achieved when the cells were fed with fresh medium containing both erythropoietin and BPA at day 7 and harvested on day 12.

M/E- and E-BPA were originally found in conditioned medium from cultures of lectin- or antigen-stimulated spleen cells. The presence of this activity in conditioned medium from continuous cultures of the WEHI-3 macrophage cell line is documented in Figure 1. In common with other mouse sources of BPA, WEHI-3 conditioned medium also stimulates

TABLE 1. Generation of new hemopoietic precursors in liquid cultures of bone marrow

| Precursor | Experiments | Mean total precursors/culture[1] Input | Output 0 | Ep[2] | BPA[3] | Ep + BPA |
|---|---|---|---|---|---|---|
| BFU-M/E + BFU-E[4] | 6 | 16 | 3.8 (0.24) | 5.5 (0.34) | 146 (9.1) | 162 (10) |
| GM-CFC[4] | 5 | 50 | 20 (0.40) | 29 (0.58) | 94 (1.9) | 98 (2.0) |
| CFU-E[5] | 3 | 107 | 0 (0) | 103 (0.96) | 11.3 (0.11) | 413 (3.9) |

[1]One milliliter cultures, 8 days. Figures in parentheses indicate ratio of output to input.
[2]Erythropoietin 0.5 units/ml.
[3]From lectin-stimulated spleen cell conditioned medium, partially purified on concanavalin A-Sepharose, Sephadex G-150, and DEAE-Sepharose. In some experiments BPA had been purified only by the latter 2 steps. The BPA was used at concentrations optimum for mixed myeloid/erythroid colony formation in methyl cellulose.
[4]Initial seeding density 8 × $10^4$ nucleated cells/ml.
[5]Initial seeding density 4 × $10^4$ nucleated cells/ml.

colony formation by GM-CFC (also shown in Fig. 1) as well as the early phase of megakaryocyte colony formation. Because of the ease of growing large numbers of these cells in very low concentrations of serum and their apparently constitutive release of activity, they are an ideal source of starting material for characterization and purification of the active molecules.

Figure 1 also shows that in the culture conditions used in this study, colony formation was totally dependent on the addition of active material, not only in the case of GM-CFC, but also in the case of erythroid and mixed myeloid/erythroid BFU. Achievement of low background erythroid burst formation was dependent on substitution of most of the serum requirement by defined materials. This allowed use of a low concentration (4%) of a fetal calf serum batch which had been selected for its low background property. Of further help was the use of marrow cells from the mouse strain C3H·HeJ, which is deficient in its response to bacterial lipopolysaccharide (Watson and Riblet, 1974). In these conditions, the erythroid bursts obtained with crude or purified WEHI-3-derived activity were indistinguishable in number or size from those obtained in more conventional conditions with higher serum concentrations. Figure 2 further indicates that the dose-response relationships of the various colony populations were not identical. In particular, the relationship between factor concentration and numbers of pure erythroid (including erythroid/megakaryocyte) colonies was often erratic and of shallower slope. Further, it was frequently observed that small numbers of such colonies were stimulated at factor concentrations insufficient for granulocyte/macrophage or mixed myeloid/erythroid colony formation.

In order to gain information on the number of molecular species that might be involved in the variety of colony-stimulating properties shown by WEHI-3 conditioned medium, a large batch was subjected to sequential purification. The initial step was size exclusion chromatography on a large column of Sephacryl S-200. Figure 2 shows results of a run at smaller scale on a column of Sephadex G-75 Superfine. The activity profiles were essentially similar to those obtained on Sephacryl S-200. Either column separated the colony-stimulating activities away from the bulk of the contaminating higher molecular weight proteins but not from each other.

Although size exclusion chromatography was used as the initial step in the present sequence of purification, it has the disadvantage of relatively limited protein capacity. Because of its higher capacity, ion exchange would be preferable as an initial step, provided it could give an equivalent degree of separation of the activity away from the major contaminating proteins. A preliminary result of chromatography on DEAE-Sepharose at pH 7.0 is shown in Figure 3. BPA and GM-CSA both eluted at one included column volume without apparent retardation in these conditions, compared with transferrin (pI 5.9), which was significantly retarded along with most of the contaminating protein. The result suggests that the active material has an isoelectric point near or above pH 7.0. The slight difference in the GM and E profiles was not consistently observed. The step achieved useful separation and will be used in modified form as the initial step in subsequent purification protocols. Previous experience

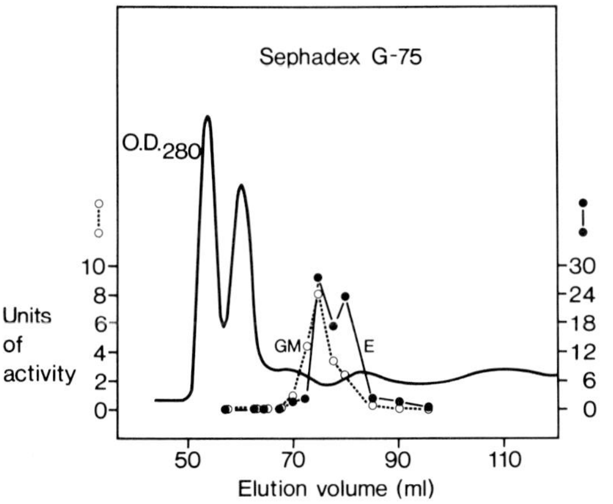


Fig. 2. Gel filtration of crude WEHI-3 conditioned medium. ○----○ GM-CSA, ●——● M/E + E-BPA. Total GM or M/E + E colony counts were transformed to activity units per culture plate as described in Methods. Equivalent of 400 ml conditioned medium applied to column, 4 μl assayed per plate.

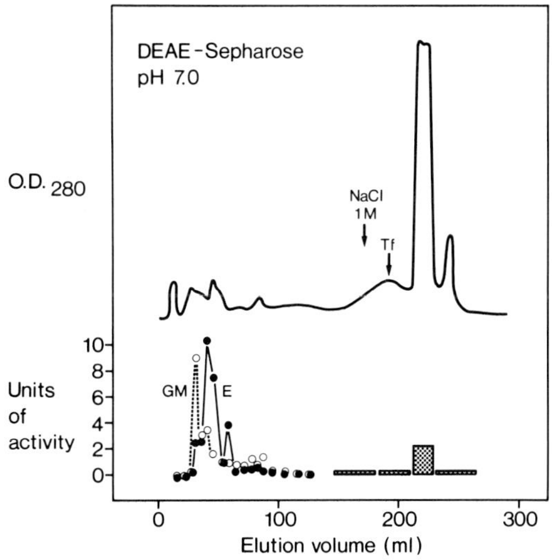


Fig. 3. Ion exchange chromatography of crude WEHI-3 conditioned medium with isocratic elution at pH 7.0, 25 mM imidazole HCl, and 50 mM NaCl. Symbols as in Figure 2. Shaded blocks represent total activity on GM-CFC (GM), M/E- and E-BFU (E) in the corresponding pools. Arrows indicate the point at which the eluting medium was changed to 1 M NaCl, and the first eluted portion of the transferrin, identified by its salmon pink color. Equivalent of 2 L of conditioned medium applied to column in a 3 ml volume, 1.6 μl assayed per plate.

with DEAE-Sephadex suggests that the unretarded material is also likely to contain Mk CSA (Williams et al., 1982), and to be free of CSF-1 (Das et al., 1980), which is present as a contaminant in unprocessed WEHI-3 conditioned medium.

The pooled active fractions from Sephacryl S-200 were next applied to a column of phenyl Sepharose for crude initial separation on the basis of hydrophobicity (Fig. 4). Approximately one third of the recovered activity eluted from the column ahead of the bulk of the interactive material. A high degree of purification relative to UV-absorbing material was thus achieved in peak II, although at the sacrifice of a considerable amount of activity remaining in peak III. The reduction of the amount of contaminating protein in the preparation down to a few milligrams permitted the application of a very large amount of activity to the subsequent high performance columns.

Peak II from the phenyl-Sepharose separation was next applied to a high performance reverse-phase column of $C_{18}$-silane and eluted with an acid n-propanol gradient (Fig. 5). BPA and GM-CSA eluted as a single peak ahead of most of the UV-absorbing material. In a separate run on the same column (not shown), the Mk-CSA profile was coincident with that of BPA + GM-CSA.

The active fractions from the reverse-phase column were pooled and chromatographed on a high performance size exclusion column (Fig. 6). BPA, GM-CSA, and Mk-CSA eluted together as a narrow peak, with additional separation from contaminating UV-absorbing material. The activity had an apparent molecular weight of 33,000, relative to globular protein standards, and was coincident with a peak of UV-absorbing material.

Active fractions from the size exclusion column were pooled and again applied to the reverse-phase column (Fig. 7). BPA and GM-CSA again eluted together, although they were now separated away from almost all the detectable UV-absorbing material. Fractions immediately preceding or following the activity peak were devoid of activity even when tested in combination. Mk-CSA was not monitored on this separation.

The partially purified BPA preparation (step III) obtained from the initial separation on $C_{18}$-silane (Fig. 5) stimulated formation of granulocyte, macrophage, and mixed granulocyte/macrophage colonies in a pattern very similar to that obtained with exhaustively purified GM-CSF from mouse lung-conditioned me-

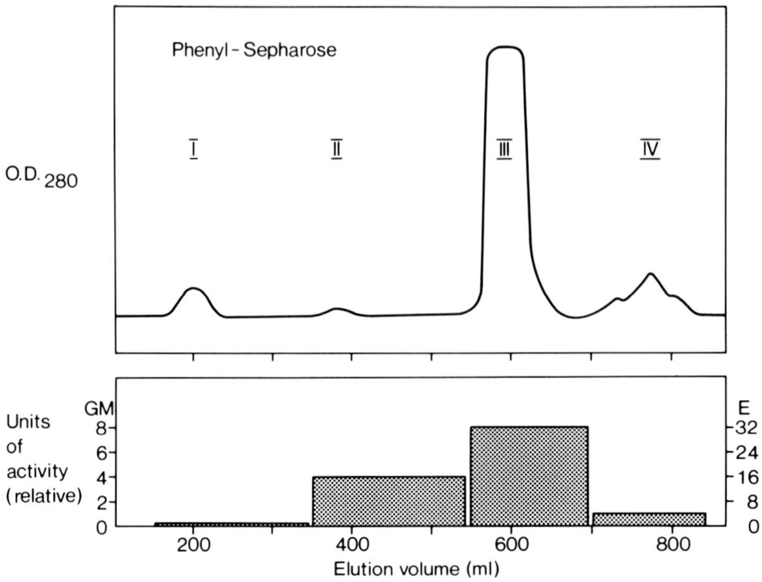


Fig. 4. Chromatography of Sephacryl S-200-purified WEHI-3 activity on phenyl-Sepharose. Shaded blocks indicate the relative total activities of the indicated pools on GM-CFC (GM) and total bursts (E). Activity scale represents units per culture plate. Equivalent of 18 L starting conditioned medium applied to column, concentrated pool volumes I 2.0 ml, II 2.0 ml, III 3.3 ml., IV 2.5 ml, relative units estimated from counts given by 0.4 μl per culture.

dium (Sheridan and Metcalf, 1972; Burgess et al., 1977), but quite distinct from the pattern of purely macrophage colony development ob-

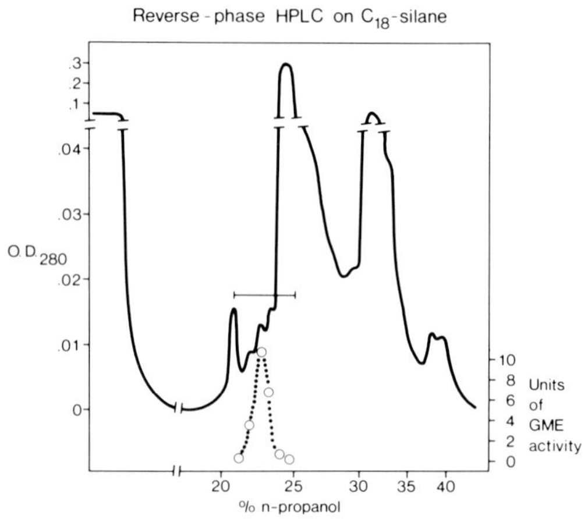


Fig. 5. Reverse phase HPLC on Lichrosorb RP-18 of pool II from phenyl-Sepharose. ○----○ activity units per culture plate for total GM-CSA and BPA. Equivalent of 4 L of starting conditioned medium applied to column, fraction size 2.7 ml, 0.8 μl assayed per plate. Horizontal bar indicates pool applied to subsequent purification step.

tained with pure CSF-1 obtained from mouse L cells (Guilbert and Stanley, 1980; Stanley and Heard, 1977). An additional similarity to GM-CSF was observed when effects on macrophage colony development were examined in methyl cellulose cultures containing maximum effective amounts of CSF-1. In serum-containing cultures, any one of GM-CSF, CSF-1, or the partially purified BPA preparation was sufficient to generate macrophage colonies having at least 100 cells per colony. However, in serum-free conditions, the macrophage colonies produced by single factors were smaller and could be further reduced in size below the scoring criterion by reducing the concentration of albumin and lipid. In such conditions (Table 2), the growth of large macrophage colonies in cultures containing pure CSF-1 required the additional presence of either pure GM-CSF or the partially purified BPA preparation. Other experiments similar to that in Table 1 showed smaller synergistic effects on numbers of colonies. However, the enhancing effect of BPA-containing preparations on the size of CSF-1-stimulated macrophage colonies was always seen.

In view of the similarity of action on GM-CFC displayed by pure GM-CSF and partially

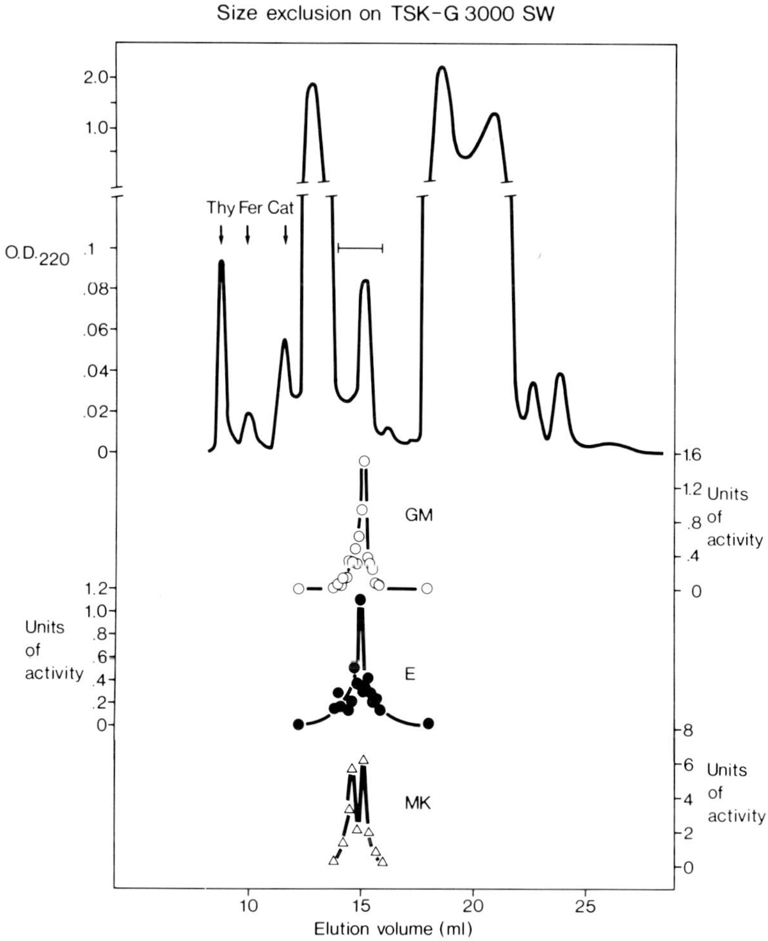


Fig. 6. Size exclusion chromatography on active fractions from reverse-phase HPLC. Arrows indicate O.D. profiles of 10 μg each of thyroglobulin, ferritin, and catalase. Profiles show GM-CSA (GM), M/E + E-BPA (E), and Mk-CSA (MK) expressed as activity units per culture plate. Equivalent of 2.3 L of starting conditioned medium applied, 0.05 μl assayed per plate. Horizontal bar indicates pool applied to subsequent purification step.

TABLE 2. Synergy of BPA[1] or GM-CSF[2] with CSF-1[3] in serum-free cultures

| Stimulus | Concentration[4] | Macrophage colonies/plate[5] | |
|---|---|---|---|
| | | Reduced BSA/lipid[6] | Optimum BSA/lipid |
| CSF-1 | 20 | 0 | 354 |
| BPA | 20 | 4 | 196 |
| GM-CSF | 30 | 9 | not tested |
| CSF-1 + BPA | 20/20 | 176 | 336 |
| CSF-1 + GM-CSF | 20/30 | 161 | not tested |
| CSF-1 + BPA | 2/2 | 26 | 236 |
| CSF-1 + GM-CSF | 2/3 | 37 | not tested |

[1]From WEHI-3, partially purified on Sephadex G-75 and reverse-phase HPLC.
[2]From mouse lung, exhaustively purified, step VI (Burgess et al., 1977).
[3]From L cells, exhaustively purified (Stanley and Heard, 1977).
[4]Microliter per culture, previously determined as optimal.
[5]One milliliter cultures containing 1 × 10$^5$ C3H HeJ marrow cells. Macrophage colonies containing 100 cells or more.
[6]One third of optimum.

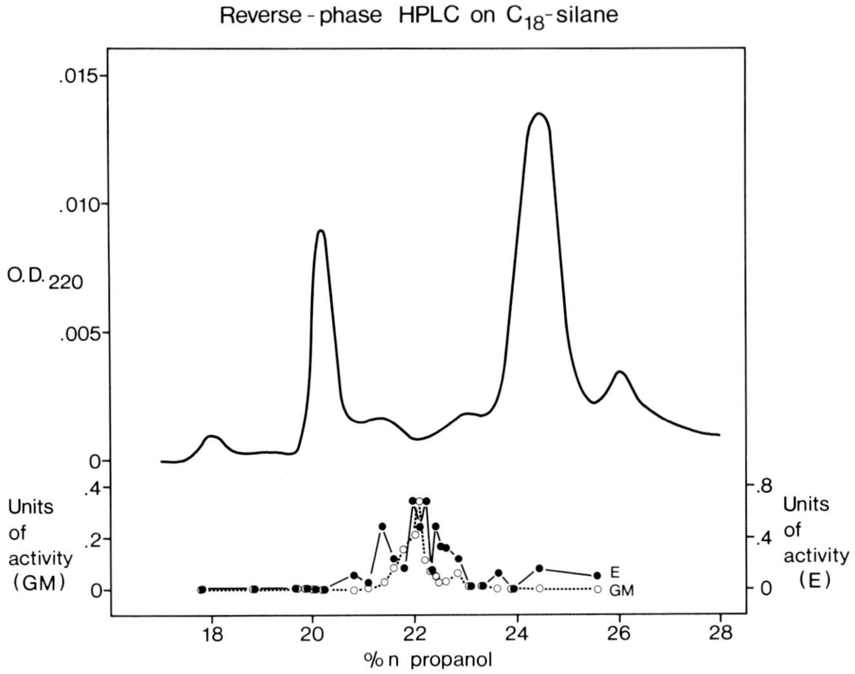


Fig. 7. Reverse-phase HPLC on Lichrosorb RP-18 of active pool from size exclusion chromatography. Activity units per plate are indicated. Equivalent of 2 L of starting conditioned medium applied, fraction size 0.43 ml, 3 μl assayed per plate.

purified BPA, the question arises as to the possible identity of these two principles. Figure 8 shows the results of parallel titration of pure GM-CSF and BPA partially purified on Sephacryl S-200 in bone marrow cultures containing erythropoietin. Over a concentration range in which both preparations gave comparable numbers of granulocyte/macrophage colonies, only the BPA preparation was effective in stimulating erythroid burst formation. An additional difference is documented in Table 3: whereas the material from WEHI-3 conditioned medium was able to stimulate colony formation by both granulocyte-macrophage and megakaryocyte precursors, GM-CSF lacked activity on megakaryocyte precursors.

TABLE 3. Test of purified GM-CSF for ability to support megakaryocyte colony formation[1]

| Stimulus | Colonies/plate[2] | |
|---|---|---|
| | Mk | GM |
| None | 2.7 ± 1.2 | 9 ± 2 |
| WEHI-3[3] | 21 ± 1.0 | > 100[4] |
| P388D1[5] | 3 ± 1.0 | 10[4] |
| WEHI-3 + P388D1 | 42 ± 5.2 | > 100[4] |
| GM-CSF[6] | 4.0 ± 0 | 58 ± 7 |
| GM-CSF + P388D1 | 4.3 ± 0.6 | 62 ± 8 |

[1]Data pooled from 2 experiments, one with 4% FCS, the other serum-free.
[2]Agar cultures (1.3 ml), $10^5$ C3H HeJ marrow cells, factors at optimal concentrations.
[3]WEHI-3 conditioned medium after passage over Sephadex G-50 (Williams et al., 1982).
[4]Estimated without counting.
[5]Source of Mk-Potentiator, crude conditioned medium 50 μl (Williams et al., 1981).
[6]Pure step VI (Burgess et al., 1977).

## DISCUSSION

These experiments show that molecules secreted by WEHI-3 cells are active on pluripotential myeloid/erythroid and more restricted erythroid progenitors in addition to earlier documented activity on granulocyte, macrophage, and megakaryocyte precursors. Further, the results show that these activities do not separate from one another during high resolution chromatography on the basis of either hydrophobicity or size, and at lower resolution on the basis of charge. This result is compatible with observations made earlier on the activities in lectin-stimulated spleen cell conditioned medium (Burgess et al., 1980). The ob-

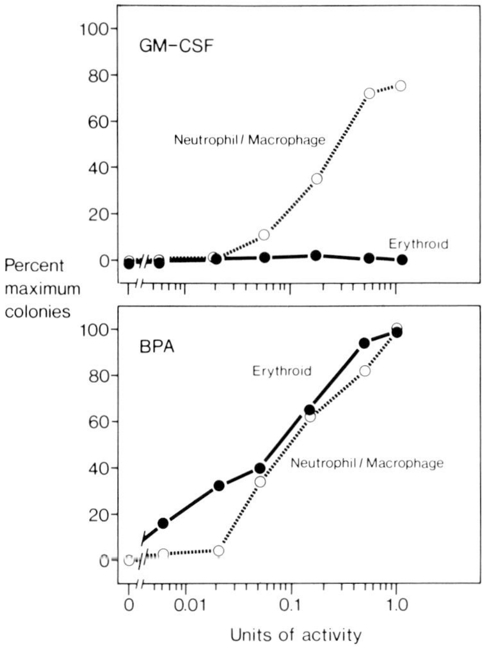


Fig. 8. Comparison of ability to stimulate GM colonies (○----○) and M/E and E bursts (●——●) by an exhaustively purified preparation of GM-CSF (Burgess et al., 1977) (upper panel) and by WEHI-3 conditioned medium partially purified on Sephadex G-75 ("BPA" preparation, lower panel). In this figure, "units of activity" signifies only the relative volume of material included in the assay cultures. Maximum colony numbers were GM 136 and M/E plus E bursts 31 per $10^5$ marrow cells.

servations are consistent with the concept that a single molecular class may be able to support growth of uncommitted as well as early committed cells of some if not all of the hemopoietic pathways.

It is also shown that partially purified BPA from WEHI-3 conditioned medium has activity on granulocyte/macrophage precursors that appears indistinguishable from that shown by exhaustively purified GM-CSF from mouse lung conditioned medium. Nonetheless, GM-CSF differs from partially purified BPA in not having activity on early committed erythroid and megakaryocyte precursors. If all the activities in the partially purified WEHI-3 material are properties of a single class of molecules, then these molecules can not be identical to lung-derived GM-CSF. It remains possible that the WEHI-3 preparations contain more than one active principle, of which one is GM-CSF. However, the extensive degree of purification of the WEHI-3 material using independent methods of high resolving power now makes the latter alternative less likely.

The phenomenon of synergy (Table 2) between different sources of GM-CSA has been documented before (Williams and Jackson, 1977; Bol et al., 1979; Wagemaker, 1980; Bertoncello et al., 1981). In general, these reports describe synergy between preparations likely to contain CSF-1 (Stanley, 1979; Das et al., 1980) and preparations containing GM-CSF or BPA. The observations are consistent with the view that the action of CSF-1 may be confined to late stages of macrophage differentiation, while other principles are required for support of earlier stages. The present results suggest that BPA and GM-CSF may be interchangeable in their capacity to augment macrophage growth in the presence of CSF-1.

Our experiments also indicate that preparations containing BPA can support net increase in the number of mixed BFU-M/E, BFU-E and GM-CFC in liquid cultures. This action may be related to the capacity of similar preparations to induce DNA synthesis in quiescent BFU-E and in spleen colony-forming cells (Wagemaker, 1980).

A model which integrates some of these observations is shown in Figure 9. It is proposed that responsiveness of hemopoietic precursor cells to hemopoietic growth factors reflects the presence of factor-specific structures in or on the cells. Since such structures are likely to be surface membrane receptors, they will be referred to here as "receptors" although details of their nature and location are immaterial to the model. Pluripotential precursors have receptors that recognize BPA. Immediate progeny of such cells would also possess these receptors, perhaps because they continue to be synthesized. However, even if their synthesis were to cease, they would be likely to persist with half times of disappearance typical of most cell proteins, i.e., 12–36 hr. This would imply responsiveness to BPA for some time after synthesis of the receptors stopped. Thus, given that pluripotential cells are responsive to a particular factor, it would be reasonable to expect similar responsiveness in their immediate progeny, including some that have already taken steps along chosen pathways.

A second fundamental element of the model is that receptors conferring responsiveness to late-acting factors such as erythropoietin, CSF-1, or possible granulocyte-specific factors, begin to be synthesized at significant rates

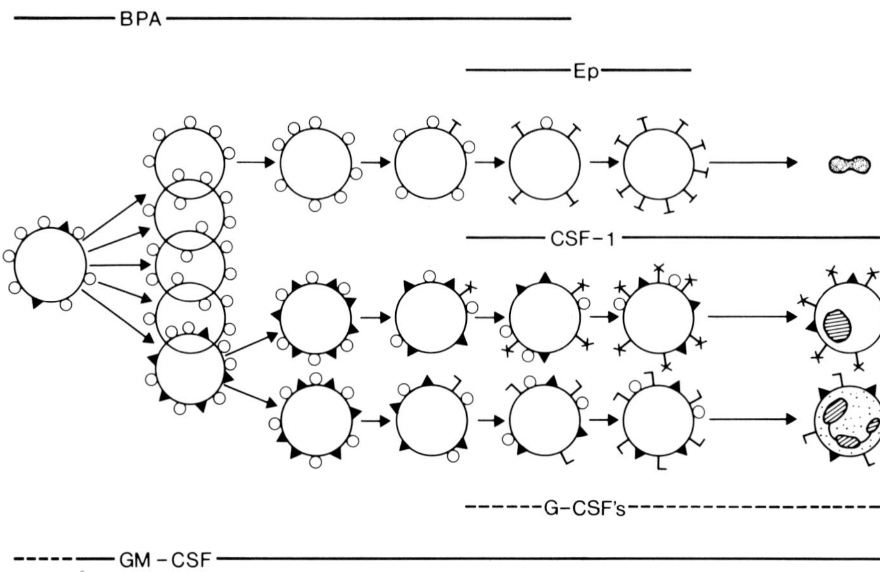


Fig. 9. A proposed distribution of "receptors" for various hemopoietic growth factors. A pluripotential cell (far left) gives rise first to partially and later totally committed cells. Receptors for erythropoietin (T), for CSF-1 (X), and for possible granulocyte-specific factors (˥) develop only as cells mature along the red cell, mononuclear phagocyte, and granulocytic pathways, respectively. Pluripotential cells and all of their early committed progeny have receptors for BPA (○). Receptors for GM-CSF (▲) are present early and late but only in the granulocyte and mononuclear phagocyte pathways.

only as part of an evolving differentiation program. The points in a particular pathway at which BPA receptor synthesis stops and at which receptors for late-acting factors begin to appear, the particular rates involved, and the relation to rate of cell proliferation (receptor dilution) might differ between pathways or between mammalian species. Such differences could lead to discrepancies in apparent biological properties of factors that may be more apparent than fundamental.

A third postulate of the model is that if a particular cell has different receptors recognizing different factors, receptor occupancy will have identical consequences within the cell regardless of which type of receptor is occupied. In the model, early granulocyte and macrophage precursors have receptors for both BPA and GM-CSF, and therefore can respond to either factor. In contrast, early erythroid precursors and megakaryocyte precursors (not shown) have receptors for BPA but lack receptors recognizing GM-CSF. The presence of GM-CSF receptors in pluripotential precursors is suggested by experiments in which purified GM-CSF appeared to initiate mixed myeloid/erythroid colony formation at a low efficiency (Metcalf et al., 1980).

Although omitted from Figure 9 in the interest of simplicity, the model would include BPA-responsiveness early in the megakaryocyte pathway, and receptors for Mk-Potentiator appearing later in the pathway.

The essential observation in the present study has been the failure of E- and M/E-BPA to dissociate from GM-CSA during the course of extensive purification of WEHI-3-conditioned medium. Assays for Mk-CSA, synergy with CSF-1, induction of DNA synthesis in short term culture and of net progenitor increase in longer term cultures still remain to be performed on highly purified material at the present and subsequent stages of purification. It therefore remains to be seen whether or not this array of activities can be definitively attributed to a single molecular class, and whether the molecules from WEHI-3 cells are the same or different from those released from lymphocyte populations, or from those that may operate on hemopoietic precursors in their normal environment (Frindel et al., 1976; Wagemaker, 1978; Porter et al., 1980; Lord, 1981) and in the normal steady state. In addition to chemical purification, the use of monoclonal antibodies to factors and receptors, and of methods involving gene cloning and translation are certain to play major roles in the resolution of these issues.


## ACKNOWLEDGMENTS

The authors wish to thank M. Schweizer, M. Barnes, and H. Jackson for excellent technical assistance, H.-P. Stahlberger and M. Pazdera for the graphic work, S. Buser for secretarial assistance, and Drs. C. Paige and K. Smith for critical readings of the manuscript. GM-CSF was a generous gift from Dr. A. W. Burgess, and CSF-1 was kindly contributed by Dr. E. R. Stanley.


## LITERATURE CITED


Alvarez, V.L., Roitsch, C.A., and Henriksen, O. (1981) Purification of H-2[a] heavy chain and $\beta_2$-microglobulin by reverse phase high-performance liquid chromatography. Anal. Biochem *115:*353–358.

Axelrad, A.A., McLeod, D.L., Suzuki, S., and Shreeve, M.M. (1978) Regulation of the population size of erythropoietic progenitor cells. In: Differentiation of Normal and Neoplastic Hematopoietic Cells, Book A. B. Clarkson, P.A. Marks, and James E. Till, eds. Cold Spring Harbor Conferences on Cell Proliferation, Cold Spring Harbor Laboratory, Vol. 5, pp. 155–163.

Aye, M.T. (1977) Erythroid colony formation in cultures of human marrow: Effect of leukocyte conditioned medium. J. Cell Physiol., *91:*69–78.

Bertoncello, I., Bradley, T.R., and Hodgson, G.S. (1981) Characterization and enrichment of macrophage progenitor cells from normal and 5-fluorouracil-treated mouse bone marrow by unit gravity sedimentation. Exp. Hematol., *9:*604–610.

Bol, S., Visser, J., and van der Engh, G. (1979) The physical separation of three subpopulations of granulocyte/macrophage progenitor cells from mouse bone marrow. Exp. Hemat., *7:*541–553.

Burgess, A.W., Camakaris, J., and Metcalf, D. (1977) Purification and properties of colony-stimulating factor from mouse lung-conditioned medium. J. Biol. Chem., *252:*1998–2003.

Burgess, A.W., Metcalf, D., Russell, S.H.M., and Nicola, N.A. (1980) Granulocyte/macrophage-, megakaryocyte-, eosinophil-, and erythroid- colony-stimulating factors produced by mouse spleen cells. Biochem. J., *185:*301–314.

Das, S.K., Stanley, E.R., Guilbert, L.J., and Forman, L.W. (1980) Discrimination of a colony-stimulating factor subclass by a specific receptor on a macrophage cell line. J. Cell Physiol., *104:*359–366.

Fauser, A.A., and Messner, H.A. (1978) Granuloerythropoietic colonies in human bone marrow, peripheral blood, and cord blood. Blood, *52:*1243–1248.

Frindel, E., Croizat, H., and Vassort, F. (1976) Stimulating factors liberated by treated bone marrow: In vitro effect on CFU kinetics. Exp. Hematol., *4:*56.

Guilbert, L.J., and Stanley, E.R. (1980) Specific interaction of murine colony-stimulating factor with mononuclear phagocytic cells. J. Cell Biol., *85:*153–159.

Hara, H., and Ogawa, M. (1978) Murine hemopoietic colonies in culture containing normoblasts, macrophages, and megakaryocytes. Am. J. Hematol., *1:*23–31.

Iscove, N.N. (1978) Erythropoietin-independent stimulation of early erythropoiesis in adult bone marrow cultures by conditioned media from lectin-stimulated mouse spleen cells. In: Hematopoietic Cell Differentiation. D.W. Golde, M.J. Cline, D. Metcalf, and C.F. Fox, eds. Academic Press, New York, pp. 37–52.

Iscove, N.N., and Guilbert, L.J. (1978) Erythropoietin-independence of early erythropoiesis and a two-regulator model of proliferative control in the hemopoietic system. In: In Vitro Aspects of Erythropoiesis. M.J. Murphy, Jr., ed. Springer-Verlag, New York, pp. 3–7.

Iscove, N.N., Guilbert, L.J., and Weyman, C. (1980) Complete replacement of serum in primary cultures of erythropoietin-dependent red cell precursors (CFU-E) by albumin, transferrin, iron, unsaturated fatty acid, lecithin, and cholesterol. Exp. Cell Res., *126:*121–126.

Iscove, N.N., and Schreier, M.H. (1979a) Clonal growth of cells in semisolid or viscous medium. In: Immunological Methods. I. Lefkovits and B. Pernis, eds. Academic Press, New York, pp. 379–385.

Iscove, N.N., and Schreier, M.H. (1979b) Involvement of T cells and macrophages in generation of burst-promoting activity. Exp. Hematol., *7:*4.

Johnson, G.R., and Metcalf, D. (1977) Pure and mixed erythroid colony formation in vitro stimulated by spleen conditioned medium with no detectable erythropoietin. Proc. Natl. Acad. Sci. U.S.A., *74:*3879–3882.

Lord, B.I. (1981) The relationship of $G_0$ to the cell cycle of haemopoietic spleen colony-forming cells. Cell Tissue Kinet., *14:*425–431.

Metcalf, D., MacDonald, H.R., Odartchenko, N., and Sordat, B. (1975) Growth of mouse megakaryocyte colonies in vitro. Proc. Natl. Acad. Sci. U.S.A., *72:*1744–1748.

Metcalf, D., Johnson, G.R., and Burgess, A.W. (1980) Direct stimulation by purified GM-CSF of the proliferation of multipotential and erythroid precursor cells. Blood, *55:*138–147.

Nathan, D.G., Chess, L., Hillman, D.G., Clarke, B., Breard, J., Merler, E., and Housman, D.E. (1978) Human erythroid burst-forming unit: T-cell requirement for proliferation in vitro. J. Exp. Med., *147:*324–339.

Parker, J.W., and Metcalf, D. (1974) Production of colony-stimulating factor in mitogen-stimulated lymphocyte cultures. J. Immunol., *112:*502–510.

Porter, P.N., Ogawa, M., and Leary, A.G. (1980) Enhancement of the growth of human early erythroid progenitors by bone marrow conditioned media. Exp. Hematol., *8:*83.

Ralph, P., Moore, M.A.S., and Nilsson, K. (1976) Lysozyme synthesis by established human and murine histiocytic lymphoma cell lines. J. Exp. Med., *143:*1528–1533.

Schreier, M.H., and Iscove, N.N. (1980) Haematopoietic growth factors are released in cultures of H-2-restricted helper T cells, accessory cells and specific antigen. Nature, *287:*228–230.

Sheridan, M.W., and Metcalf, D. (1972) A low molecular weight factor in lung-conditioned medium stimulating granulocyte and monocyte colony formation in vitro. J. Cell Physiol., *81:*11–24.

Stanley, E.R., and Heard, P.M. (1977) Factors regulating macrophage production and growth: Purification and some properites of the colony-stimulating factor from medium conditioned by mouse L cells. J. Biol. Chem., *252:*4305–4312.

Stanley, E.R. (1979) Colony-stimulating factor (CSF) radioimmunoassay: Detection of a CSF subclass stimulating macrophage production. Proc. Natl. Acad. Sci. U.S.A., *76:*2969–2973.

Wagemaker, G. (1978) Cellular and soluble factors influencing the differentiation of primitive erythroid progenitor cells (BFU-e) in vitro. In: In Vitro Aspects of Erythropoiesis. M.J. Murphy, Jr., ed. Springer-Verlag, New York, Heidelberg, Berlin, pp. 44–57.

Wagemaker, G. (1980) Early erythropoietin-independent stage of in vitro erythropoiesis: Relevance to stem cell differentiation. In: Experimental Hematology Today 1980. S.J. Braun, G.D. Ledney, and D.W. van Bekkum, eds. S. Karger, Basel, pp. 47–60.

Watson, J., and Riblet, R. (1974) Genetic control of responses to bacterial lipopolysaccharides in mice. I. Evidence for a single gene that influences mitogenic and immunogenic responses to lipopolysaccharides. J. Exp. Med., *140:*1147–1161.

Williams, N., and Jackson, H. (1977) Analysis of populations of macrophage-granulocyte progenitor cells stimulated by activities in mouse lung-conditioned medium. Exp. Hematol., *5:*523–534.

Williams, N., Eger, R.R., Moore, M.A.S., and Mendelsohn, N. (1978) Differentiation of mouse bone marrow precursor cells into neutrophil granulocytes by an activity separated from WEHI-3 cell-conditioned medium. Differentiation, *11:*59–63.

Williams, N., Jackson, H., Ralph, P., and Nakoinz, I. (1981) Cell interactions influencing murine marrow megakaryocytes: Nature of the potentiator cell in bone marrow. Blood, *57:*157–163.

Williams, N., Eger, R.R., Jackson, H.M., and Nelson, D.J. (1982) Two-factor requirement for murine megakaryocyte colony formation. J. Cell Physiol *110:*101–104.

# Polycythemia Vera: Studies of Hemopoiesis in Continuous Long-Term Culture of Human Marrow

**JERRY S. POWELL,* PHILIP J. FIALKOW, AND JOHN W. ADAMSON**

*Divisions of Hematology (J.S.P., J.W.A.) and Medical Genetics (P.J.F.), Department of Medicine, University of Washington School of Medicine, and the Hematology Research Laboratory (J.S.P., J.W.A.) and Section of Medical Genetics (P.J.F.), Medical Service, Veterans Administration Medical Center, Seattle, Washington 98195*

Long-term cultures of marrow cells from ten normal subjects and three patients with polycythemia vera were established to compare normal and neoplastic hemopoiesis in vitro. Suspended cells were removed periodically from the cultures and assayed for their content of various colony-forming cells, including erythroid colony- and burst-forming cells (CFU-E and BFU-E), granulocyte/macrophage colony-forming cells (CFU-C), and "mixed cell" colony progenitors (CFU-GEMM). To determine if mixed cell colonies arise from a single progenitor, we used the cellular mosaicism conferred by X-chromosome inactivation. The isoenzymes of glucose-6-phosphate dehydrogenase (G-6-PD) were used as markers of the mosaicism. Preliminary results suggest that these colonies are clonal only at low plating densities. The G-6-PD system was also used to determine whether selection or "drift" occurs in continuous long-term cultures. The ratios of G-6-PD isoenzyme types in pooled colonies from cultures of two normal heterozygotes remained similar, indicating stable cultures. Long-term cultures of normal marrow and marrow from the patients with polycythemia vera maintained BFU-E for a mean of 8.7 ($\pm$ 0.6) and 12.5 ($\pm$ 0.5) weeks ($P = 0.03$), respectively. The fractions of total BFU-E detected as endogenous erythroid colonies remained similar over the culture period. These results demonstrate that 1) hemopoiesis in polycythemia vera can be analyzed in long-term culture; 2) polycythemia vera marrow grows as well or better than normal in long-term culture; and 3) the proportion of the neoplastic clone in polycythemia vera represented by endogenous erythroid colony growth is unchanged over time, suggesting no reemergence of normal stem cell progeny in this system.

Polycythemia vera is a clonal disease of hemopoietic stem cells, typically characterized by increased red cells, platelets, and granulocytes (Adamson et al., 1976). Although this disease is a neoplasm, at least erythropoiesis retains some degree of responsiveness to normal physiologic regulators. Consequently, phlebotomy results in increases in plasma iron turnover and reticulocyte production in patients who are not iron deficient (Adamson, 1970) and, in vitro, erythroid colony growth by polycythemia vera marrow cells is enhanced by erythropoietin, glucocorticoids, and other hormones (Golde et al., 1977; Prchal et al., 1978; Mladenovic and Adamson, 1979). A distinctive feature of polycythemia vera marrow or peripheral blood cells is their formation in vitro of erythroid colonies in the absence of added erythropoietin. These so-called "endogenous erythroid colonies" may reflect the growth of a population of colony-forming cells that is extremely sensitive to the very low levels of erythropoietin in the serum used as part of the culture medium (Zanjani et al., 1977). While all endogenous erythroid colony-forming cells are members of the neoplastic clone, they represent only a fraction of the abnormal population (Adamson et al., 1980).

The proliferation and differentiation of populations of hemopoietic progenitors are dy-

*To whom correspondence/reprint requests should be addressed.

namic processes of development which occur continually. While clonogenic assays have been employed for a number of years and have provided much useful information about progenitor populations, such assays have two major limitations: 1) sampling is restricted to a single time and thus reflects hemopoiesis only at one point, and 2) interactions between different cell populations over time are difficult to study. Dexter et al. (1977) developed a system for the continuous long-term culture of murine marrow, which has been adapted by several groups to studies of human marrow (Gartner and Kaplan, 1980; Hocking and Golde, 1980; Greenberger et al., 1979; Moore et al., 1980; Toogood et al., 1980). We have employed a similar technique to study the regulation of in vitro hemopoiesis in marrow from normal subjects and from patients with polycythemia vera and to compare the maintenance of progenitor cells, their frequency, and the frequency of endogenous erythroid colony-forming cells with time.

## METHODS

### Source of marrows

Marrow samples used in this study were obtained with informed consent from ten normal volunteers and three patients with polycythemia vera. Each patient met the diagnostic criteria of the Polycythemia Vera Study Group (Berlin, 1975) and had been treated by phlebotomy alone; none had received myelosuppressive therapy.

### Long-term culture system

Marrow cells were aspirated from the posterior iliac crest into sterile heparinized syringes and the cells were immediately diluted at least ten-fold with culture medium. A buffy coat was prepared twice by spinning the marrow cells at 200g for 10 minutes. Nucleated cells were counted using a hemacytometer and $20 \times 10^6$ trypan blue dye-excluding cells in culture medium added to Corning T-25 flasks.

Gartner and Kaplan (1980) reported successful long-term culture of granulocyte precursors from human rib marrow cells using medium supplemented with hydrocortisone and equal amounts of fetal calf serum (FCS) and horse serum.

After preliminary experiments, we modified their techniques for optimal growth of erythroid progenitors from marrow cells obtained by aspiration. The culture medium consisted of α-medium, 12.5% FCS, 12.5% horse serum, $10^{-4}$ M β-mercaptoethanol, $10^{-6}$ M hydrocortisone, 20 units/ml penicillin, 20 μg/ml streptomycin, 2 mM L-glutamine, 5 mg vitamin C per 100 ml, and 100 units insulin per 100 ml. The FCS used was selected for its ability to support optimal human erythroid colony growth. Differences in the sources of horse serum were made insignificant with the addition of hydrocortisone, as previously shown by Greenberger, et al. (1979).

After being seeded with cells the flasks were gassed with 5% $CO_2$ and cultured at 33°C. Four to five days later excess red cells were removed by centrifugation of the suspended cells over Lymphoprep® (1.077 specific gravity). The nucleated cells that did not enter the gradient were washed three times, placed in fresh medium, and put back into the flasks. Each marrow was set up in four to ten culture flasks. The numbers of suspended cells in the medium were counted each week at the time of feeding.

### Methylcellulose assays

At weekly intervals the flasks were gently rocked, half of the medium removed, and an equal volume of medium replaced. The cells recovered at the times of feeding were cultured for various colony-forming cells. The methylcellulose technique as described by Iscove et al. (1974) and modified for this laboratory was utilized for erythroid colony growth (Prchal et al., 1978). The suspension of cells removed was centrifuged, washed, and plated at a nucleated cell concentration of $10^5$ per ml in methylcellulose, α-medium, 30% FCS, $10^{-4}$ M β-mercaptoethanol, 20 units/ml penicillin, 20 μg/ml streptomycin, and 1% deionized bovine serum albumin. Erythropoietin (1 unit) and phytohemagglutinin-stimulated leukocyte conditioned medium (5%) were added to most plates. Other plates had no erythropoietin added in order to permit observation of endogenous erythroid colony growth. Plates for granulocyte/macrophage progenitors (CFU-C) were prepared by the addition of 10% conditioned medium.

## RESULTS

### Cell recovery

With removal of 50% of the medium each week, the nonadherent cell population remaining in the flask was reduced. To correct for depopulation, a first approximation would be to multiply the number of cells obtained at each feeding by $2^{n-1}$, where n = the number of times the cultures were depleted by feeding. Rates of cell regeneration vary with cell density, interaction between cell populations, and other factors, but this correction provides an estimate of the proliferative capacity of the marrow in culture.

The cell numbers recovered per week are compared in Figure 1. The total numbers of nonadherent cells removed each week were not significantly different between the cultures of polycythemia vera and normal marrow cells. For both groups, the number of cells recovered dropped off rapidly to a nadir at 3–4 weeks followed by a variable period of stable proliferation. Depletion of the cultures occurred when weekly removal of nonadherent cells exceeded the proliferative capacity of the cultures. There was no statistically significant difference in the longevity of the cultures.

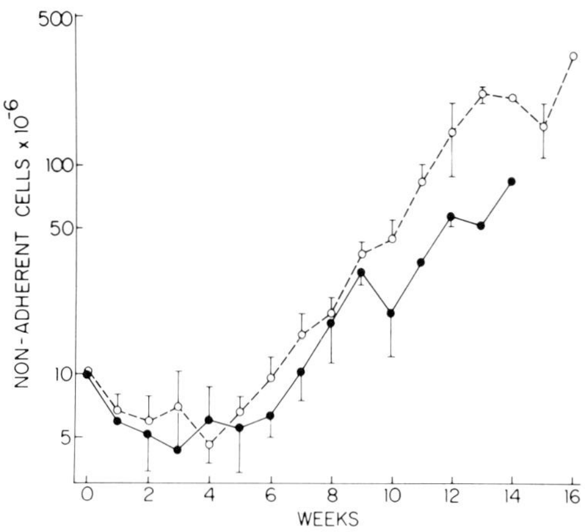


Fig. 1. The numbers of nonadherent cells removed each week per culture flask. The results represent the means ± SEM for cultures of marrow from polycythemia vera (●-●) and normal (○---○) subjects and have been corrected for depopulation.

### Adherent layer

Each week the culture flasks were examined by inverted light microscopy. Adherent layers formed in all marrow specimens and reached confluence by 2–3 weeks. The adherent layers formed equally rapidly in polycythemia vera marrow cultures and in normal cultures and appeared similar over the culture period. Lipid-laden macrophages (fat cells) were seen usually beginning about 5–6 weeks. Their appearance in great numbers preceded the rapid decline of the culture by 1–2 weeks. Cultures completely lacking these cells, however, had a shorter survival. Therefore, some lipid-laden macrophages were necessary for a functional adherent layer. Morphologically, the adherent layer consisted of endothelial cells and a few fibroblasts. Areas of small round cells were seen clustered together on the surface of the adherent layer. When these cells were removed and stained, they were found to be largely myelocytes, promyelocytes, and blast cells. These areas decreased as the cultures declined over time. Individual marrows varied in the number of macrophages and other cell types seen, but there were no distinctive differences in the cultures of polycythemia vera marrow when compared to normal.

### Assays for committed colony-forming cells

Erythroid progenitor cells were assayed each week as has been described previously. Colonies derived from erythroid colony-forming cells (CFU-E) were enumerated on day 7 as clusters of 8 to 32 hemoglobinized cells. These more differentiated erythroid progenitors were not detectable beyond 2–3 weeks in cultures of normal marrow. In cultures of polycythemia vera marrow, endogenous erythroid colonies, scored on day 10–12 as hemoglobinized colonies in the plates without added erythropoietin, were detected for the duration of the culture period and represented a relatively stable fraction of the total number of erythroid burst-forming cells (BFU-E) (Table 1). The numbers of BFU-E recovered per week are compared in Figure 2. BFU-E were detectable from some normal cultures for 16 weeks but the average number of weeks that normal cultures yielded BFU-E was significantly less than the average for the cultures of polycythemia vera marrow cells (Table 2). Although the numbers of BFU-E were somewhat greater from the polycythemia vera marrows, the differences were not statistically significant. The frequencies of BFU-E per $10^5$ cells are shown in Figure 3. No consistent difference is observed when the fre-

TABLE 1. Endogenous erythroid colonies in long-term culture[1]

| Patient | Initial value | Mean | (± SEM) | Weeks of study |
|---|---|---|---|---|
| 1 | 42% | 30% | (± 4.7) | 9 |
| 2 | 9% | 17% | (± 3.3) | 13 |
| 3 | 34% | 25% | (± 5.2) | 15 |

[1]Percent of total erythroid colonies represented by the number of endogenous colonies in long-term marrow cultures from three patients with polycythemia vera. The initial value is that obtained with marrow cells at the time the long-term cultures were established. There were no statistically significant differences between initial values and values obtained at any subsequent week.

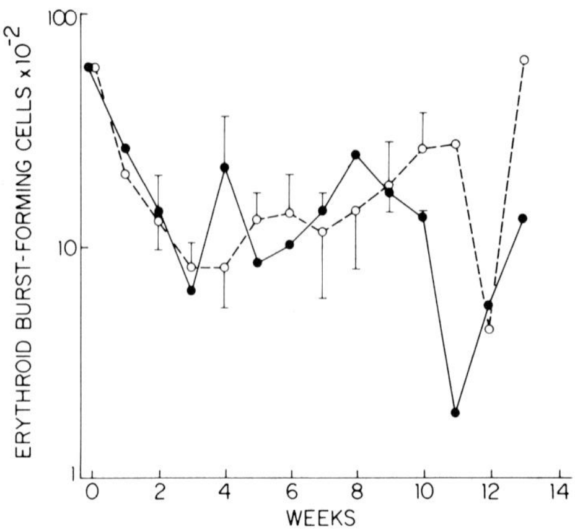


Fig. 2. Total BFU-E recovered each week in continuous long-term marrow culture. The results represent the means ± SEM for cultures of marrow from polycythemia vera (●-●) and normal (○---○) subjects and have been corrected for depopulation.

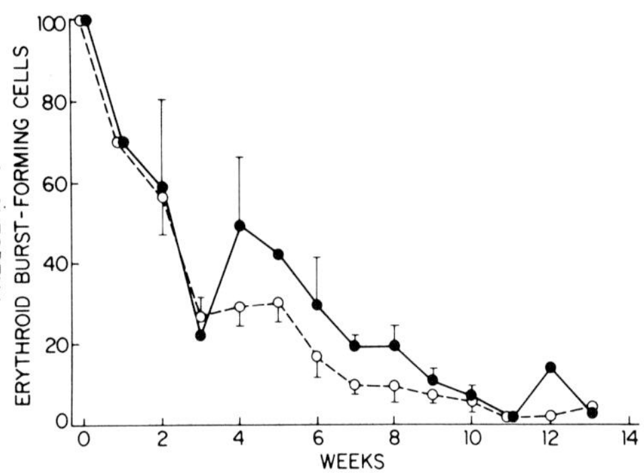


Fig. 3. The frequency of BFU-E per $10^5$ nonadherent cells assayed each week from long-term cultures of polycythemia vera (●-●) and normal (○---○) subjects. The results represent the means ± SEM expressed as percent of BFU-E per $10^5$ mononuclear marrow cells plated on day 0.

TABLE 2. Duration of recovery of BFU-E from long-term culture

| | Mean | ± SEM |
|---|---|---|
| Normal | 8.7 weeks* | ± 0.6 |
| Polycythemia vera | 12.5 weeks* | ± 0.5 |

*P = 0.03.

quencies for normal and polycythemia vera marrow cultures are compared.

From both polycythemia vera and normal marrow cultures, CFU-C regularly were detected for a longer period than erythroid precursors. However, no consistent difference was observed in the numbers or frequencies of CFU-C.

## Assay for mixed cell colony-forming cells (CFU-GEMM)

Colonies containing multiple cell types ("mixed cell" colonies) may represent the growth in culture of a multipotent stem cell as described by Fauser and Messner (1978). These colony-forming cells occur at low frequencies (1–10 per $10^5$ light density mononuclear cells) in normal marrows. Such colonies were identified from the long-term cultures of both normal and polycythemia vera marrows for 4–5 weeks. However, the frequencies were too low to allow statistical comparisons after the first week.

To determine whether these mixed cell colonies arose from single pluripotent stem cells, we used the cell marker system provided by the isoenzymes of glucose-6-phosphate dehydrogenase (G-6-PD). This marker system is particularly useful because early in embryogenesis a portion of one of the two X chromosomes containing the structural locus for G-6-PD undergoes inactivation in somatic cells of females. Following inactivation, only one structural gene for G-6-PD remains active in that cell and in all its daughter cells. Thus, females heterozygous at the G-6-PD locus have cellular mosaicism. In one population of cells, the usual type B enzyme is synthesized and in another population of cells the electrophoretic variant, A or A$^-$, is produced. Therefore, colonies arising from a single pluripotent progenitor will show only one isoenzyme type (e.g., A or B).

In initial experiments, mixed cell colonies from the marrow of a normal G-6-PD heterozygote were tested. Light density (< 1.077 g/ml) mononuclear cells were prepared over Lymphoprep® and then separated by unit gravity velocity sedimentation using a modification (Singer and Adamson, 1976) of the method of Miller and Phillips (1969). Fractions with cells having sedimentation velocities of 4–8 mm/hr contained the majority of the progenitor cells. All doublets sedimented rapidly and thus were excluded. These single cells were then pooled and plated at several cell concentrations ranging from $2 \times 10^4$ to $3 \times 10^5$ cells/ml in the methylcellulose assay.

Mixed cell colonies were counted on day 14 of culture and a number analyzed for G-6-PD (Fig. 4). At the lowest cell concentration, only

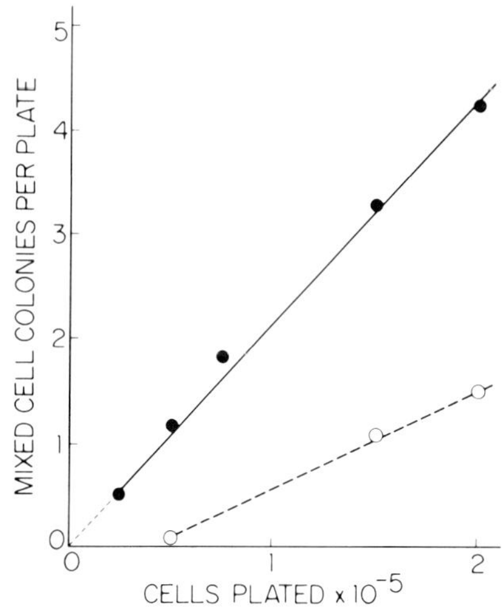


Fig. 4. Mixed cell colonies identified in cultures of marrow plated at several cell concentrations from a G-6-PD heterozygote. Each point represents the average number per plate and at each concentration $2 \times 10^6$ cells were plated. Closed circles represent total colonies; open circles represent the fraction of mixed cell colonies showing two isoenzyme types.

a small fraction of the mixed cell colonies tested displayed double isoenzyme activities, thus supporting the single cell origin of these colonies. In contrast, when more than 125 total colonies (CFU-C-, BFU-E-, and CFU-GEMM-derived) were observed per plate, a substantial fraction of the mixed cell colonies revealed double-enzyme activities, indicating that at higher cell concentrations a significant number of apparent mixed cell colonies arose from two or more progenitors. These preliminary data are consistent with results for granulocyte-macrophage colonies previously reported by Singer et al. (1979). They found that when CFU-C-derived colonies exceeded 30 per plate, between 13% and 34% of the colonies displayed two isoenzyme types and therefore arose from more than one cell.

### Tests for selection or "drift" in long-term cultures

Observations on precursor cell populations in long-term culture have the potential to be skewed by factors related to culture conditions leading to growth of one population more than another. Such in vitro phenomena have been reported by Zauala et al. (1978) in established fibroblast cultures carried beyond 3 months, and by Behcet et al. (1974) in EBV-transformed lymphoid cell lines. To interpret studies comparing normal and neoplastic progenitor growth in vitro, it must be determined whether such drift and/or selection occur, particularly when the same patient is the source of both normal and neoplastic cells. To study this question, the marker system provided by the isoenzymes of G-6-PD was employed. If culture conditions favored the emergence of one stem cell line over time, the ratios of activity that characterized the two isoenzymes of G-6-PD at the beginning of the cultures would be altered. We cultured marrow cells from two hematologically normal G-6-PD heterozygotes and assayed recovered cells and colonies. As shown in Table 3, the original ratios did not change significantly for erythroid or for granulocytic progenitors over time in long-term culture. Therefore, observations on precursor cell populations in this system are not skewed owing to culture conditions favoring one population over another during the 15-week period of study.

## DISCUSSION

This study compares hemopoiesis in continuous long-term marrow cultures from normal subjects and patients with polycythemia vera. In most instances, the neoplastic hemopoiesis of polycythemia vera and normal hemopoiesis were quantitatively similar. Thus, both types of marrow were characterized by an initial, rapid decline in recoverable nonadherent cell numbers and BFU-E. A nadir was reached at 3–4 weeks and was followed by a period of regeneration.

In contrast, BFU-E were recoverable from cultures of polycythemia vera marrow for a mean of 12.5 ($\pm$ 0.5) weeks compared to 8.7 ($\pm$ 0.6) weeks for normal cultures ($P = 0.03$). In addition, the frequency of BFU-E from the polycythemia vera cultures was somewhat higher at weeks 4 and 5, and the total number of BFU-E obtained from these cultures was greater than the total number obtained from normal cultures. These data suggest slightly greater regeneration of the erythroid precursors of the neoplastic clone.

Progenitors giving rise to mixed cell colonies were recovered for only 4–5 weeks from the long-term cultures. The initial results of the studies of mixed cell colonies using the cell marker system of G-6-PD isoenzymes confirmed the single cell origin of these colonies in vitro if the concentration of colony-forming cells per plate is low. However, at the higher colony-forming cell concentrations there were significant numbers of mixed cell colonies that

TABLE 3. Tests for clonal selection in long-term culture

| | Subject 1 | | Subject 2 | |
|---|---|---|---|---|
| | Erythrocytes (A⁻) | Granulocytes 0.50[1] | Erythrocytes 0.60 | Granulocytes 0.60 |
| | BFU-E | CFU-C | BFU-E | CFU-C |
| Week 1 | 0.50 | 0.40 | 0.50 | 0.60 |
| 4 | 0.40 | 0.80 | 0.50 | 0.50 |
| 5 | N.T. | N.T. | 0.75 | 0.50 |
| 6 | N.T. | N.T. | 0.75 | 0.50 |
| 7 | N.T. | N.T. | 0.75 | 0.65 |
| 8 | 0.65 | 0.50 | 0.75 | 0.50 |
| 13 | N.T. | 0.35 | N.T. | N.T. |
| 15 | N.T. | 0.35 | N.T. | N.T. |
| Mean(±SEM) | 0.50(±0.06) | 0.50(±0.07) | 0.65(±0.05) | 0.55(±0.03) |

[1]G-6-PD isoenzyme ratios for erythroid and granulocyte/macrophage colonies from long-term cultures of two normal heterozygous females. Ratios are expressed as the fraction of total colonies having isoenzyme type B. There was no detected A-like G-6-PD activity in the red cells of Subject 1, who is an A⁻/B heterozygote. Unlike nucleated erythroid cells, mature erythrocytes are deficient in G-6-PD type A⁻. N.T. = not tested.

displayed both isoenzyme types, indicating these colonies arose from at least two cells. Since colonies with two isoenzyme activities arise from a minimum of two progenitors (e.g., one type A and one type B), the actual frequency of colonies arising from multiple cells is more than twice the fraction of colonies with double isoenzyme activity. Although these results are preliminary, they clearly indicate that studies employing high cell concentrations for enumerating mixed cell colonies must be carefully reviewed to confirm that such colonies are indeed of single cell origin.

One of the distinctive features of polycythemia vera is the growth in vitro of endogenous erythroid colonies in the absence of added erythropoietin. This feature of the disease was reflected in continuous long-term culture by the maintenance of a population of cells with similar properties. The fraction of total erythroid precursors represented by endogenous colony-forming cells did not change over the time of the culture, indicating that the abnormal clone is stable in this system. These results also imply there is no selective reemergence of normal stem cell progeny. If there had been dramatic proliferation of normal relative to neoplastic progenitors, then the fraction of endogenous colony-forming cells might have changed with time. That the proportion of endogenous colonies remained constant also suggests that the population of normal BFU-E, if present, remained stable as well. It is possible that no normal progenitors were placed into culture or that this population was present in such reduced numbers from the outset that its members did not influence the total colonies measured. In previous studies in G-6-PD heterozygotes, Prchal et al. (1978) found that a significant proportion of BFU-E and CFU-C were normal. With time, however, their frequency decreased (Adamson et al., 1980). It is likely that the majority, if not all, of the erythroid precursors assayed in the long-term cultures of polycythemia vera marrows were members of the abnormal clone. However, continued active suppression of normal erythropoiesis, evident from observations in vivo (Adamson et al., 1976) may have persisted in the long-term cultures as well. Whether any normal progenitors can be detected under these circumstances awaits the results of similar studies in patients with appropriate cell markers.

The mechanisms by which the neoplastic clone in polycythemia vera is responsible for the clinical phenotype of the disease are unclear. A small fraction of erythroid progenitors, which is independent of the normal negative feedback of reduced levels of erythropoietin, could account for the increased red cell mass. Such a population persisted in long-term cultures. Assuming equal numbers of pluripotent stem cells in polycythemia vera and normal marrows, an increased rate of commitment of this population to erythroid and granulocyte/macrophage progenitors would result in more BFU-E and, presumably, more CFU-C in polycythemia vera than in normal. Such a mechanism has been suggested by Messner and Fauser (1980) who reported similar frequencies of CFU-GEMM in direct cultures of polycythemia vera and normal marrow, but increased cell cycle activity of these progenitors in polycythemia vera. Alternatively, greater proliferative capacities of BFU-E and CFU-C in polycythemia vera compared to normal would yield more erythroid and granulocytic cells. While the precise mechanisms remain to be determined, these initial results provide evidence that the long-term marrow culture system may be directly applicable to studies of regulatory abnormalities in human hematologic diseases.

## ACKNOWLEDGMENTS

The excellent technical assistance of Cathy Wise and Laura Steinmann and the secretarial assistance of Maxine Cormier and Patricia Jenkins are gratefully acknowledged. This work was supported by grants AM 19410, CA 16448, CA 18029, CA 21765, and CA 20180 from the National Institute of Arthritis, Metabolism, Digestive and Kidney Diseases, the National Cancer Institute, National Institutes of Health, and designated research funds of the Veterans Administration. Dr. Powell is the recipient of an Associate Investigator Award of the Veterans Administration.

## LITERATURE CITED

Adamson, J.W. (1970) The regulation of erythropoiesis in polycythemia vera and related myeloproliferative disorders. In Clarke, W.J., Howard, E.B., Hackett, P.L. eds.: Myeloproliferative Disorders of Animals and Man. United States Atomic Energy Commission, Washington, D.C.: United States Government Printing Office, pp. 440–452.

Adamson, J.W., P.J. Fialkow, S. Murphy, J.F. Prchal, and L. Steinmann (1976) Polycythemia vera: Stem cell and probable clonal origin of the disease. New Engl. J. Med., *295*:913–916.

Adamson, J.W., J.W. Singer, P. Catalano, S. Murphy, N. Lin, L. Steinmann, C. Ernst, and P.J. Fialkow (1980) Polycythemia vera: Further in vitro studies of hematopoietic regulation. J. Clin. Invest., *66*:1363–1368.

Behcet, J.M., P.J. Fialkow, K. Nilsson, G. Klein, and S. Singh (1974) Immunoglobulin synthesis and G-6-PD as cell markers in human lymphoblastoid cell lines. Exp. Cell Res., *89*:275–282.

Berlin, N.I. (1975) Diagnosis and classification of the polycythemias. Semin. Hematol., *12*:339–351.

Dexter, T.M., T.D. Allen, and L.G. Lajtha (1977) Conditions controlling the proliferation of hemopoietic stem cells in vitro. J. Cell. Physiol., *91*:335–344.

Fauser, A.A., and H.A. Messner (1978) Granuloerythropoietic colonies in human bone marrow, peripheral blood, and cord blood. Blood, *52*:1243–1248.

Gartner, S., and H. Kaplan (1980) Long-term culture of human bone marrow cells. Proc. Natl. Acad. Sci. U.S.A., *77*:4756–4759.

Golde, D.W., N. Bersch, and M.J. Cline (1977) Polycythemia vera: Hormonal modulation of erythropoiesis in vitro. Blood, *49*:399–406.

Greenberger, J.S., M. Sakakeeny, and L. Parker (1979) In vitro proliferation of hemopoietic stem cells in long marrow cultures: Principles in mouse applied to man. Exp. Hematol., *7* (suppl 5):135-141.

Hocking, W., and D.W. Golde (1980) Long-term human bone marrow cultures. Blood, *56*:118-124.

Iscove, N.N., F. Sieber, and K.H. Winterhalter (1974) Erythroid colony formation in cultures of mouse and human bone marrow: Analysis of the requirement for erythropoietin by gel filtration and affinity chromatography on agarose-concanavalin A. J. Cell. Physiol., *83*:309–315.

Messner, H.A., and A.A. Fauser (1980) Culture studies of human pluripotent hemopoietic progenitors. Blut, *41*:327–333.

Miller, R.G., and R.A. Phillips (1969) Separation of cells by velocity sedimentation. J. Cell. Physiol., *73*:191–202.

Mladenovic, J., and J.W. Adamson (1979) Adrenergic modulation of normal and neoplastic erythroid colony growth. Blood, *54*:142a.

Moore, M.S., H.E. Broxmeyer, A.P. Sheridan, P.A. Meyers, N. Jacobsen, and R.J. Winchester (1980) Continuous human bone marrow culture: Ia antigen characterization of probable pluripotential stem cells. Blood, *55*:682–690.

Prchal, J.F., J.W. Adamson, S. Murphy, L. Steinmann, and P.J. Fialkow (1978) Polycythemia vera: The in vitro response of normal and abnormal cell lines to erythropoietin. J. Clin. Invest., *61*:1044–1047.

Singer, J.W., and J.W. Adamson (1976) Steroids and hematopoiesis. II. The effect of steroids on in vitro erythroid colony growth: Evidence for different target cells for different classes of steroids. J. Cell. Physiol., *88*:135–144.

Singer, J.W., P.J. Fialkow, L.W. Dow, C. Ernst, and L. Steinmann (1979) Unicellular or multicellular origin of human granulocyte-macrophage colonies in vitro. Blood, *54*:1395–1397.

Toogood, I.R.G., T.M. Dexter, T.D. Allen, T. Suda, and L.G. Lajtha (1980) The development of a liquid culture system for the growth of human bone marrow. Leuk. Res., *4*:449–461.

Zanjani, E.D., J.D. Lutton, R. Hoffman, and L.R. Wasserman (1977) Erythroid colony formation by polycythemia vera bone marrow in vitro: Dependence on erythropoietin. J. Clin. Invest., *59*:841–848.

Zauala, C., G. Herner, and P. Fialkow (1978) Evidence for selection of cultured diploid fibroblast strains. Exp. Cell Res., *117*:137–141.

JOURNAL OF CELLULAR PHYSIOLOGY SUPPLEMENT 1:87–94 (1982)

# Stromal Cell Associated Haemopoiesis

T.M. DEXTER*

*Paterson Laboratories, Christie Hospital and Holt Radium Institute, Withington, Manchester M20 9BX, United Kingdom*

The role of marrow stromal cells in haemopoietic cell development has been a matter of speculation for many years. Are such cells merely acting as a supportive matrix providing a privileged site for haemopoiesis to occur, or do the stromal cells themselves play an instructive role in determining proliferation and cell lineage development within the haemopoietic system? A corollary to the latter part of the question is to ask to what extent the production and release of diffusible factors (as opposed to specific cell to cell interaction) play a role in any regulatory pathway.

## STROMAL:HAEMOPOIETIC CELL ASSOCIATIONS IN VIVO

Evidence in vivo indicating specific association between stromal cells and developing foci of haemopoiesis comes from several sources. Sorrell and Weiss (1980), in studies of chick bone marrow, showed that erythroid cells were associated with the sinusoidal endothelium, while granulopoiesis occurred in association with the extravascular reticulum cells; Westen and Bainton (1979) reported the presence of an alkaline phosphatase positive reticulum cell (in rodent marrow), generally found contiguous with early granulocytes, and a macrophage-like, acid phosphatase positive cell in proximity with erythroid precursors. Intimate associations between macrophages and developing erythroblasts have also been described by Ben-Ishay and Yoffey (1972) and Bessis (1973). Indeed, while it is generally believed that the fixed macrophages of the bone marrow are themselves derived from the haemopoietic stem cell (via a committed progenitor cell), the evidence for this is scanty, and it could well be that (in the adult) such macrophages are independently derived. Mitosis have been observed in such cells (Westen and Bainton, 1979), indicating that marrow macrophages are capable of proliferation and, as such, may be viewed as an integral component of the marrow stroma. Adipocytes are also characteristic cells in marrow stroma. Such cells are thought to be derived from adventitial reticular cells, which accumulate lipid (Sorrell and Weiss, 1980) and are often found in areas of mature granulopoiesis (Weiss, 1980). Similar associations between fat cells, macrophages, reticular cells, and active haemopoiesis have been observed on implanted cellulose ester membranes (Knospe et al., 1978) and in the marrow of humans recovering from intensive chemotherapy for acute leukaemia (Islam et al., 1980). Furthermore, marrow implanted in the omentum shows active granulopoiesis in association with the omental fact (Meck et al., 1973).

However, while these data indicate consistent and specific cellular association, they do not prove that the stromal cells *directly* influence haemopoietic cells. It could be, for example, that the surface coat of the stromal cells (the external cell matrix and membrane associated glycoconjugates) merely provide a suitable surface (Edwards, 1978) for the attachment of appropriate haemopoietic cell types, which are subsequently exposed to other inductive signals. A genetically anaemic mouse strain (*Sl*/*Sl*$^{d}$) has proved to be fairly important in this context. These animals have about half the number of haemopoietic stem cells (CFU-S) as compared with their normal littermates, although the concentration of CFU-S/$10^5$ cells is the same (McCulloch et al., 1965). However, they do have a macrocytic anaemia that cannot be rescued by injection of normal marrow cells, and, when irradiated, such mice cannot support the growth of CFU-S (McCulloch et al., 1965), i.e., the spleen "microenvironment" is defective for stem cell growth. Further light on this defect was shed by the studies of Wolf (1978, 1979 review) who showed that when spleen or marrow from nor mal mice was implanted or joined to *Sl*/*Sl*$^{d}$ spleen, the implanted normal stroma became actively erythropoietic, while the adjoining *Sl*/

*To whom correspondence/reprint requests should be addressed.

$Sl^d$ stroma remained relatively quiescent. These data and those of McCulloch et al. (1965), using parabiosed mice, indicate that the defect is fixed, is not "cured" by a migrating cell, and that the stromal cells exert a local influence on haemopoiesis. Nonetheless, the precise nature of this defect has yet to be determined.

One approach to the analysis of stromal:haemopoietic cell interactions is to develop in vitro systems where the characteristics and function of the various cell types can be examined. Using cellular and molecular cloning techniques, it may then be possible not only to define the individual cellular elements and their role in haemopoiesis, but also to determine the nature of the inductive process—be it cell contact mediated or via the production and release of diffusible regulatory molecules.

## STROMAL CELLS IN VITRO

Several authors have reported the development of fibroblast-like colonies derived from adherent cells of bone marrow from many species (Friedenstein et al., 1970; Wilson et al., 1976; Friedenstein et al., 1974; Wilson and O'Grady, 1976). These cells can be grown for long periods in vitro following subculture, and implantation of such cells under the renal capsule results ultimately in the formation of osseous stroma and associative haemopoiesis (Friedenstein et al., 1974). However, it is likely that such fibroblasts represent only one component of the marrow environment since evidence shows that fibroblastic cells of this type do not support stem cell proliferation in vitro (Lubennikova and Domaratskay, 1976), although they do produce the granulocyte/macrophage colony-stimulating factor (Wilson et al., 1974). The finding that the Friedenstein fibroblast-like cells will not support in vitro CFU-S growth may be very significant. If the same rule applies in vivo, then the success of haemopoiesis in fibroblast grafts under the kidney capsule may be due primarily to infiltrating host stromal elements (macrophages?). In other words, the fibroblasts may merely be providing the supportive matrix for the enlodgment and development of such cells. Only then can haemopoiesis become established. This raises several interesting questions. Since the fibroblast colony-forming cells are migrating and appear in the circulation (Werts et al., 1980), and other stromal elements are presumably also migrating (the kidney graft experiment), then a) in the $Sl/Sl^d$ model discussed earlier, why does parabiosis not lead to a permanent cure of the $Sl/Sl^d$ anaemia (McCulloch et al., 1965), and in the spleen ligation experiments (Wolf, 1978), why does the $Sl/Sl^d$ spleen not become progressively infiltrated with normal stromal elements? b) Following marrow transplantation, experiments indicate that donor stromal (fibroblastic) cells do not become established in the host (Friedenstein et al., 1978; Wilson et al., 1978). These observations may be reconciled by the finding that fibroblasts and macrophages are relatively radioresistant (and the turnover time relatively long), so that in the irradiated mouse (and in the nonirradiated $Sl/Sl^d$ mouse) the intact haemopoietic stromal tissue exerts a negative influence on the development of any donor stromal cells. Nevertheless, fibroblast stromal precursor cells are detectable in vivo and they may facilitate the development of haemopoiesis in the kidney capsule.

In summary, the arguments presented above indicate that the fibroblasts may represent a mandatory but not an exclusive component of the haemopoietic stroma.

## LONG-TERM MARROW CULTURES

Under appropriate conditions, adherent cells derived from bone marrow will support the maintenance of proliferation and differentiation of CFU-S for many months (Till and McCulloch, 1961; Dexter et al., 1977a). Unlike the Friedenstein-type cultures, this culture system contains adherent cells with widely differing phenotypes, ranging from cells with an endothelial appearance, adipocytes, macrophages, and reticular cells (Allen and Dexter, 1976; Dexter et al., 1977a; Wilson et al., 1981; Bentley and Foidart, 1980). When such adherent layers are inoculated with fresh marrow cells, the CFU-S that are present in the marrow inoculum migrate into the adherent layer and establish sites of active haemopoiesis (Dexter et al., 1978; Dexter, 1979). Within the adherent layer, continuous stem cell proliferation and differentiation results in the long-term production of commited granulocyte/macrophage progenitor cells (GM-CFC) (Dexter et al., 1977a; Williams et al., 1977), primitive erythroid progenitor cells (BFU-E) (Eliason et al., 1979; Gregory and Eaves, 1978), and megakaryocyte precursors (Williams et al., 1978). All these cells, including CFU-S, are continuously being released into the overlying growth medium. This means that when the cultures are fed (by demi- or total depopulation of the growth medium) these cells can be assayed using conventional in vivo (CFU-S) or in vitro clonogenic assay systems, and events occurring in the cultures can be monitored for the duration of haemopoiesis. Using such assays, combined with other experimental protocols, it has been shown that the stem cells and committed progenitor cells are very similar to, if not identical with, the corresponding cells produced in vivo (Dexter et al., 1980a) and strengthens our contention that the long-term culture system is representative of haemopoiesis in vivo.

Cells with lymphoid potential are also being produced in such long-term cultures (Dexter et al., 1977a; Jones-Villenueve and Phillips, 1980a; Schrader and Schrader, 1978). Thus, injection of the cultured cells into haemopoietically ablated mice results in reconstitution of both the myeloid and lymphoid elements

In terms of mature cell production, extensive granulopoiesis is occurring such that for every one CFU-S there are, on average, 10 GM-CFC, and for every GM-CFC there is the production of 500–1,000 mature granulocytes, i.e., the same order of magnitude as that seen in vivo (Dexter et al., 1980a). Moreover, the mature granulocytes produced in long-term cultures appear to be identical with their counterparts produced in vivo and show a greater degree of maturation than granulocytes produced from GM-CFC in soft-gel cultures (Williams et al., 1977). While mature (platelet producing) megakaryocytes are also present for many months, their proportion in relation to other cell types has so far not been determined. Of major interest is that erythroid development is normally "blocked" at the primitive day 10 and day 14 BFU-E stages (Eliason et al., 1979), although terminal maturation can be induced in appropriate circumstances (Dexter et al., 1981). This will be discussed in more detail later. Finally, B-lymphocytes and mature T-cells are not present after the first week of culture, although Thy-1 bearing cells can be derived from progenitor cells in some circumstances (Jones-Villenueve et al., 1980b). However, the functional significance of such Thy-1 positive cells is unclear at present since granulocytic cell lines bearing Thy-1 and lacking Ly-1 and Ly2.3 can also be established from long-term cultures (Dexter et al., 1980b). Nonetheless, in normal conditions of long-term marrow culture, development of alloreactive T-cells does not occur, which means that the cultures can supply a source of stem cells for marrow transplantation in the absence of graft-versus-host disease (Dexter and Spooncer, 1978).

## ROLE OF THE ADHERENT LAYER IN LONG-TERM HAEMOPOIESIS

In the absence of an adherent cell layer, sustained stem cell proliferation will not occur, e.g., using siliconised culture flasks to inhibit cell attachment or in short-term suspension cultures prior to adherent layer formation. Moreover, only adherent cell layers derived from marrow stroma appear to be effective in supporting prolonged haemopoiesis; adherent cells derived from foetal liver, spleen, and embryonic fibroblasts do not maintain stem cell proliferation and differentiation (Reimann and Burger, 1979b), although CFU-S obtained from spleen tissue will readily grow on a marrow adherent cell layer (G. Keller, personal communication, June 1981). In other words, only the marrow adherent cells can support haemopoiesis in vitro, an effect which is presumably mediated by cell/cell interaction and/or regulator production by specific stromal elements. In support of the role of the adherent marrow cells in the maintenance of haemopoiesis is the observation that the microenvironmental and CFU-S defects observed, respectively, in *Sl*/$Sl^d$ and *W*/$W^v$ mice (discussed earlier) can be reproduced in the long-term marrow culture system (Dexter and Moore, 1977).

## CELLULAR COMPOSITION OF THE ADHERENT CELLS

Within 2–3 weeks after establishing the primary cultures (normally by inoculating the cellular contents of a single femur into 25 cm$^2$ TC flasks containing Fischer's medium supplemented with horse serum), several morphologically distinct cell types can be recognised. These include flattened cells which form a pavement-like structure (putative endothelial cells), large branching dendritic cells (reticular cells?), and foci of lipid laden adipocytes (Dexter, Allen and Lajtha, 1977a; Allen and Dexter, 1976; Dexter, 1979). The latter are highly characteristic "giant" cells with a distinctive ultrastructural morphology. The absence of phagocytic inclusion allows them to be readily separated from lipid-accumulating macrophages (Dexter, 1979). These three cell types usually occur in association with phagocytic mononuclear cells (macrophages) and often form a complex multilayer with several cell types in association. It is in these areas that haemopoiesis occurs, and membrane functional complexes between the stromal cells and the developing haemopoietic cells indicates extensive metabolic cooperation.

When foetal calf serum is substituted for horse serum in the culture medium, the adherent layer formed is obviously different. While putative endothelial-fibroblastic and macrophage cells are present, adipocytes are notably absent (Dexter et al., 1980a). Sustained haemopoiesis will not occur on such adherent layers, maybe associated with this lack of adipocytes. Indeed, if hydrocortisone is added to foetal-calf supplemented cultures, adipocytes rapidly appear and the adherent cell layers then become competent to support stem cell proliferation and differentiation (Dexter et al., 1980a; Greenberger, 1980). Of some interest is that if hydrocortisone addition is delayed for 2 or 3 weeks after establishing the cultures, adipogenesis is markedly reduced, and although the other stromal cell types are present the cultures are generally defective for maintaining haemopoiesis. This indicates that the adipocytes (or their precursors) are a cell population distinct from the endothelial and fibroblastic cells.

A similar association between endothelial cells, adipocytes, and sustained haemopoiesis is seen also when marrow cells are plated in collagen gels (Lanotte et al., 1981). These cultures, like the long-term marrow cultures in liquid medium, emphasize the clone similarities between interaction of stromal cells/haemopoietic cells seen in vitro and similar interactions reported by Weiss (1980) in vivo.

## MAINTENANCE OF STEM CELLS (CFU-S)

Stem cell proliferation in long-term cultures occurs with all mouse strains so far tested, although there are major differences in the longevity of CFU-S production associated with the different genotypes (Dexter et al., 1977b); Reimman and Burger, 1979a; Greenberger, 1980). That stem cell maintenance requires close range cell interactions has been inferred from studies showing that the major site of stem cell production is from within the adherent cell multilayer (Dexter et al., 1980a) and more directly from studies by Bentley (1981), who showed that CFU-S in diffusion chambers placed on preformed marrow adherent layer were not maintained. In fact, when medium conditioned by growth of marrow stromal cells is added to fresh marrow, CFU-S short-term survival is increased, but there is no evidence for stem cell proliferation—after 7 days incubation CFU-S have disappeared (Blackburn and Patt, 1980). Thus, if the stem cell maintenance is determined by the production of diffusible regulatory molecules, such activities are in very low concentrations in conditioned medium or are labile molecules.

However, there is evidence that stromal cell conditioned medium does contain factors that can modify the proliferative activity of stem cells. This work stems from the observation that CFU-S cycling in long-term cultures (as measured by $^3$HTdR killing) follows a cycling pattern related to the feeding regime (Dexter et al., 1977c). Shortly after culture depopulation, CFU-S are actively cycling (high $^3$HTdR kill), whereas several days later, CFU-S are in a proliferatively quiescent stage (low $^3$HTdR kill). Subsequently, it was demonstrated that this oscillation was associated with the production, into the growth medium, of molecules that could inhibit or stimulate DNA synthesis in CFU-S (Toksoz et al., 1980), similar to the ones described by Lord for fresh bone marrow (Lord et al., 1976, 1977). Shortly after feeding the cultures (when CFU-S cycling is "high"), the concentration of the stimulatory material (30,000–50,000 MW) increases relative to the inhibitor (50,000–100,000 MW); several days later, the reverse situation occurs and the concentration of the inhibitory material increases relative to the stimulator. Also, the cycling status of CFU-S in the long-term cultures can be modified by addition of excess exogenous factor, i.e., addition of partially purified stimulatory material to cultures containing low-cycling CFU-S will lead, within hours, to a rapid onset of DNA synthesis in such stem cells. The stimulus for production of these opposing activities appears not to be related to absolute cell numbers or stem cell density (Dexter et al., 1980a). Rather, it appears to be related to the proximity of CFU-S with the adherent cell layer. This led to the formulation of a model whereby the adherent layer contains local cellular milieu (perhaps fixed in number) in which the presence of a stem cell would signal its presence and lead to the production of an "inhibitor," and displacement of a stem cell would have the opposite effect, leading to stimulator production. Certainly, the inhibitor and stimulator-producing cells are themselves adherent and comprise part of the adherent cell layer (Wright et al., 1980). The benefit of this model is that it suggests a way in which stem cell numbers can be rigidly controlled in vivo (homeostasis) based upon the number of "sites" available. In many ways, this is an extension of the stem cell "niche" hypothesis proposed by Schofield (1978).

## THE REGULATION OF GRANULOPOIESIS IN VITRO

The elegant experiments of Bradley and Metcalf (1966) and Pluznik and Sachs (1966) (reviewed in Metcalf, 1977) established that the growth of granulocyte/macrophage progenitor cells (GM-CFU) in soft-gel media required the continual presence of colony-stimulating factor (CSF). In the presence of GM-CSF, clonal proliferation of GM-CFC resulted in the formation of discrete colonies containing mature neutrophils and/or macrophages. In the absence of GM-CSF, proliferation did not occur and the progenitor cells rapidly disappeared. However, while these clonal assays in soft-gel media are continuing to supply useful information regarding the characteristics of the colony-forming cells and the various stimulatory activities necessary for their development, such systems have the disadvantage that the events occurring prior to the production of the committed GM progenitor cells cannot be easily investigated. Neither can the regulation of homeostasis in the granulocyte compartment be examined.

In long-term marrow cultures, however, the complete developmental progression from stem cells, through the GM progenitor cells, to the mature functional end cells can be readily followed for several months. An interesting finding to emerge from this system is the observation that the extensive granulopoiesis in the long-term cultures is apparently occurring in the absence of GM-CSF; i.e., conditioned medium removed from actively haemopoietic long-term cultures was unable to stimulate the growth of normal marrow GM-CFC (Dexter et al., 1977a; Williams et al., 1977; Reimann and Burger, 1979b; Bentley and Foidart, 1980). In fact, addition of highly purified CSF derived from L-cells or from WEHI-myelomonocytic leukaemia cells to the long-term cultures had no effect upon stem cell proliferation, GM-CFC production, or the development of granulocytes and macrophages (Dexter and Shadduck, 1980; Williams and Burgess, 1980). Also, addition of antiserum raised against L-cell type CSF did not interfere with granulocyte/macrophage development (Dexter and Shadduck, 1980).

What does this mean? The absence of biodetectable GM-CSF in the conditioned medium suggests that granulopoiesis is occurring independently of CSF. This is strengthened by the observation that addition of antiserum has no significant effect; if L-cell type CSF *were* being produced, locally, at concentrations sufficient to stimulate GM development, the antibodies may be expected to bind and biologically inactivate the CSF molecule. The lack of effect of added (exogenous) CSF suggests that this granulopoiesis is occurring at an optimal level, which is not influenced by addition of excess "regulator" molecules. Can this be reconciled with the known requirement of GM-CFC for GM-CSF in soft-gel clonal assays? A striking observation in the long-term cultures is that developing granulocytes are usually found in association with the adherent layer, and granulopoiesis is especially intense in areas of the adherent layer that are undergoing adipogenesis (Allen and Dexter, 1976; Dexter, 1979). In these areas, membrane complexes are often seen between the adherent cells and the developing haemopoietic cells. This suggests a mechanism for the regulation of granulocyte/macrophage development via intimate cell/cell contact and the transmembrane passage of appropriate stimulating factors (CSF?). Whether this is mediated by the GM-CSFs which have so far been described and purified must await the development of suitable probes.

These data do, however, stress the importance of cell/cell interactions in the regulation of granulopoiesis. Further experiments that shed some light on this phenomenon have been described by Zipori and Sassoon (1980). When marrow GM-CFC were plated (in the presence of GM-CSF) on a preformed marrow adherent cell layer, colony formation was reduced to about 10% of the control level. This effect is seen only when the target cells and the marrow adherent cell are in close proximity; if the cells are separated by a thin layer of agar, no inhibitory effect is seen. Thus, marrow adherent cells can modify the response of GM-CFC and this suggests the mechanism (in vivo) whereby excess levels of GM-CSF in the serum do not lead to a prolonged, exuberant granulopoiesis. Also intriguing was their finding that the few colonies that did form in the presence of marrow adherent cells contained significantly more GM-CFC (after 7 days growth) than did colonies developing in the absence of adherent cells, i.e., adherent cells not only modify the response of GM-CFC to CSF but also maintain the proliferative progeny.

## ERYTHROPOIESIS IN LONG-TERM CULTURES

As stated previously, erythroid cell development in long-term cultures normally proceeds only to the level of the primitive progenitor cells—the day 10 and day 14 BFU-E.

Addition of partially purified erythropoietin to the cultures will facilitate the progression of the BFU-E to more mature cells (the day 4 BFU-E and day 2 BFU-E) but will not ordinarily lead to the production of haemoglobinising cells (Eliason et al., 1979). However, if the cultures are mechanically shaken throughout the period of erythropoietin treatment, mature, nonnucleated RBC are produced. In more recent work (Dexter et al., 1981), we have circumvented the requirement for mechanical agitation by simply adding anaemic mouse serum (AMS) to the cultures. In these circumstances, extensive erythropoiesis is induced (including mature erythrocytes) and the duration of erythropoiesis is a reflection of the treatment time with AMS. For example, a single addition of AMS will lead to a fairly well defined peak of erythropoiesis commencing 10–14 days later (i.e., the time interval taken for progression of the BFU-E through the various stages to erythrocyte formation). Twice weekly treatment with AMS leads to a prolonged period of erythropoiesis. It is of interest that AMS-stimulated erythropoiesis in long-term cultures promotes the production of adult-type haemoglobin—comparable with the situation in vivo.

In terms of stromal cell interaction in such erythropoietically active cultures, several points are worthy of comment. Firstly, erythropoiesis occurs in synchronous cohorts, and all morphologically recognisable stages occur in association with a central macrophage, i.e., clusters of 4–16 normoblasts in association with a central macrophage and 26–64 reticulocytes in close juxtaposition with a macrophage are regularly seen. Such an intimate membrane contact between developing red cells and macrophages has also been observed in vivo (Bessis, 1973). Secondly, the increase in the production of erythroid cells occurs concommitantly with a *decrease* in granulocyte production. This effect is not associated with inhibition of the production of the GM-CFC (i.e., there is no "competition" at the stem cell or progenitor cell level), but rather seems to be due to a selective inhibition of GM-CFC development into mature neutrophils and macrophages. However, AMS is not toxic to GM-CFC; neither does AMS prevent development of GM-CFC in soft-agar cultures stimulated with GM-CSF. Rather, the effect of AMS in granulopoiesis in long-term cultures appears to be mediated indirectly through the stromal environment. Shortly after adding AMS, adipocytes undergo lipolysis and the fat cell areas became indistinct as the lipid is discharged into the growth medium. At the same time, such areas, which were actively granulopoietic prior to the addition of AMS, became hypocellular; and other areas of the adherent layer, which were essentially nonhaemopoietic before addition of AMS, became actively erythropoietic. The stromal cells in these areas consist of sheets of endothelial-like cells associated with macrophages.

A lipolysis associated with increased erythropoietic demand has also been observed in vivo following phenylhydrazine (PHZ), induced erythroid aplasia (Bathija et al., 1978). The effect of PHZ appears to be specifically directed to marrow fat; extramedullary adipose tissue is not affected. Also, although there is a reduction in marrow fat cell volume, there is no reduction in esterification ability, which is unlike the effect of, e.g., starvation on extramedullary lipid when both volume and esterification capacity decrease. The data obtained in long-term cultures are also intriguing in that they indicate that addition of excess lineage specific regulator(s) may indirectly influence development of another cell lineage without affecting committment at the progenitor cell level. In some respects, this is analogous to the situation reported by Smith et al. (1980a,b), who showed that hypertransfusion of sublethally irradiated mice led to enhanced recovery of the megakaryocytic and granulocytic cell compartments, i.e., a *suppression* of erythropoiesis leads to a "stimulation" of other haemopoietic lineages. Our data in the long-term cultures suggest that this may be mediated by alterations in the stromal environment, maybe by influencing the functional activity of the various cell components. The data also raise the controversial possibility that the marrow enviornment may be structured in discrete areas for the development of granulocytes, erythrocytes, and megakaryocytes, respectively, each area facilitating the development of the appropriate committed progenitor cell. Such a model has important differences from the hypothesis proposed by Trentin, Wolf, and co-workers (reviewed in Wolf, 1979). Primarily, the long-term culture work suggests that such cellular niches (microenvironments) play a role in progenitor cell development rather than stem cell commitment.

## CONCLUSIONS AND FUTURE TRENDS

Evidence in vivo and in vitro indicates that stromal cells have a role to play in haemopoiesis. Data obtained from the long-term cultures indicate that their function is not simply a passive one, providing a cellular architecture in which haemopoiesis occurs, but rather that the stromal cells play a determinative role in haemopoiesis, from the stem cell through the committed progenitor cells to the mature end cells. Some examples have been cited which suggest that regulation is mediated by the intimate cell contact, rather than by the production of diffusible regulators, and that cell lineage development within the haemopoietic system occurs in association with discrete stromal elements. An outstanding question, however, is the extent to which the putative regulators involved in colony growth (in soft-gel media), of the committed progenitor cells, plays a similar role in stromal cell-mediated haemopoiesis in long-term cultures. Only when appropriate probes are available can some of the anomalies be resolved. Furthermore, neither analysis of haemopoiesis in vivo nor in vitro (liquid culture or soft-gels) has provided clues as to how a pluripotential stem cell becomes restricted in its developmental potential. Is differentiation caused by some random event in the genome of a stem cell, or is it determined by cell/cell interactions or receptor binding of specific molecules that commit a stem cell to one or the other lineages? Obviously, stem cell selfrenewal and differentiation can occur in long-term cultures and in clonal assay systems (Metcalf et al., 1979). One possible approach to this problem, therefore, would be to define the conditions necessary for self-renewal versus differentiation, i.e., to produce stem cell "lines" in which commitment can be investigated. Accessory to this problem, of course, is the purification of molecules such as GM-CSF, burst-promoting activity, eosinophil, and megakaryocyte-stimulating factor, etc.; the production of monoclonal antibodies against such factors; and the analysis of cloned lines of marrow stroma and their associative function in haemopoiesis. Such studies are already in an advanced stage of development.

## ACKNOWLEDGMENTS

The author is a Fellow of the Cancer Research Campaign (United Kingdom).

## LITERATURE CITED

Allen, T.D., and T.M. Dexter (1976) Cellular interrelationships during in vitro granulopoiesis. Differentiation, *6*:191–194.

Bathija, A., S. Davis, and S. Trubowitz (1978) Marrow adipose tissue: Response to erythropoiesis. Am. J. Hematol., *5*:315–321.

Ben-Ishay, Z., and J.M. Yoffey (1972) Ultrastructural studies of erythroblastic islands of rat bone marrow. II. The resumption of erythropoiesis in erythropoietically depressed rebound marrow. Lab. Invest., *26*:637–648.

Bentley, S.A. (1981) Close range cell: cell interaction required for stem cell maintenance in continuous bone marrow culture. Exp. Hematol., *9*:308–312.

Bentley, S.A., and J.M. Foidart (1980) Some properties of marrow derived adherent cells in tissue culture. Blood, *56*:1006–1012.

Bessis, M. (1973) Living Blood Cells and Their Ultrastructure. New York: Springer.

Blackburn, M.J., and H.M. Patt (1980) Influence of a marrow stromal factor on survival of hemopoietic stem cell in vitro. Exp. Hematol., *8*:77–82.

Bradley, T.R., and D. Metcalf (1966) The growth of mouse bone marrow cells in vitro. Aust. J. Exp. Biol. Med. Sci., *44*:297–300.

Dexter, T.M., and M.A.S. Moore (1977) In vitro duplication and "cure" of haemopoietic defects in genetically anaemic mice. Nature, *269*:412–414.

Dexter, T.M., T.D. Allen, and L.G. Lajtha (1977a) Conditions controlling the proliferation of haemopoietic stem cells in vitro. J. Cell Physiol., *91*:335–344.

Dexter, T.M., M.A.S. Moore, and A.P.C. Sheridan (1977b) Maintenance of hemopoietic stem cells and production of differentiated progeny in allogeneic and semi-allogeneic bone marrow chimeras in vitro. J. Exp. Med., *145*:1612–1616.

Dexter, T.M., E.G. Wright, F. Krizsa, and L.G. Lajtha (1977c) Regulation of haemopoietic stem cell proliferation in long-term bone marrow cultures. Biomedicine, *27*:344–349.

Dexter, T.M., T.D. Allen, L.G. Lajtha, F. Krisza, N.G. Testa, and M.A.S. Moore (1978) In vitro analysis of self renewal and commitment of hematopoietic stem cells. In Clarkson, B., Marks, P.A., Till, J.E. eds.: Differentiation of Normal and Neoplastic Hematopoietic Cells. Cold Spring Harbor, New York: Cold Spring Harbor Publications, pp. 63–80.

Dexter, T.M., and E.M. Spooncer (1978) Loss of immunoreactivity in long-term bone marrow culture. Nature, *275*:135–136.

Dexter, T.M. (1979) Cell Interaction In Vitro. Clin Haematol., *8*:453–468.

Dexter, T.M., E. Spooncer, D. Toksoz, and L.G. Lajtha (1980a) The role of cells and their products in the regulation of in vitro stem cell proliferation and granulocyte development. J. Supramol. Struct., *13*:513–524.

Dexter, T.M., J. Garland, D. Scott, E. Scolnick, and D. Metcalf (1980b) Growth of factor-dependent hemopoietic precursor cell lines. J. Exp. Med., *152*:1036–1047.

Dexter, T.M., and R.K. Shadduck (1980) The regulation of haemopoiesis in long-term bone barrow cultures. I. Role of L-cell CSF. J. Cell Physiol. *102*:279–286.

Dexter, T.M., N.G. Testa, T.D. Allen, T. Rutherford, and E. Scolnick (1981) Molecular and cell biological aspects of erythropoiesis in long-term bone marrow cultures. Blood, *58*:699–707.

Edwards, P.A. (1978) Differential cell adhesion may result from nonspecific interactions between cell surface glycoproteins. Nature, *271*:248–249.

Eliason, J.F., N.G. Testa, and T.M. Dexter (1979) Erythropoietin-stimulated erythropoiesis in long-term bone marrow cultures. Nature, *281*:382–384.

Friedenstein, A.J., R.K. Chailakhjan, and K.S. Lalykina (1970) The development of fibroblast colonies in monolayer cultures of guinea pig bone marrow and spleen cells. Cell Tissue Kinet., *3*:393–403.

Friedenstein, A.J., R.K. Chailakhjan, N.V. Latsinik, A.F. Panasynk, and I.V. Keiliss-Borok (1974) Stromal cells responsible for transferring the microenvironment of hemopoietic tissues. Transplantation, *17*:331–340.

Friedenstein, A.J., A.A. Ivanov-Smolenski, R.K. Chailakhjan, V.F. Gorskaya, A.I. Kuralesova, N.W. Latzinik, and V.W. Gerasimow (1978) Origin of bone marrow stromal mechanocytes in radiochimeras and heterotopic transplants. Exp. Hematol., *6:*440–444.

Greenberger, J.S. (1980) Self-renewal of factor-dependent hemopoietic progenitor cell lines derived from long-term bone marrow cultures demonstrates signficiant mouse strain genotypic variation. J. Supramol. Struct., *13:*501–511.

Gregory, C.J., and A.C. Eaves (1978) In vitro studies of erythropoietic progenitor cell differentiation. In Clarkson, B., Marks, P.A., Till, J.E. eds.: Differentiation of Normal and Neoplastic Hematopoietic Cells. Cold Spring Harbor, New York: Cold Spring Harbor Publications, pp. 179–192.

Islam, A., D. Catovsky, and D.A.G. Galton (1980) Histological study of bone marrow regeneration following chemotherapy for actue myeloid leukaemia and chornic granulocytic leukaemia in blast transformation. Br. J. Haematol., *45:*535–540.

Jones-Villenueve, E.V., and R.A. Phillips (1980a) Potentials for lymphoid differentiation by cells from long-term cultures of bone marrow. Exp. Hematol., *8:*65–69.

Jones-Villenueve, E.V., J.J. Rusthoven, R.G. Miller, and R.A. Phillips (1980b) Differentiation of Thy-1 bearing cells from progenitors in long-term bone marrow cultures. J. Immunol., *124:*597.

Knospe, W.H., S. Husseini, and F.E. Trobaugh (1978) Hematopoiesis on cellulose ester membranes (CEM). II. Enrichment of the hematopoietic microenvironment by the addition of selected cellular elements. Exp. Hemat., *6:*601–612.

Lanotte, M., S. Schor, and T.M. Dexter (1981) Collagen gels as a matrix for haemopoiesis. J. Cell Physiol., *106:*269–277.

Lord, B.I., K.J. Mori, E.G. Wright, and L.G. Lajtha (1976) An inhibitor of stem cell proliferation in normal bone marrow. Br. J. Haematol., *34:*441–445.

Lord, B.I., K.J. Mori, and E.G. Wright (1977) A stimulator of stem cell proliferation in regenerating bone marrow. Biomed. Express (Paris), *27:*223–226.

Lubennikova, E.I., and E.I. Domaratskay (1976) Changes in the number of stem cells during cultivation of mouse bone marrow on a bed of fibroblast-like cells. Byull. Eksp. Biol. Med. *81:*718–719.

McCulloch, E.A., L. Siminovitch, J.E. Till, E.S. Russell, and S.E. Bernstein (1965) The cellular basis of the genetically determined hemopoietic defect in anemic mice of genotype *Sl/Sl*$^{d}$. Blood, *26:*399–410.

Meck, R.A., J.E. Haley, and G. Brecher (1973) Hematopoiesis versus osteogenesis in ectopic bone marrow transplants. Blood, *42:*611–669.

Metcalf, D. (1977) Hemopoietic Colonies in Vitro: Cloning of Normal and Leukemic Cells. Berlin, Heidelberg: Springer, p. 227.

Metcalf, D., G.R. Johnson, and T.E. Mandel (1979) Colony formation in agar by multipotential hemopoietic cells. J. Cell Physiol., *98:*401–412.

Pluznik, D.H., and L. Sachs (1966) The induction of colonies of normal "mast" cells by a substance in conditioned medium. Exp. Cell Res., *45:*553–563.

Reimann, J., and H. Burger (1979a) In vitro proliferation of haemopoietic cells in the presence of adherent cell layers. I. Culture conditions and strain dependence. Exp. Hematol., *7:*45–51.

Reimann, J., and H. Burger (1979b) In vitro proliferation of haemopoietic cells in the presence of adherent cell layers. II. Differential effect of adherent cell layers derived from various agars. Exp. Hematol., *7:*52–58.

Schofield, R. (1978) The relationship between the spleen colony-forming cell and the haemopoietic stem cell: A hypothesis. Blood Cells, *4:*7–25.

Schrader, J.W., and S. Schrader (1978) In vitro studies on lymphocyte differentiation. I. Long-term in vitro culture of cells giving rise to functional lymphocytes in irradiated mice. J. Exp. Med., *148:*823–828.

Smith, P.J., C.W. Jackson, L.W. Dow, C.C. Edwards, and M.A. Whidden (1980a) Effect of hypertransfusion on bone marrow regeneration in sublethally irradiated mice. I. Enhanced granulopoietic recovery. Blood, *56:*52–57.

Smith, P.J., C.W. Jackson, M.A. Whidden, and C.C. Edwards (1980b) Effects of hypertransfusion on bone marrow regenerated in sublethally irradiated mice. II. Enhanced recovery of megakaryocytes and platelets. Blood, *56:*58–63.

Sorrell, J.M., and L. Weiss (1980) Cell interactions between hematopoietic and stromal cells in the embryonic chick bone marrow. Anat. Rec., *197:*1–19.

Till, J.E., and E.A. McCulloch (1961) A direct measurement of the radiation sensitivity of normal mouse bone marrow cells. Radiat. Res., *14:*213–222.

Toksoz, D., T.M. Dexter, B.I. Lord, E.G. Wright, and L.G. Lajtha (1980) The regulation of hemopoiesis in long-term bone marrow cultures. II. Stimulation and inhibition of stem cell proliferation. Blood, *55:*931–936.

Weiss, L. (1980) The haemopoietic microenvironment of bone marrow: An ultrastructural study of the interactions of blood cells, stroma and blood vessels. In Blood Cells and Vessel Walls: Functional Interactions. Elsevier/North Holland: Ciba Foundation series 71, pp. 3–19.

Werts, E.D., D.P. Gibson, S.A. Knapp, and R.I. DeGowin (1980) Stromal cell migration precedes hemopoietic repopulation of the bone marrow after irradiation. Radiat. Res., *81:*20–30.

Westen, H. and D.F. Bainton (1979) Association of alklaine-phosphatase positive reticulum cells in bone marrow with granulocytic precursors. J. Exp. Med., *150:*919–937.

Williams, N., H. Jackson, and E. Rabellino (1977) Proliferation and differentiation of normal granulopoietic cells in continuous bone marrow cultures. J. Cell Physiol., *93:*435–440.

Williams, N., H. Jackson, A.P.C. Sheridan, M.J. Murphy, A. Elste, and M.A.S. Moore (1978) Regulation of megakaryopoiesis in long-term murine bone marrow cultures. Blood, *51:*245–255.

Williams, W., and A.W. Burgess (1980) The effect of mouse lung granulocyte-macrophage colony stimulating factor and other colony stimulating activities on the proliferation and differentiation of murine bone marrow cells in long-term cultures. J. Cell Physiol., *102:*287–295.

Wilson, F.D., L. O'Grady, M.C. McNeill, and S.L. Munns (1974) The formation of bone marrow derived fibroblastic plaques in vitro: Preliminary results contrasting these population to CFU-C. Exp. Hematol., *2:*343–354.

Wilson, F.D., and L. O'Grady (1976) Some observations of the hematopoietic status in vivo and in vitro on mice of genotype *Sl/Sl*$^{d}$. Blood, *48:*601–608.

Wilson, F.D., B.R. Greenberg, P.N. Konrad, A.K. Klein, and P.A. Walling (1978) Cytogenetic studies on bone marrow fibroblasts from a male-female hematopoietic chimeras. Transplantation, *25:*87–88.

Wilson, F.D., M. Tavossili, B.R., Greenberg, D. Hinds, and A.K. Klein (1981) Morphological studies on "adherent cells" in bone marrow cultures from humans, dogs, and mice. Stem Cells, *1:*15–29.

Wolf, N.S. (1978) Dissecting the hematopoietic microenvironment. III. Evidence for a positive short-range stimulus for cellular proliferation. Cell Tissue Kinet., *11:*335–345.

Wolf, N.S. (1979) The haemopoietic microenvironment. In Lajtha, L.G. ed.: Clinics in Haematology. London, Philadelphia, Toronto: W.B. Saunders Co., vol. *8:*469–500.

Wright, E.G., G. Garland, and B.I. Lord (1980) Specific inhibition of haemopoietic stem cell proliferation: Characteristics of the inhibitor producing cells. Leukemia Res., *4:*537–545.

Zipori, D., and T. Sassoon (1980) Adherent cells from mouse bone marrow inhibit the formation of colony stimulating factor (CSF)-induced myeloid colonies. Exp. Hematol., *8:*816–817.

JOURNAL OF CELLULAR PHYSIOLOGY SUPPLEMENT 1:95-97 (1982)

# Human Leukaemia: What Next?

R. SCHOFIELD*
*Paterson Laboratories, Christie Hospital and Holt Radium Institute, Withington, Manchester M20 9BX, United Kingdom*

During the past 20 years considerable advance has been made in our knowledge of haemopoiesis in the normal mouse. This has been made possible directly as a consequence of the recognition by Till and McCulloch (1961) of the significance of the colonies formed on the spleens of mice by injecting bone marrow into animals whose haemopoietic tissues had been ablated by exposure to ionising radiation. Despite this impetus and the work that it has made possible during the intervening period there are many outstanding questions about normal murine haemopoiesis still awaiting answers and even more puzzles exist when we look at aberrant situations in the mouse.

The research that has been carried out was not undertaken, however, with the ultimate objective of understanding how the mouse produces its blood cells or of curing murine disorders. Any information we obtain is in the anticipation that it will prove useful in understanding and dealing with human disease. The indirect approach, i.e., studies in another mammal, is used because the study of disease in the human is beset by such enormous difficulties and restrictions. We do not have syngeneic humans with identical diseases; no one would think of initiating disease in a human in order to study its whole course; any experimentation must necessarily be secondary to an attempt at influencing the disease for the benefit of the patient.

What then is the most profitable approach to the study of leukaemia in the human animal? It is possible to ask fundamental questions, such as, which cell is the target for the leukaemogen? What are the changes that occur in order to convert the normal cell to a leukaemic stem cell? At the other end of the spectrum one can ask questions that are more immediately practical, such as, are there any detectable features within a single type of leukaemia, say acute myeloid leukaemia, which would give predictive information about response to treatment or life span of the patient? I do not know the answers; there is more than one way to skin a rabbit and I am sure that different approaches would sooner or later lead in the same general direction. Nor is any one approach exclusive.

I am neither a clinician treating patients with leukaemia nor a patient having, or expecting to have in the near future, leukaemia. Since these two groups of people are ultimately the ones with the greatest practical interest in the results of research, I have considered what I would want the professional researchers to give priority to if I were a member of either of these groups.

At the top of my list, I would want a system to be developed in which any human leukaemia could be studied in a variety of ways. There is no mouse leukaemia that is comparable to the human leukaemias, and transplanted leukaemias, in any case, do not provide good models for the study of spontaneously occurring conditions. Therefore the best system one can visualise, and which is probably achievable, is a long-term culture similar to the one that already exists for normal mouse bone marrow (Dexter et al., 1977). The advantage of such a culture would be that it could be set up before treatment was begun and maintained for study throughout a considerable course of the disease no matter what therapy was being given to the patient. There is already evidence that the AML of any given patient retains at least some of its original characteristics virtually unchanged throughout the disease (McCulloch et al., 1981). However, the disease in one patient varies considerably from that in another patient. The experimental freedom such a system would provide could reasonably be expected to permit a range of questions, both of immediate practical and more fundamental interest, to be asked. It should be possible by this means to investigate the role of the environment, the nature of the stem cell, and the effect of cytotoxic agents, to monitor the state of remission. It could profitably be combined with recently developed technology, such as the use of monoclonal antibodies, perhaps to separate out leukaemic

*To whom correspondence/reprint requests should be addressed.

0021-9541/82/SUPPL. 1-0095$01.50 

cells and grow up the normalised marrow for grafting. If it behaved in the same way as cultures of mouse marrow it would fail to support lymphoid cells and such cultured haemopoietic cells may be used for grafting without invoking graft-versus-host reaction. It may even be possible to grow normal marrow from patients with lymphatic leukaemia and use it for autografting.

What is the expectation that such a system could be developed? I do not know. I do know that until 20 years ago the ability to study the behaviour of the normal mouse stem cell population awaited a major advance, and that came as a result of acute observation and hard work. About 8 years ago we moved a step nearer when the beginnings of the present system for long-lasting cultures of mouse bone marrow were produced. These have subsequently been developed to provide a valuable tool for the study of many aspects of haemopoiesis. The conditions used in this murine culture system are certainly not directly transferable to human marrow cultures and so far attempts being made in Manchester and elsewhere to establish such cultures on cellular monolayers have met with only limited success (Moore and Sheridan, 1979; Toogood et al., 1980; Gartner and Kaplan, 1980). Further attempts to produce cultures using a matrix provided by collagen fibres in a gel have shown foci of haemopoietic cells growing in association with cells that appear to provide a suitable environment (Lanotte et al., 1981). It is probably a fair bet that a system for long-term culture of normal human bone marrow will be developed in the near future. Whether the step from this to culture of marrow from patients with leukaemia is a big one or a small one is unknown at present. But the rewards are large enough to warrant investment in the attempts.

The second question I would like to ask is whether the most effective treatments in existence are indeed available to the clinician. At the present time the major weapons are cytotoxic chemicals used alone or in combination. Of the vast array of compounds synthesized very few have come to clinical trial because they have failed to satisfy certain criteria. The candidacy of a compound for trial as an antileukaemic agent is invariably ultimately judged by its activity against a number of transplanted mouse tumours such as the L1210, B16 melanoma, and the Lewis lung carcinoma. It is probably quite fortuitous if a drug that satisfies these criteria also has a selective action against the human leukaemic stem cells. For many years Fox and his colleagues in Manchester have synthesized various closely related alkylating agents. When administered to rodents they have shown surprising selectivity of action against cells at different stages of development in the spermatogenic epithelium (Fox and Fox, 1967; Fox, 1975). In a few experiments, investigating their activity in the mouse haemopoietic tissues, different drugs with the same functional groups have been seen to have strikingly different effects. For example, busulphan and isopropylmethane sulphonate, both alkylmethane sulphonates, affect the haemopoietic populations totally differently. It is possible that selectivity of this sort is what we should be seeking in our cytotoxic chemical.

At present Fox and Connors, along with other experimental and medical oncologists in the United Kingdom, have organised a Trials Committee under the sponsorship of the Cancer Research Campaign. The objective of this group is to select novel agents for testing in patients on the basis of parameters that do not necessarily include activity against rodent tumours.

And, of course, attempts to augment the activity of certain drugs by manipulation of stem cell proliferation are already in progress experimentally. Ameliorating the adverse effects of cytotoxic agents on stem cells by increasing amplification of the intermediate precursors using specific proliferation stimulators would make therapy more manageable. But these things also are already in the experimental pipeline.

In making this short introduction to the session on human leukaemia I have selected two particular aspects in which I am more or less connected and which represent some of the thoughts leading to experimental approaches in the Paterson Laboratories in Manchester. Of course it is virtually impossible to predict the events that will produce the next major advance, and a multipronged approach is absolutely essential. These are just two of the prongs that we are using.

## LITERATURE CITED

Dexter, T.M., T.D. Allen, and L.G. Lajtha (1977) Conditions controlling the proliferation of haemopoietic stem cells in vitro. J. Cell Physiol., *91*:335–344.

Fox, B.W., and M. Fox (1967) Biochemical aspects of the action of drugs on spermatogenesis. Pharmacol. Rev., *19*:21–57.

Fox, B.W. (1975) Mechanism of action of methanesulphonates. In Sartorelli and Johns eds.: Handbuch der Experimental Pharmacologie. *38*:35–46.

Gartner, S., and H. Kaplan (1980) Long-term culture of human bone marrow cells. Proc. Soc. Natl. Acad. Sci. U.S.A., *77:*4756–4759.

Lanotte, M., S. Schor, and T.M. Dexter (1981) Collagen gels as a matrix for haemopoiesis. J. Cell Physiol., *106:*269–277.

McCulloch, E.A., R.N. Buick, J.E. Curtis, H.A. Messner, and J.S. Senn (1981) The heritable nature of clonal characteristics in acute myeloblastic leukemia. Blood, *58:*105–109.

Moore, M.A.S., and A.P.C. Sheridan (1979) Pluripotent stem cell replication in continuous human, prosiminian, and murine bone marrow culture. Blood Cells, *5:*1297–xxx.

Till, J.E., and E.A. McCulloch (1961) A direct measurement of the radiation sensitivity of normal mouse bone marrow cells. Radiat. Res., *14:*213–222.

Toogood, I.R.G., T.M. Dexter, T.D. Allen, T. Suda, and L.G. Lajtha (1980) The development of a liquid culture for the growth of human bone marrow. Leuk. Res., *4:*449–461.

JOURNAL OF CELLULAR PHYSIOLOGY SUPPLEMENT 1:99–102 (1982)

# Bone Marrow Transplantation for Hematologic Malignancies

KARL G. BLUME* AND STEPHEN J. FORMAN
*City of Hope Bone Marrow Transplant Group, City of Hope National Medical Center, Duarte, California 91010*

Allogeneic and syngeneic bone marrow transplantation has developed into an important treatment for hematologic malignancies such as acute myelogenous or lymphoblastic leukemia, chronic granulocytic leukemia in accelerated phase, acute malignant myelosclerosis, and various forms of lymphoma. So far, more than 500 patients with these disorders have been transplanted at centers throughout the world. Marrow transplantation is still surrounded by several significant clinical problems. However, optimal results are obtained when marrow grafting is performed early during the clinical course of the respective disorders.

Based on the rapid growth of knowledge of stem cells accumulated during the last 30 years, clinical bone marrow transplantation (BMT) has developed into an important treatment modality for patients with serious hematologic disorders. During the last decade, close to 1,000 patients have undergone BMT, more than half of them suffering from hematological malignancies. In the vast majority of patients, the marrow was collected from allogeneic sibling donors whereas a small proportion of patients received marrow grafts from syngeneic donors. With rapidly growing clinical experience, the indications for BMT have widened (see Table 1).

In principle, the technique for marrow transplantation for hematologic malignancies consists of pretransplant high-dose antileukemic chemotherapy and radiation followed by intravenous infusion of the marrow inoculum (Thomas et al., 1975). It needs to be emphasized that marrow graft rejection is an extremely rare phenomenon in patients who undergo BMT for hematologic malignancies. During the early posttransplant phase, the recipients are susceptible to bacterial and fungal infections because of pancytopenia. The risk decreases after functional engraftment has been attained at the end of the first month. This is followed by a phase during which acute graft-versus-host disease may become clinically relevant. Also, viral infections such as cytomegalovirus associated interstitial pneumonia, herpes zoster infection, and generalized herpes simplex infection occur during this time. After approximately six months following transplantation, chronic graft-versus-host disease may be a significant problem. Also, leukemic recurrence represents a major risk factor; it mostly occurs during the first two years after marrow transplantation.

This presentation deals with the results obtained with patients suffering from acute myelogenous leukemia (AML), acute lymphoblastic leukemia (ALL), chronic granulocytic leukemia (CGL), and acute malignant myelosclerosis (AMMS).

## ACUTE LEUKEMIA

Beginning in 1970, patients with end-stage leukemia refractory to conventional chemotherapy have been transplanted after preparation with marrow ablative radiochemotherapy. It was possible to demonstrate that marrow engraftment could be accomplished. However, graft-versus-host disease, infections, and leukemic recurrence were the major obstacles to long-term success. Nevertheless, 10–15% of such patients have become disease-free long-term survivors who have been followed now for up to ten years. This is a remarkable result considering the grave prognosis for patients with end-stage leukemia (Thomas et al., 1977). It then was a logical step to utilize BMT for patients with leukemia who were early in their clinical course, namely, in complete remission, while the leukemic clone had not yet become resistant to antileukemic

*To whom correspondence/reprint requests should be addressed.

0021-9541/82/SUPPL. 1-0099$01.50 

drugs. Since 1976, several centers have performed BMT for patients with AML who were in complete first remission. Actuarial disease-free survival for such patients is in the order of 60% with a follow-up time reaching now five years (Thomas et al., 1979a, Blume et al., 1980; Powles et al., 1980). The incidence of leukemic recurrence for those patients transplanted in first remission is approximately 10% with most of the relapses occurring during the first post-transplant year. Patients in second or subsequent complete remission and patients who are in partial remission or "early" relapse seem to have a 20–40% chance to be free of leukemia and hopefully cured of their disease (Thomas et al., 1979b; Blume et al., 1981).

The influence of the pretransplant hematological status on the outcome of BMT is shown in Figure 1. A follow-up of our initial evaluation of the first 33 consecutive patients is presented in the lower part of the figure. We have now extended our study to more than 100 patients with acute leukemia. The previous results have proven to be reproducible.

TABLE 1. Clinical uses of bone marrow transplantation

| |
|---|
| Acute leukemia (AML, ALL) |
| Chronic granulocyte leukemia |
| Acute malignant myelosclerosis |
| Lymphoma |
| Aplastic anemia |
| Severe combined immunodeficiency |
| Osteopetrosis |
| Wiskott-Aldrich syndrome |
| Severe hemoglobin abnormalities |
| Congenital leukocyte dysfunction syndromes |

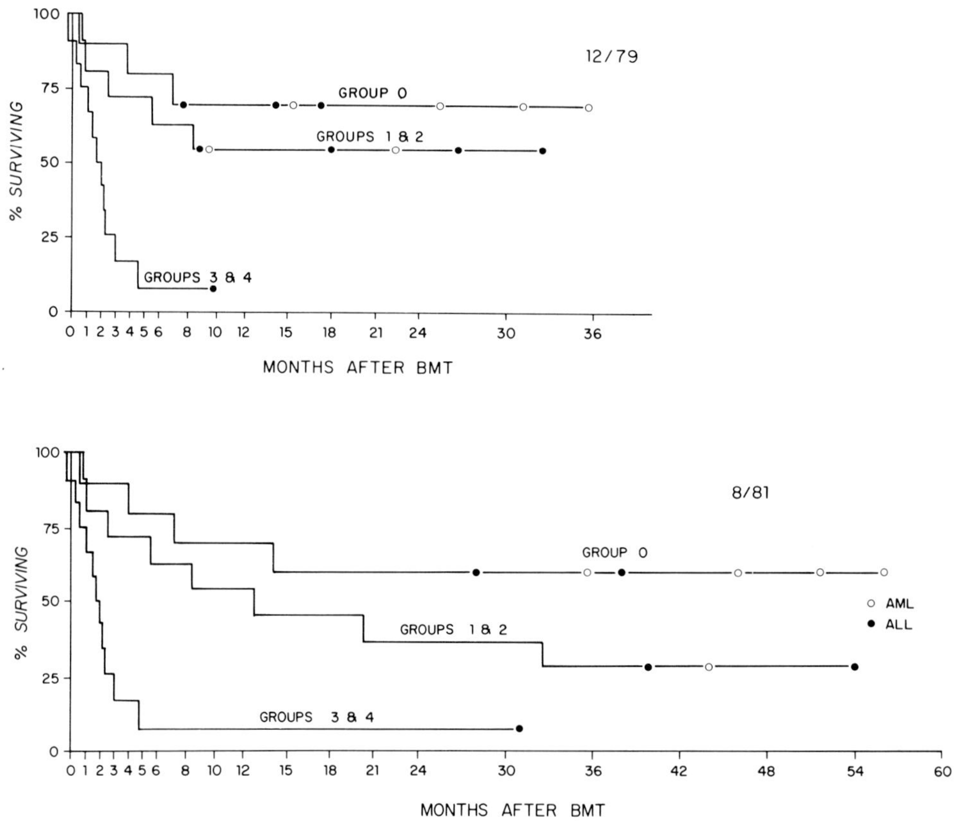


Fig. 1. Survival analysis of 33 patients with acute leukemia undergoing bone marrow transplantation between May 1976 and May 1979. The upper panel shows the evaluation in December 1979 and the lower panel that in August 1981. Ten patients were in complete first, second, or third remission (Group 0), 11 were in partial remission or "early" relapse (Groups 1 and 2), and 12 were in advanced relapse (Groups 3 and 4) at the time of bone marrow transplantation (BMT). Open circles indicate living patients with acute myelogenous leukemia (AML) and solid circles those with acute lymphoblastic leukemia (ALL).

## CHRONIC GRANULOCYTIC LEUKEMIA

Attempts to utilize allogeneic marrow transplantation for patients with CGL in the acute blastic phase have been generally unsuccessful (Doney et al., 1978). Graft failure, interstitial pneumonitis, and recurrent leukemia have been the major reasons for failure. The first promising results using marrow transplantation in CGL was a series of patients who were grafted from identical twin donors while they were still in the chronic phase of their disease (Fefer et al., 1979). The Philadelphia chromosome was abrogated in the majority of patients and no further therapy has been required. This series has now been extended to more than 15 patients and relapse-free survival is over 75%.

It appears to be an acceptable concept at the present time to subject CGL patients who show signs of clinical acceleration of their disease or who have initially other stigmata of poor prognosis to ablative radiochemotherapy followed by allogeneic marrow grafting. Such criteria are: change in cytogenetic pattern, absence of the Philadelphia chromosome, blasts and promyelocytes over 20% in the marrow or blood, peritrabecular infiltration with blasts on marrow biopsy, development of resistance to the drugs busulfan or hydroxyurea, or other clinical signs of progressive CGL such as anemia, thrombocytopenia, fever, and increasing splenomegaly. Also patients with juvenile CGL and patients who have been successfully treated for the acute phase of CGL are candidates for BMT. Currently, several transplant centers are actively involved in pursuing this therapeutic concept. However, it will take years to evaluate the results because of the chronic nature of the underlying disease.

## ACUTE MALIGNANT MYELOSCLEROSIS

AMMS is probably the hematologic malignancy with the worst prognosis of all hematologic disorders. A response to chemotherapy is almost never attained and the patients die within weeks or months after the diagnosis of AMMS has been established. It appears to be a reasonable concept to utilize BMT as primary treatment for patients with AMMS without any prior attempt with conventional chemotherapy. Six such patients have undergone allogeneic BMT and three patients have become disease-free, long term survivors (Smith et al., 1981; Wolf et al., 1982; McMillan et al., 1982). Our patient with AMMS is now 18 months after BMT and continues to be in complete remission. The initially massive myelosclerosis has completely resolved as demonstrated by serial bone marrow biopsies. Based on previous fibroblast culture studies it can be assumed that the fibrosis/sclerosis of the marrow is a reactive phenomenon secondary to the proliferation of the abnormal hemopoietic cell clone.

## LYMPHOMA

Thus far, only twenty patients with non-Hodgkin lymphoma who were refractory to conventional management have been treated with marrow ablative therapy followed by marrow grafting. Nine patients have survived with out recurrent lymphoma for one to ten years (Lutcher et al., 1980; O'Leary et al., 1980; Appelbaum, 1981; Goldman et al., 1981). This again is a remarkably good result considering the prognosis of patients with aggressive forms of lymphoma who have failed chemotherapy.

## DISCUSSION

As indicated in the introductory paragraphs, several areas in bone marrow transplantation require further intensive clinical research.

1. So far, successful transplantation in patients over the age of 40 years is rare. With the current concept of preparation for marrow transplantation by high-dose radiochemotherapy and the limited therapeutic success in the treatment of acute graft-versus-host disease in older patients, it is not likely that the results will be improved upon without major changes being implemented.

2. The limited number of patients with a histocompatible donor still restricts wider applicability of BMT. However, very encouraging results have been reported recently when partly compatible donors were used (Clift et al., 1981). These studies need to be extended and if successful may lead to an important change in the management of patients with hematologic malignancies.

3. The incidence and clinical relevance of acute and chronic graft-versus-host disease makes this an area that demands further studies. The use of methotrexate, prednisone, and antithymocyte globulin has led to prevention or control of graft-versus-host disease in a significant number of patients. However, the morbidity and mortality from this syndrome is still significant. The use of cyclosporin A and the manipulation of the marrow inoculum with specific antisera are attractive concepts that we are now just beginning to understand.

4. Interstitial pneumonia occurs in a large

number of marrow transplant recipients. Its development is poorly understood and the causative role of the cytomegalovirus is not well established. In many patients the interstitial pneumonia has been diagnosed as being idiopathic. For the pneumonias that are associated with positive viral findings, the use of new antiviral agents may lead to improved results. The more judicious use of granulocyte support may also decrease the incidence of interstitial pneumonia. Finally, new modes of total body irradiation or preparation for transplantation without radiation may lead to a decreased incidence of interstitial pneumonia.

5. As outlined above, the recurrence of leukemia is still a major obstacle to long-term survival for many patients. The step to earlier marrow transplantation has led to a decrease in the relapse rate; however, about 10% of the patients transplanted in first complete remission still suffer a leukemic relapse. Better preparative regimens and the use of graft-versus-leukemia effect may lead to a successful ablation of the leukemic clone (Weiden et al., 1981).

In conclusion, bone marrow transplantation has become a very powerful mode of therapy for patients with hematologic malignancies. Long-term disease-free survival and hopefully cure has been achieved in a significant number of patients who had an otherwise poor prognosis.

## ACKNOWLEDGMENTS

This work was supported by National Cancer Institute grant CA 30206-01.

## LITERATURE CITED

Appelbaum, F.R., M.A. Cheever, A. Fefer, C.D. Buckner, P.D. Greenberg, H.G. Kaplan, R. Storb, and E.D. Thomas (1981) Treatment of non-Hodgkins lymphoma with marrow transplantation in identical twins. Blood *58:*509–513.

Blume, K.G., E. Beutler, K.J. Bross, R.K. Chillar, O.B. Ellington, J.L. Fahey, M.J. Farbstein, S.J. Forman, G.M. Schmidt, E.P. Scott, W.E. Spruce, M.A. Turner, and J.L. Wolf (1980) Bone-marrow ablation and allogeneic marrow transplantation in acute leukemia. N. Engl. J. Med., *302:*1041–1046.

Blume, K.G., W.E. Spruce, S.J. Forman, J.L. Wolf, M.J. Farbstein, E.P. Soctt, and J.L. Fahey (1981). Bone-marrow transplantation for acute leukemia. N. Engl. J. Med., *305:*101–103.

Clift, R.A., J.A. Hansen, and E.D. Thomas for the Seattle Marrow Transplant Team. (1981) The role of HLA in marrow transplantation. Transplant. Proc., *13:*234–236.

Fefer, A., M.A. Cheever, E.D. Thomas, C. Boyd, R. Ramberg, H. Glucksberg, C.D. Buckner, and R. Storb (1979) Disappearance of Ph[1]-positive cells in four patients with chronic granulocytic leukemia after chemotherapy, irradiation, and marrow transplantation from an identical twin. N. Engl. J. Med., *300:*333–337.

Doney, K., C.D. Buckner, G.E. Sale, R. Ramberg, C. Boyd, and E.D. Thomas (1978) Treatment of chronic granulocytic leukemia by chemotherapy, total body irradiation, and allogeneic bone marrow transplantation. Exp. Hematol., *6:*738–747.

Goldman, J.M., S.A. Johnson, D. Catovsky, G. Agnarsdottir, A.W.G. Goolden, and D.A.G. Galton (1981) Identical twin marrow transplantation for patients with leukemia and lymphoma. Transplantation., *31:*140–141.

Lutcher, C.L., G.B. Faquet, R.R. Moores, and C.D. Buckner (1980) Ten year unmaintained remission from diffuse poorly differentiated lymphoma following bone marrow transplantation. Clin. Res., *28:*887A.

McMillan, R., and E. Beutler Bone marrow transplantation in acute malignant myelosclerosis. (in preparation).

O'Leary, M., N.K.C. Ramsay, M.E. Nesbit, W. Krivit, P.F. Coccia, T.H. Kim, and J.H. Kersey (1980) Bone marrow transplantation (BMT) for childhood non-Hodgkin's lymphoma (NHL). Am. Assoc. Clin. Res., *21:*173.

Powles, R.L., H.M. Clink, G. Bandini, J.G. Watson, D. Spence, B. Jameson, H.E.M. Kay, G. Morgenstern, D. Hedley, H. Lumley, D. Lawson, A. Barrett, S. Lawler, and T.J. McElwain (1980) The place of bone-marrow transplantation in acute myelogenous leukaemia. Lancet, *1:*1047–1050.

Smith, J.W., H.M. Shulman, E.D. Thomas, A. Fefer, and C.D. Buckner (1981) Bone marrow transplantation for acute myelosclerosis. Cancer *48:*2198–2203.

Thomas, E.D., R. Storb, R.A. Clift, A. Fefer, F.L. Johnson, P.E. Neiman, K.G. Lerner, H. Glucksberg, and C.D. Buckner (1975) Bone-marrow transplantation. N. Engl. J. Med., *292:*832–843, 895–902.

Thomas, E.D., C.D. Buckner, M. Banaji, R.A. Clift, A. Fefer, N. Flournoy, B.W. Goodell, R.O. Hickman, K.G. Lerner, P.E. Neiman, G.E. Sale, J.E. Sanders, J. Singer, M. Stevens, R. Storb, and P.L. Weiden (1977) One hundred patients with acute leukemia treated by chemotherapy, total body irradiation, and allogeneic marrow transplantation. Blood, *49:*511–533.

Thomas, E.D., C.D. Buckner, R.A. Clift, A. Fefer, F.L. Johnson, P.E. Neiman, G.E. Sale, J.E. Sanders, J.W. Singer, H. Shulman, R. Storb, and P.L. Weiden (1979a) Marrow transplantation for acute nonlymphoblastic leukemia in first remission. N. Engl. J. Med., *301:*597–599.

Thomas, E.D., J.E. Sanders, N. Flournoy, F.L. Johnson, C.D. Buckner, R.A. Clift, A. Fefer, B.W. Goodell, R. Storb, and P.L. Weiden (1979b). Marrow transplantation for patients with acute lymphoblastic leukemia in remission. Blood, *54:*468–476.

Weiden, PL., K.M. Sullivan, N. Flournoy, R. Storb, E.D. Thomas, and the Seattle Marrow Transplant Team. (1981) Antileukemic effect of chronic graft-versus-host disease. Contribution to improved survival after allogeneic marrow transplantation. N. Engl. J. Med., *304:*1529–1533.

Wolf, J.L., W.E. Spruce, R.M. Bearman, S.J. Forman, E.P. Scott, J.L. Fahey, M.J. Farbstein, H. Rappaport, and K.G. Blume (1982) Reversal of acute malignant myelosclerosis by allogeneic bone marrow transplantation. Blood *59:*191–193.

JOURNAL OF CELLULAR PHYSIOLOGY SUPPLEMENT 1:103–111 (1982)

# Renewal and Determination in Leukemic Blast Populations

E.A. McCULLOCH,* C.A. IZAGUIRRE, L.J-A. CHANG, AND L.J. SMITH
*Institute of Medical Science, School of Graduate Studies, University of Toronto and The Ontario Cancer Institute, Toronto, Ontario M4X 1K9, Canada*

Genetic evidence is available that a number of hematological diseases originate in pluripotent stem cells (for reviews see Fialkow, 1980; McCulloch, 1979). These diseases include Polycythemia vera (P vera), idiopathic myelofibrosis (IMF), chronic myeloblastic leukemia (CML), and acute myeloblastic leukemia (AML). In each of these disorders a single stem cell, carrying a heritable lesion, gives rise to a dominant clone that comes to occupy the total hemopoietic tissue in each patient. Other clones fail to expand in vivo, although culture evidence is available that their capacity for growth and differentiation is not lost (Adamson et al., 1976; Cunningham et al., 1979; Fialkow et al., 1981). With time these coexisting normal stem cells may become extinct.

We have suggested that disorders with the characteristics described above may be grouped together as clonal hemopathies (McCulloch et al., 1977). The diagnostic features of each of the entities reflect the cellular composition of the dominant clone; the natural history of each disease is determined by the biological characteristics of the clone including growth rate and capacity for differentiation. The stem cell origin of the clonal hemopathies is usually displayed as differentiation along granulopoietic, erythropoietic, and megakaryocytic pathways. However, in one disease, AML, the important population shows little or no evidence of differentiation; because of their morphology, these cells have been called blasts.

The abnormal stem cells of the clonal hemopathies are genetically unstable. With time, growth patterns change and transitions between disease entities are seen. For example, patients with P vera or CML may develop features of myelofibrosis (Nelson et al., 1968); more commonly, progression towards AML occurs with the appearance of blast cells (Dameshek et al., 1968).

Thus, the generation of blasts is common in the clonal hemopathies, and, since treatment is not highly effective, the appearance of blasts usually heralds the beginning of the terminal phase of each illness. Therefore, blast populations have significance beyond AML itself; their appearance within abnormal clones is a further unifying property of the clonal hemopathies.

We have proposed a model, at the cellular level, of leukemic blast populations (McCulloch et al., 1981b). These are considered as lineages, in which events analogous to those of normal differentiation progressively reduce the capacity of cells for proliferation. However, these changes in blast cells are not associated with morphological evidence of specialization. Although originally derived from pluripotent stem cells, the earliest progenitors of the blast lineage are considered to be capable of self-renewal, providing them with the capacity to maintain blast populations without further recruitment from stem cells. We have suggested that the "differentiation" programmes of blast populations are assembled abnormally from components that properly belong to normal programmes, and, further, that this abnormal assemblage is the consequence of abnormal gene expression.

In this paper the experimental evidence for these hypotheses will be reviewed with particular emphasis on those findings that may provide links between cellular and molecular events.

## BLAST PROGENITOR PROPERTIES

Blast cells from AML patients form colonies in culture (Buick et al., 1977); in our hands the technical requirements are methylcellulose to provide a viscid supporting structure and stimulatory substances released into culture media by normal leukocytes incubated in the presence of phytohemagglutinin (PHA-LCM). The colonies consist of cells morphologically similar to those present in the peripheral blood of patients from whom they were derived. Evi-

*To whom correspondence/reprint requests should be addressed.

dence of orderly differentiation along myelopoietic pathways is absent, although expression of certain components of myelopoietic differentiation programmes may be seen.

The leukemic nature of the progenitors of blast colonies was first inferred from the morphological characteristics of the cells within them. Support was obtained when it was observed that blast colonies could be grown from blood or marrow of patients with leukemia (Minden et al., 1979) or preleukemia (Senn et al., 1981), but not from normals. In patients where chromosomal abnormalities were present in freshly obtained blast cells the same markers were also found in metaphases from colonies (Izaguirre et al., 1978), a finding that provided a cytogenetic link with leukemic clones in those patients.

The lectin requirement for the proliferation of blast progenitors appeared to be indirect; mitogenically stimulated cells released bioactive molecules that stimulated blast cell growth (Aye et al., 1975). PHA is the common ingredient in a number of techniques developed for the growth of AML blasts. Dicke et al. (1976) preincubated leukemic marrow for 24 hours in the presence of PHA and obtained colonies when the cells were plated subsequently in soft agar even if further sources of stimulation were not provided. Park et al. (1980) noted an increased plating efficiency of leukemic marrow when cultures were fed repeatedly with PHA-LCM. Lowenberg et al. (1978) incorporated PHA in an underlayer of normal leukocytes; they obtained an increased plating efficiency when leukemic marrow was plated over such feeders. Recently, they have reported (Swart et al., 1981) a detailed comparison between their method and that devised by Robinson et al. (1971), in which leukocytes, but not lectin, provide the source of colony stimulation. They concluded that the addition of PHA to the feeders induced the formation of a separate subpopulation of leukemic cells than that observed in the Robinson technique.

We have searched for a relationship between the progenitors of blast colonies in culture and other hemopoietic populations. Since the biologically active materials in PHA-LCM have not been purified, stimulator specificity cannot provide evidence. Indeed, the crude culture supernatants are effective stimulators of granulopoietic progenitors (CFU-C) (Minden et al., 1979), the early stages of erythropoiesis (Aye, 1977), pluripotent stem cells (CFU-GEMM) (Fauser et al., 1979), and T-lymphocyte progenitors (Price et al., 1978). When PHA-LCM is used to promote blast colony formation, the method is improved in sensitivity and specificity when the cell suspensions are depleted exhaustively of T-lymphocytes prior to plating (Minden et al., 1979). However, the kinetics of blast colony growth are different from those of myelopoietic growth; blast cell cultures consisting of single cells after plating are observed to contain doublets and 4 cell clusters after as little as 24 hrs. Maximum size is reached after between 5 and 7 days and thereafter the colonies quickly degenerate.

In contrast, nascent myelopoietic colonies cannot be confidently identified during the first week and reach maturity only after 12–14 days of growth. Thus, the cells of blast colonies divide rapidly and do not undergo terminal granulopoietic differentiation even in the presence of a stimulator adequate for granulopoiesis; nor does the addition of erythropoietin lead to the appearance of erythropoietic cells. Indeed, in cultures of leukemic marrow, myelopoietic colonies identified to be leukemic by the presence of characteristic chromosomal markers can be observed after blast colonies have appeared and degenerated. From these data, we postulated (McCulloch et al., 1978) that blast cells did not represent a stage in any known myelopoietic lineage, but rather a novel abnormal population.

Replating experiments provided direct evidence that the blast population is organized as a lineage (Buick et al., 1979). In these, either cells from individual colonies or from a pool of colonies were replated in microwells with methylcellulose and PHA-LCM. In most experiments colonies in the wells were similar in colonial and cellular morphology to the blast colonies from which their cells of origin were obtained. This observation, showing that new cells with colony-forming ability were generated during blast colony formation, providing evidence that blast progenitors possess the property of self-renewal. However, the plating efficiency of blast cells from colonies, though variable, was low, usually less than 1% and never greater than 10%. Therefore, the majority of new cells of blast morphology generated during colony formation lacked proliferative potential; they might be considered to have arisen as a consequence of a determination step analogous to the process in normal myelopoiesis in which pluripotent stem cells enter a specific differentiation programme and lose their ability to renew themselves. Additional support for this view was obtained when the distribution of new blast progenitors among blast colonies was determined by replating individual colonies; marked heterogeneity was observed, similar to that found when individual spleen colonies were tested for their content of new stem cells (Siminovitch et al., 1963). Such asymmetrical distributions may be generated if renewal and determination occur at random during the expansion of a clone from a single cell (Till et al., 1964).

## HETEROGENEITY OF BLAST PROGENITOR CELLS

Marked individual variation was found when the blast colony method was used to study AML patients (McCulloch et al., 1980). This heterogeneity was evident when the property of blast cell self-renewal was measured (Buick et al., 1979; McCulloch et al., 1981a). The values of the plating efficiencies of blast cell populations obtained by pooling colonies (secondary plating efficiency or PE2) varied between indetectable and in excess of 500 colonies/$10^4$ cells. Figure 1 depicts the cumulative distribution of such values where PE2 was measured by plating $2 \times 10^4$ cells per microwell. Not only is the great range evident, but also the asymmetrical nature of the distribution; blast progenitors from many patients exhibited little renewal capacity, while those of a few were capable of extensive self-replication.

Such extensive variation might represent fluctuations in blast progenitor physiology occurring at random during the course of the disease. Alternatively, renewal capacity might be a stable blast progenitor property. Support for the latter view was obtained in studies where individual patients were examined repeatedly during the course of their disease (McCulloch et al., 1981a). Values of PE2 were obtained at intervals from a month to more than a year; in spite of extensive chemotherapy between measurements, and in two instances complete remission, PE2 values were very similar when measured repeatedly. Such stable properties of blast progenitors might be more likely to be determinants of the course of the disease in individual patients than those varying with time.

Measurements of the sensitivities of blast progenitors to the chemotherapeutic drugs adriamycin (Adria) and cytosine arabinoside (Ara-C) were also found to vary from patient to patient and, like PE2, repeated measurements showed that drug sensitivities were stable properties of leukemic blast progenitors (McCulloch et al., 1982).

We asked whether measurements of self-renewal and drug sensitivity were risk factors, contributing to clinical outcome (McCulloch et al., 1981c), using data from AML patients entered in sequential clinical trials. Preliminary evidence, indicating that self-renewal was a significant attribute contributing to the success of remission-induction, was confirmed; a weak association was found between sensitivity to Ara-C, but not adriamycin, and successful treatment with regimens that included both drugs. Table 1 contains the results of a multivariate analysis, using logistic regression, and assessing both PE2 and Ara-C sensitivity as they entered regression succes-

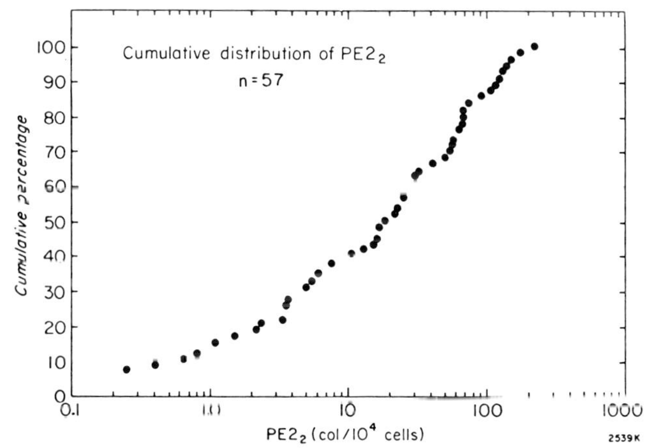


Fig. 1. The cumulative distribution of values of PE2, when measured at $2 \times 10^4$ cells per well. Data are from 57 AML patients. The semilogarithmic display is used because the values are widely spread.

TABLE 1. Contribution of blast progenitor properties to remission[1]

| Attribute | $\chi^2$ | Cumulative percent of variance | P |
|---|---|---|---|
| PE2 | 5.4 | 17 | 0.02 |
| $D_{10}$ Ara-C | 2.3 | 23 | 0.13 |

[1]Remission-induction attempted in 30 AML patients studied on protocols including Ara-C.

sively. It is apparent that PE2 made a significant contribution accounting for 17% of the variance; the additional contribution to variance from Ara-C sensitivity did not reach significance. When the more stringent test of length of survival after diagnosis was used, PE2 continued to be highly significant; indeed, its contribution was greater than that of such known risk factors as age and percentage blasts in the marrow, although these were also significant.

These clinical studies provide evidence that the property of blast progenitor renewal, as determined in culture, reflects important events in vivo. Further, these events are related to the biological behavior of the blast population. The reason underlying the lack of association between drug sensitivity measured in culture and clinical outcome is not evident. Pharmacological factors, such as drug distribution and metabolism, may be of greater importance than intrinsic sensitivity. But it is also possible that presently available chemotherapeutic agents, including Ara-C and Adria, are not sufficiently effective to be both powerful and specific attributes contributing to remission induction. This hypothesis is consistent with observations that show a treatment regimen consisting of cyclophosphamide, arabinosyl cytosine, and vincristine was not significantly more effective in inducing remissions or prolonging survival than more toxic treatments using adriamycin and Ara-C (Curtis et al., 1979) or the two drugs in combination with thioguanine (Curtis, 1981).

## BLAST CELL DIFFERENTIATION PROGRAMMES

The capacity of blast progenitors to renew themselves provides them with the capacity to maintain blast cell populations without requiring recruitment from pluripotent stem cells. The clinical evidence, cited above, provides evidence for the biological relevance of the cell culture demonstration of blast progenitor replication. Only negative evidence, the failure of the majority of cells in colonies to grow on replating, has been presented for a concurrent process, initiated by a determination event and continued as an abnormal differentiation programme (McCulloch et al., 1981b). Positive evidence of maturation after a determination event might provide insights of two kinds. First, its existence would support strongly the cellular model of blast cell proliferation, presented here and elsewhere. Second, the nature of the differentiation programmes followed by maturing blasts might reflect the molecular lesions that underlie blast cell production.

A number of putative blast differentiation markers are available for use in tracing programmes; myeloperoxidase (Kaplow, 1965) and NASD chloracetate esterase (Yam et al., 1971) are well known. Recently two monoclonal antibodies have been developed both specific for normal granulopoiesis and both found on blast cells; these are anti-My-1, obtained by Civin et al. (1981) from fusions of plasmacytomas and spleen cells of mice immunized with HL60, and anti-OKTM1 prepared by Breard et al. (1980) by fusion with human mononuclear cells immunized B-cells c/j mouse spleen cells. The monoclonal has specificity for monocytes and granulocytes and has been reported to stain human myelomonocytic and AML cells.

To seek for evidence of blast cell maturation in kinetics, anti-My-1 and NASD esterase were measured during blast cell colony formation (Marie et al., 1981a). While My-1 positive cells usually increased in culture, those containing the esterase declined. In contrast, when normal granulopoietic colonies were followed, both markers increased together until day 9, after which both declined together as the colonies became predominantly macrophage or monocytic in composition, rather than granulopoietic.

These experiments showed maturation events occurring during blast cell proliferation. In addition, the observation that the kinetics My-1 and chloracetate esterase were similar during normal granulopoiesis but different in blast cell growth supports the concept that blast programmes are abnormal. More convincing evidence would be the presence in single leukemic cells of markers of different lineages, since such an observation would violate the usual finding that commitment to one lineage excludes the expression of markers of other lineages. The first examples of doubly marked cells were found in continuous lines in culture. Fioritoni et al. (1980) identified spectrin, a marker of erythropoiesis, and NASD chloracetate esterase, a marker of granulopoiesis, in a high proportion of cells in a cloned line of Friend cells. A similar approach was applied to the human cell line K-562 (Marie et al., 1981b); this line was derived from a patient with CML in blast crisis (Lozzio et al., 1979) and has often been considered erythroleukemic because both spectrin and glycophorin have been identified within the cells. However, approximately 50% of K-562 cells (McCulloch et al., 1981a) were found to be positive for My-1; of these many contained both erythropoietic and granulopoietic markers. The finding is illustrated in Figure 2.

A beginning has been made on the search for similarly doubly marked cells in freshly obtained or cultured blast populations (Smith et al., 1981). In two patients, blasts were identified that were positive for both spectrin and My-1. In two other cases, spectrin and OK-M1 positive cells appeared during culture; in both instances a small number of doubly marked cells were identified. Similar abnormal differentiation marker associations have been found in two patients with preleukemia. Doubly marked cells have also been found after culture in the peripheral blood or marrow of a patient with myelofibrosis.

Controls consisted of freshly obtained marrow, regenerating marrow in patients with transplants and colonies in culture (including colonies of erythropoietic, granulopoietic, and megakaryocytic composition) taken at various times of incubation. Doubly marked cells have not been found in any of these controls.

Doubly marked cells have not been identified in blasts from eight other AML patients, either when freshly obtained or after culture. If the concept of abnormally assembled differentiation programmes is correct, such negatives are to be anticipated, particularly since only a small number of markers of different lineages are available for study. During differentiation certain markers might be expected to be expressed for only brief time periods; the simultaneous appearance at any point of time of two markers might be expected to be an unusual event. From the same point of view, the failure to identify doubly marked cells in normal hemopoietic populations should not be taken to mean that such cells do not exist. Indeed, Till (1976) has suggested that stem cells may express a variety of markers and that differentiation may be associated with simplification of the cell surface. Nonetheless, the finding of doubly marked cells in marrow or blood of AML patients provides the best evidence available in support of the concept of abnormal differentiation programmes in the blasts of AML. The identification of similar cells in preleukemia and myelofibrosis provides additional links between these entities and AML.

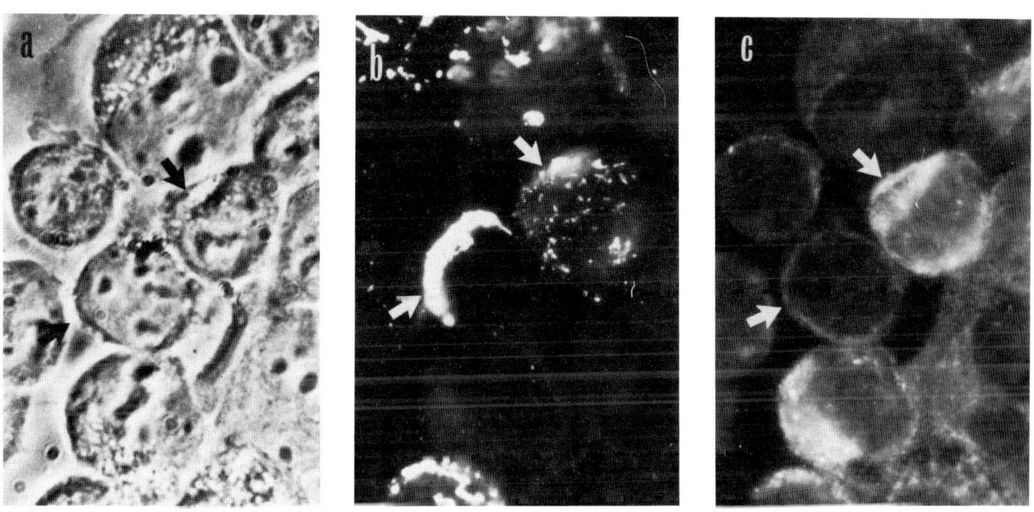


Fig. 2. A sample of K-562 cells (Lozzio et al., 1979) prepared to demonstrate both My-1 and spectrin. Panel (a) shows a field with phase microscopy. Panels (b) and (c) show the presence of My-1 and spectrin, respectively. Cells with both markers are indicated with arrows in the three frames. (Reprinted from Marie et al., 1981b, by permission of the publisher).

## MOLECULAR APPROACHES

The infrequency of doubly marked cells limits their value in establishing the nature of abnormal differentiation programmes. A more powerful approach might be developed if it were possible to study directly the expression of the genes that code for macromolecular differentiation markers. A requirement for such an approach is a method that permits the recognition within DNA of intact sequences with coding specificity. Observations indicating that genes may be expressed transiently following transfection suggested a way to develop a convenient assay (Chang et al., 1982). The technique exploits the sensitivity and specificity of monoclonal antibodies by using them to detect gene product following transfection. By choosing recipients incapable of expressing a gene product of interest, a system may be developed that is free of background and therefore very sensitive.

Such a method was developed using genomic human DNA for transfection and mouse LTA cells as recipients. In the initial experiments two monoclonal antibodies were used, anti-My-1 and OK-T3. The method is shown in Figure 3; it is based on the calcium phosphate coprecipitation technique of Graham et al. (1973), as modified in the transfection methods of Wigler et al. (1979). For convenience, coverslips are added to the petri dishes containing the recipient cells. These can be removed periodically (usually at daily intervals up to 72 hrs) and prepared for immunofluorescence.

Following transfection of DNA either from AML blast or from HL60 cells My-1 was detected on the surface of recipient LTA cells. When blast cell DNA was used, positive recipients were infrequent, never reaching 10% and often being less than 1%; however, when HL60 DNA was used, positive cells were seen in number usually exceeding 10% of the total.

Four lines of evidence support the conclusion that the presence of My-1 in recipient cells is the consequence of expression of the My-1 gene in the transfected DNA. First controls, consisting of transfer of salmon sperm DNA, LTA DNA, and calf-thymus DNA, or sham transfers without DNA never produced positive recipient cells. Second, positive cells were not seen for 48 hrs, and usually declined after reaching a peak 2 to 4 days after transfer. An example of the kinetics is shown in Figure 4. The delay means the direct effect of the transferred DNA is unlikely; the decline demonstrates the transient nature of the effect, a finding to be expected in the absence of selection. Third, when transcription and translation were inhibited by actinomycin and cyclohexamide, respectively, My-1 positive recipient cells were greatly reduced in number or eliminated. Finally, a specific endonuclease digestion pattern was observed. My-1 positive recipients were seen when DNA digested by *Eco* R1 or *Bam* H1 was transferred, but not when *Hin*d III or *Sal* 1 were used for the digestion. OK-T3 positive recipient LTA cells were seen when DNA from an OK-T3 positive T-cell leukemia was used for transfer. This activity was sensitive

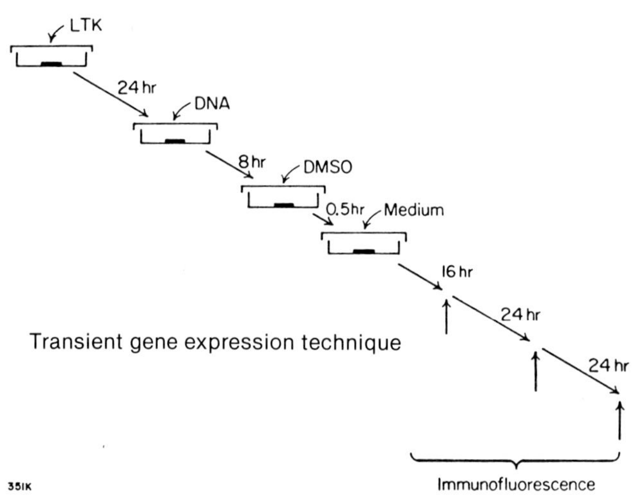


Fig. 3. A diagram illustrating the technique for obtaining transient gene expression.

to digestion with *Eco* R1 and *Bam* H1, but was not abolished by digestion with *Hind* III, a pattern different from that seen for the expression of My-1.

In the preliminary transient expression experiments, DNA for transfection was obtained from leukemic cells that expressed the differentiation marker to be searched for in the recipient cells. However, markers not expressed in donors were found after transfection. Thus, My-1 positive AML blasts or HL60 cells provided DNA that could transfer both My-1 and OK-T3, while DNA from OK-T3 positive ALL blast cells transferred My-1 as well as the pan T-cell marker. The example of kinetics shown in Figure 4 illustrates this observation. The DNA was obtained from OK-T3 negative AML blasts; the gene product observed in the LTA recipients was that of OK-T3. This observation is interpreted to mean that, at least in hemopoiesis, differentiation is not associated with irreversible inactivation of the genes of alternative programmes. The observation is compatible with the finding of differentiation markers of different lineages in the same leukemic cell; however, the mechanism of the abnormal gene expression remains unexplained.

## CONCLUSIONS

The themes of this paper all contribute to the model of blast cell growth. That blast cells possess the capacity to renew themselves is a central component of the model. The finding of a clinical correlation for the cell-culture measurements of self-renewal provides convincing evidence that the process is important in vivo. The observation that differentiation markers change during blast colony formation supports the second major tenet of the model—that determination occurs and is followed by maturation; the finding of markers of granulopoiesis and erythropoiesis together in the same cell strongly supports the concept that blast cell differentiation programmes contain abnormally assembled sequences. The phenomenon of transient gene expression following DNA transfection is the basis of an assay that may permit the model to be tested at the molecular level. For example, the model predicts that abnormal genes characteristic of leukemia will not be found. As leukemic specificity is proposed for newly described markers, this can be tested by searching for genes coding for such markers in cells that do not express them and in normal cells.

Each of the paper's themes has significance beyond the model. Analysis of blast cell properties failed to provide evidence that sensitivity of colony formation to chemotherapeutic agents is related to outcome. This finding casts doubt on the predictive value of clonogenic assays in leukemia. Rather, approaches designed to limit drug toxicity by staging patients on the basis of risk factors appear to be in order; therapeutic innovation might properly be directed towards the use of agents, such as interferon (Taetle et al., 1980), that affect self-renewal.

A beginning has been made on the problem of defining blast cell maturation. It is already evident that marked patient-to-patient variation is present in this, as in other components of blast cell growth. Analysis of this heterogeneity may contribute to the staging of patients. The mechanisms for the generation of heterogeneity, when uncovered, may have implications for the origins of leukemia and perhaps other cancers.

The molecular chapter of the biology of differentiation in human leukemia has just begun. It is already evident that genes coding for differentiation antigens can be characterized in respect to their sensitivity to restriction endonucleases. It seems likely that molecular clones for many differentiation genes will soon be available. With those as probes, it should be possible to determine whether abnormal gene expression is the consequence of rearrangements within the genomes of cells or lesions in genes that regulate the ordering of components within differentiation programmes.

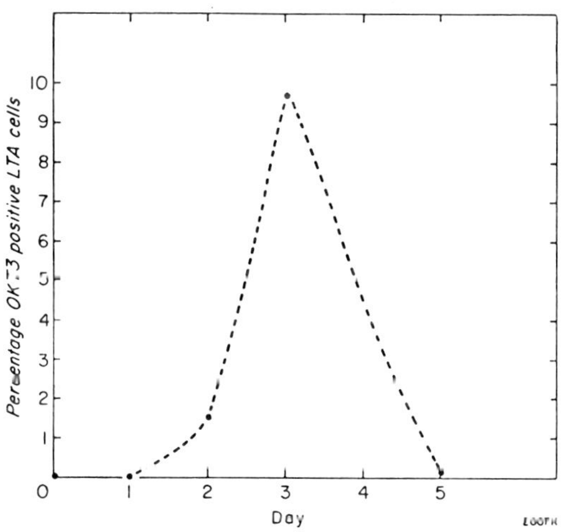


Fig. 4. The kinetics of expression of OK-T3 on the surface of LTA cells after transfection of DNA from OK-T3 negative AML blasts.

## ACKNOWLEDGMENTS

This work was supported by grants from the Medical Research Council of Canada, the National Cancer Institute of Canada, and the Ontario Cancer Treatment and Research Foundation. The authors are grateful to Dr. T. Mak and Miss C. Gamble for collaboration in the molecular aspects of the work.

## LITERATURE CITED

Adamson, J.W., P.J. Fialkow, S. Murphy, J.F. Prchal, and L. Steinmann (1976) Polycythemia vera: Stem cell and probable clonal origin of the disease. N. Engl. J. Med., *295:*913–916.

Aye, M.T., J.E. Till, and E.A. McCulloch (1975) Interacting populations affecting proliferation of leukemic cells in culture. Blood, *45:*485–493.

Aye, M.T. (1977) Erythroid colony formation in cultures of human marrow: Effect of leukocyte conditioned medium. J. Cell Physiol., *91:*69–77.

Breard, J., E.L. Reinherz, P.C. Kung, G. Goldstein, and S.F. Schlossman (1980) A monoclonal antibody reactive with human peripheral blood monocytes. J. Immunol., *124:*1943–1948.

Buick, R.N., J.E. Till, and E.A. McCulloch (1977) Colony assay for proliferative blast cells circulating in myeloblastic leukemia. Lancet, *1:*862–863.

Buick, R.N., M.D. Minden, and E.A. McCulloch (1979) Self-renewal in culture of proliferative blast progenitor cells in acute myeloblastic leukemia. Blood, *54:*95–104.

Chang, L.J-A., C.L. Gamble, C.A. Izaguirre, M.D. Minden, T.W. Mak, and E.A. McCulloch (1982) Detection of genes coding for human differentiation markers by their transient expression following DNA transfer. Proc. Natl. Acad. Sci. U.S.A. (in press).

Civin, C.I., J. Mirro, and M.L. Banquerigo (1981) My-1, a new myeloid specific antigen identified by a mouse monoclonal antibody. Blood, *57:*842–845.

Cunningham, I., T. Gee, M. Dowling, R. Chaganti, R. Bailey, S. Hopfan, L. Bowden, A. Turnbull, W. Knapper, and B. Clarkson (1979) Results of treatment of $Ph^{1+}$ chronic myelogenous leukemia with an intensive treatment regimen (L-5 protocol). Blood, *53:*375–395.

Curtis, J.E., J.E. Till, H.A. Messner, P. Sousan, and E.A. McCulloch (1979) Comparison of outcomes and prognostic factors for two groups of patients with acute myeloblastic leukemia. Leuk. Res., *3:*409–416.

Curtis, J.E., H.A. Messner, and E.A. McCulloch (1981) Clinical and biological prognostic factors in acute myelogenous leukemia (AML). Clin. Invest. Med., *4:*24B (abstr).

Dameshek, W. (1968) The myeloproliferative disorders. In Clarke, W.J., Howard, E.B., Hackett, P.L., eds.: Myeloproliferative Disorders of Animals and Man. 8th Hanford Biology Symposium. U.S. Atomic Energy Commission, Washington, D.C.: United States Government Printing Office, pp. 413–420.

Dicke, K.A., G. Spitzer, and M.J. Ahearn (1976) Colony formation in vitro by leukaemic cells in acute myelogenous leukaemia with phytohaemagglutinin as stimulating factor. Nature, *259:*129–130.

Fauser, A.A., and H.A. Messner (1979) Identification of megakaryocytes, macrophages, and eosinophils in colonies of human bone marrow containing neutrophilic granulocytes and erythroblasts. Blood, *53:*1023–1027.

Fialkow, P.J. (1980) Clonal and stem cell origin of blood cell neoplasms. In LoBue, J., ed.: Contemporary Hematology/Oncology. New York: Plenum Press, pp. 1–46.

Fialkow, P.J., J.W. Singer, J.W. Adamson, K. Vaidya, L.W. Dow, J. Ochs, and J.W. Moohr (1981) Acute nonlymphocytic leukemia: Heterogeneity of stem cell origin. Blood, *57:*1068–1073.

Fioritoni, G., L. Bertolini, G. Toriotano, and R. Revoltella (1980) Cytochemical characteristics of leukopoietic differentiation in murine erythroleukemic (Friend) cells. Cancer Res., *40:*866–872.

Graham, F.L., and A.J. van der Eb (1973) A new technique for the assay of infectivity of human adenovirus-5-DNA. Virology, *52:*456–467.

Izaguirre, C.A., and E.A. McCulloch (1978) Cytogenetic analysis of leukemic clones. Presented at the American Society of Hematology annual meeting, December 1978, New Orleans. Blood (Suppl. 1), *52:*255.

Kaplow, L.S. (1965) Simplified myeloperoxidase stain using benzidine dihydrochloride. Blood, *26:*215–219.

Lowenberg, B., and A. Hagemeijer (1978) Attempts at improving colony methods for human leukemic cells. J. Supramol. Struct. Suppl. *2:*182 (abstr 461).

Lozzio, B.B., and C.B. Lozzio (1979) Properties and usefulness of the original K-562 human myelogenous leukemia cell line. Leuk. Res., *3:*363.

Marie, J.P., C.A. Izaguirre, C.I. Civin, J. Mirro, and E.A. McCulloch (1981a) Granulopoietic differentiation in AML blasts in culture. Blood *58:*670–674.

Marie, J.P., C.A. Izaguirre, C.I. Civin, J. Mirro, and E.A. McCulloch (1981b) The presence within single K-562 cells of erythropoietic and granulopoietic differentiation markers. Blood *58:*708–711.

McCulloch, E.A., and J.E. Till (1977) Stem cells in normal early haemopoiesis and certain clonal hemopathies. In Hoffbrand, A.V., Brain, M.C., Hirsch, J. eds.: Recent Advances in Haematology-2. London: Churchill-Livingston, pp. 85–110.

McCulloch, E.A., R.N. Buick, M.D. Minden, and C.A. Izaguirre (1978) Differentiation programmes underlying cellular heterogeneity in the myeloblastic leukemias of man. In Golde, D.W., Cline, M.J., Metcalf, D., Fox, C.F. eds.: Hematopoietic Cell Differentiation. ICN-UCLA Symposium on Molecular and Cellular Biology. New York: Academic Press: Vol. X., pp. 317–333.

McCulloch, E.A. (1979) Haemopoiesis in myeloblastic leukaemia. Clin. Haematol., *8:*501–523.

McCulloch, E.A., R.N. Buick, M.D. Minden, and C.A. Izaguirre (1980) Cellular heterogeneities in acute myeloblastic leukemia. In McKinnell, R.G., DiBerardino, M.A., Blumenfeld, M., Bergad, R.D. eds.: Differentiation and Neoplasia. New York: Springer, Vol. II, pp. 217–227.

McCulloch, E.A., R.N. Buick, J.E. Curtis, H.A. Messner, J.S. Senn (1981a) The heritable nature of clonal characteristics in acute myeloblastic leukemia. Blood, *58:*105–109.

McCulloch, E.A., and J.E. Till (1981b) Blast cells in acute myeloblastic leukemia: A model. Blood Cells, *7:*63–77.

McCulloch, E.A., J.E. Curtis, H.A. Messner, J.S. Senn, and T.P. Germanson (1982) The contribution of blast cell properties to outcome variation in acute myeloblastic leukemia (AML). Blood (in press).

Minden, M.D., R.N. Buick, and E.A. McCulloch (1979) Separation of blast cell and T-lymphocyte progenitors in the blood of patients with acute myeloblastic leukemia. Blood, *54:*186–195.

Nelson, B., and R.M. Knisely, (1968) Marrow fibrosis in myeloproliferative disorders. In Clarke, W.J., Howard, E.B., Hackett, P.L. eds.: Myeloproliferative Disorders of Animals and Man. 8th Hanford Biology Symposium, U.S. Atomic Energy Commission, Washington, D.C.: United States Government Printing Office, pp. 533–555.

Park, C.H., M. Amare, M.A. Savin, J.W. Goodwin, M.M. Newcomb, and B. Hoogstraten (1980) Prediction of chemotherapy response in human leukemia using an in vitro chemotherapy sensitivity test on the leukemic colony-forming cells. Blood, *55:*595–601.

Price, G.B., and E.A. McCulloch (1978) Cell surfaces and the regulation of hemopoiesis. Semin. Hematol., *15:*283–300.

Robinson, W.A., J.E. Kurnick, and B.L. Pike (1971) Colony growth of human leukemic peripheral blood cells in vitro. Blood, *38:*500–508.

Senn, J.S., H.A. Messner, P.H. Pinkerton, L.J-A. Chang, B. Nitsch, and E.A. McCulloch (1981) Peripheral blood blast cell progenitors in human preleukemia. Blood (in press).

Siminovitch, L., E.A. McCulloch, and J.E. Till (1963) The distribution of colony-forming cells among spleen colonies. J. Cell Comp. Physiol. *62:*327–336.

Smith, L.J., and E.A. McCulloch (1981) Phenotypic classification of acute myeloblastic leukemic (AML) blasts: The presence in single cells of erythropoietic and granulocytic/monocytic markers. Blood, *59* (Suppl. 1):153a (Abstr. #521).

Swart, K., A. Hagemeijer, and R. Lowenberg (1981) Acute myeloid leukemia colony growth in vitro: Differences of colony forming cells in PHA-supplemented and standard leukocyte feeder cultures. Blood (in press).

Taetle, R., R.N. Buick, and E.A. McCulloch (1980) Effect of interferon on colony formation in culture by blast cell progenitors in acute myeloblastic leukemia. Blood, *56:*549–552.

Till, J.E. (1976) Regulation of hemopoietic stem cells. In Cairnie, A.B., Lala, P.K., and Osmond, D.J., eds.: Stem Cells. New York: Academic Press, pp. 143–145.

Till, J.E., E.A. McCulloch, and L. Siminovitch (1964) A stochastic model of stem cell proliferation, based on the growth of spleen colony forming cells. Proc. Natl. Acad. Sci. U.S.A., *51:*29–36.

Wigler, M., A. Pellicer, S. Silverstein, R. Axel, G. Urlaub, and L. Chasin (1979) DNA-mediated transfer of the adenine phosphoribosyltransferase locus into mammalian cells. Proc. Natl. Acad. Sci. U.S.A., *76:*1373.

Yam, L.T., C.Y. Li, and W.H. Crosby (1971) Cytochemical identification of monocytes and granulocytes. Am. J. Clinical Pathol., *55:*283.

JOURNAL OF CELLULAR PHYSIOLOGY SUPPLEMENT 1:113–125 (1982)

# "Target" Cells, Cellular Phenotypes, and Lineage Fidelity in Human Leukaemia

MELVYN F. GREAVES*
*Membrane Immunology Laboratory, Imperial Cancer Research Fund, Lincoln's Inn Fields, London WC2A 3PX, United Kingdom*

A detailed analysis of lymphoid leukaemia cells with a panel of monoclonal antibodies plus other markers indicates a striking conservation of phenotype that closely mimics that of normal lymphoid precursor or progenitor cells and reflects the imposition of maturation arrest. The composite phenotypes detected show some minimal deviations or asynchronies (with respect to maturation "stages") but are faithful with respect to cell lineage. These observations are suggested to shed light on the available "target" populations for lymphoid malignancy in humans and challenge the view that dedifferentiation or qualitatively aberrant gene expression is common and important in malignancy. They are further taken to support the view that normal gene products that regulate or couple proliferation and maturation are critically involved in the successive clonal selection events underlying the evolution of leukaemia. No qualitatively unique and consistent leukaemia "markers" may exist or be required. The only exception to this rule are the nonrandom chromosome changes, which, it is suggested, involve genetic loci critical to the regulation of growth and maturation in particular normal cell types.

The search for unique phenotypic features of tumour cells has been a persistent goal of cancer research, particularly with respect to cell surface properties (Wallach, 1975; Marchesi, 1976; Nicolson, 1976; Hynes and Fox, 1981; Poste, 1977). The heterogeneity of cancer cells not only thwarts design of rational therapeutic strategy but also frustrates attempts to identify consistent phenotypic properties that might be causally related to tumour cell behaviour. Indeed, this diversity, even with a "single" tumour (tissue) type, led directly to the concept that the only unique property of cancer cells is their expression of multiple independent variables (Foulds, 1969). A corollary view, fermented by both pathologists and biologists, is that cancer cells are frequently if not invariably subject to anaplasia, dedifferentiation (or "retrodifferentiation"; Coggin, 1978), or "transdifferentiation", i.e., expressing "illegitimate" products of other cell lineages or tissues.

The central issue concerning cell phenotypes in malignancy is to identify, at a molecular level, consistent, inheritable features that distinguish cancer cells from their normal counterparts; more especially we are interested in which alterations are essential and sufficient to convert a normal cell to a cancer cell. While this question is directly pertinent to leukaemia it requires rephrasing or a subsidiary enquiry in the context of other, more major cancers (e.g., epithelial carcinomas); that is, what changes are responsible for metastasis?

It is likely a priori that cancer cell phenotypes are a combination of normal, programmed, selective gene expression (i.e., differentiation) plus abnormal feature(s). Furthermore, since malignant populations are clearly subject to selective pressure and clonal evolution, the balance and relevance (biologically and clinically) of those determining influences will change with time. Figure 1 summarises in a simple fashion what are likely to be the major controls over cellular phenotype in cancer. In principle, phenotype will reflect a conservative "framework" of normal differentiation upon which clonal alterations are progressively imposed. In view of the clear monoclonality of emerging or clinically evident tumours (Fialkow, 1976), the origin of the malignant clone in terms of cell lineage is of considerable importance. Components of the

*To whom reprint requests/correspondence should be addressed.

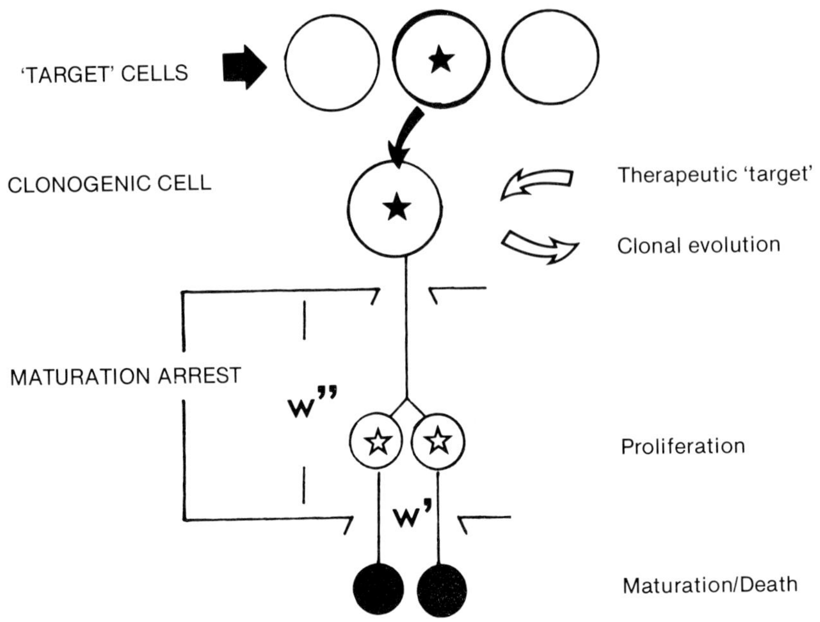


Fig. 1. Target cells: The ultimate "target" cells are those cells (probably polyclonal) subject to a rare genetic event that creates a clone with selective advantage. It is quite possible that one or more preceding rare genetic events also occurred and were essential to the eventual clonal dominance, in which case the concept of "target" populations should broaden to include all cell types involved. For example, although T-ALL clones may be "selected" from within a T lineage committed stem cell pool, it cannot be excluded that abnormal genetic events in pluripotential stem cells or in the germ line itself were essential prerequisites. Clonogenic cells: Cells with the capacity for extensive self-renewal, i.e., substantial longevity and proliferative capacity—properties of normal stem cells. These cells may cycle more frequently than normal stem cells but do not necessarily divide continuously. They provide the major challenge for effective therapy. Maturation arrest: The dominance of immature haemopoietic cells reflecting in part at least a reduced rate of maturation. Two "windows" reflecting the stringency of "arrest" are denoted (see Fig.): one (horizontally) reflecting the rate at which cells may still be able to mature (w') and another (vertically) that indicates the range of immature phenotypes in apparent arrest (some may be still in transit) (w''). Arrest (at least initially in leukaemia) is not regarded as complete or particularly efficient: neither is it necessarily irreversible. Selection: The clonogenic cells are the main focus for malignant evolution since it is in these cells that "appropriate" change can gain selective advantage. Random genetic changes may occur continuously in those cells and very rarely (i.e., in single cells) provide the basis for intraclonal selection (Nowell, 1976), e.g., via drug resistance mechanisms or altered growth requirements. Clinical "progression" of malignancy will inevitably cause further distortions of the normal cellular phenotype (e.g., additional chromosomal abnormalities) but may in addition involve changes in phenotype that reflect a more stringent maturation arrest (i.e., more "immature" phenotypes dominate) or even selection of clonogenic cells in a different lineage from that which previously was dominant (see discussion below on CGL).

lineage programme are likely to provide a dominant contribution to phenotype (Greaves and Janossy, 1978; Pierce et al., 1978). Herein lies an infrequently acknowledged difficulty. Since many or most cancers probably arise from tissue stem cell or progenitor cell populations, and may retain many "immature" features, i.e., be subject to apparent "maturation arrest", how can we interpret the significance of their phenotypes without access to the normal appropriate controls? At present this is extremely difficult for solid tumours. There is little point, for example, in comparing any biochemical feature of a monoclonal tumour with a ground-up liver or colon since the inevitable differences observed will be a) multiple and b) uninterpretable.

Leukaemias offer the distinct advantage that they are available as reasonably homogenous cell suspensions derived from a tissue much, but by no means all, of whose lineage structure and population control is understood or accessible to investigation. Over the past few years we have carried out an analysis of human leukaemic cell heterogeneity in comparison with their putative normal counterparts using a library of monoclonal antibodies and flow sorting (FACS), (Greaves, 1981a,b;

Greaves et al., 1981a,b). Monoclonal antibodies offer the advantages of diversity (as a "library"), homogeneity, and specificity, as well as providing access to the molecular structure bearing the antigenic determinant that is detected. They are clearly complementary to molecular probes for unique DNA or RNA sequences.

This study, developed as part of a routine immunodiagnostic service for leukaemia in the United Kingdom (Greaves et al., 1981d), has revealed some interesting, consistent features of leukaemic phenotypes in relation to normal haemopoietic differentiation. With respect to the central question identified above, this approach lacks the directness of molecular biology (e.g., with viral gene probes); however, it provides the opportunity to expose some popular "red herrings" and supports one particular view of what makes a cancer cell a cancer cell. Most of these observations and speculations have been discussed before (Greaves and Janossy, 1978; Greaves, 1979a,b, 1981a,b,c; Greaves et al., 1981b,f) and only salient or essential details with some more recent supporting data will therefore be reiterated here.

## MATERIALS AND METHODS

### Patients and cells

Leukaemic samples (heparinised blood and bone marrow, lymph node biopsies, cerebrospinal fluid, and peritoneal aspirates) were obtained from referring hospitals throughout the United Kingdom as part of a routine immunodiagnostic service linked principally to the Medical Research Council's UKALL trials of chemotherapy (and radiotherapy) (Greaves et al., 1981d). All samples were separated on ficoll-isopaque (1.077 gm/cc$^3$, 40 min at 400g/20°C). A panel of established leukaemic cell lines were maintained as previously described (Minowada et al., 1978).

### Monoclonal antibodies

The monoclonal antibodies used in this study, their apparent cellular specificity and, where known, the general biochemical features of the antigens they detect, are listed in Table 1.

Absorbed (but possibly polyclonal) heteroantisera used included rabbit or horse antithymus (Greaves and Janossy, 1976) and rabbit anticommon ALL/gp100 (Greaves et al., 1975; Brown et al., 1975). All leukaemic samples were also tested with an affinity purified anti-terminal deoxynucleotidyl transferase (TdT; Bollum, 1979).

### Immunofluorescence, flow cytofluorimetry and sorting

Binding of murine monoclonal antibodies was assessed with fluorochrome labeled, affinity purified IgG F(ab')$_2$ fragments of goat antimouse Ig using both fluorescence microscopy (Standard 16 Zeiss photomicroscope with epi-illuminescence) and flow cytometry (Fluorescence Activated Cell Sorter, FACS-I, Becton Dickinson, Mountain View, CA).

For intranuclear TdT staining cytospin cell preparations were first fixed in cold methanol, then incubated with rabbit antibodies specific for TdT (Bollum, 1979), followed, after washing, by fluorochrome labeled goat antibodies specific for rabbit Ig.

### E-rosette test

Twenty-five microliters of lymphocytes (at 10$^7$/ml) were mixed with 25 μl foetal calf serum (absorbed with sheep red blood cells) and 50 μl of 2% sheep red blood cells/E (pretreated with 15 U/ml neuraminidase at 37°C for 30 min). The mixture was centrifuged for 5 min at 400g (av) and left 1 hr at room temperature before gentle resuspension, dilution, and counting.

### Isolation of cell surface antigens

Membrane glycoproteins were labeled ($^{125}$I-lactoperoxidase or $^{35}$S-methionine) and isolated by detergent solubilisation, lectin affinity chromatography, immunoprecipitation, and SDS-polyacrylamide gel electrophoresis as previously described (Sutherland et al., 1978; Newman et al., 1981).

## RESULTS

### Phenotypic heterogeneity of T cell lineage malignancies

A presumptive normal sequence of antigen (and other marker) expression in T cell differentiation is given in Table 2. This scheme is based upon reported antigenic heterogeneity of cortical and medullary thymocytes and subsets of blood T cells (Bradstock et al., 1980b; Reinherz et al., 1980; Tidman et al., 1981; Greaves, 1980). The pattern differs from that which Reinherz and colleagues (1980) used to compare T-ALL phenotypes in several respects: 1) E (sheep) rosettes, OKT11A, and TdT are included and are maturation stage associated markers; 2) OKT1 and OKT3 are not regarded as exclusive to mature T cells since ~90% of thymocytes react with OKT1 (weakly) and ~30% react with OKT3, including some TdT$^+$ cells (Tidman et al., 1981); and 3) OKT9

TABLE 1. A library of monoclonal antibodies reactive with haemopoietic cells[1]

| Antibody | Specificity | Molecular weight |
|---|---|---|
| HLA-associated | | |
| W6/32, PA.2.6, BB.7.8 | HLA-ABC (monomorphic) | |
| BB5, EC3 | $\beta_2$ microglobulin | |
| DA2, OKIa-1 | HLA-DR (monomorphic) | |
| Genox 353 | HLA-DR (polymorphic, $DR_W$ 1,2,6)[2] | |
| T lineage associated[3] | | |
| OKT1, L17F12 (Leu-1, T101, UCH-$T_2$, UCH-$T_3$ | Pan T | p65–69 |
| OKT3, UCH-$T_1$ | Mature T | p19 |
| OKT4, Leu-3a | "Helper" T subset | p62 |
| OKT6, NA134 | Cortical thymocyte | p49 + p12 |
| OKT8, Leu-2a | "Suppressor", cytotoxic T subset | p30 + p32 |
| OKT11, OKT11A | Pan T (E-rosette receptor) | p50[4] |
| B lineage associated | | |
| BA-1 | | |
| FMC1 | | |
| FMC7 | | |
| Nonlymphoid | | |
| OKM-1 | Monocyte/granulocyte | |
| TG-1 | Granulocyte | |
| MI/N1 | Granulocyte (+ neural) | |
| AN51 | Platelet (glycoprotein I) | |
| LICR.LON.R10 | Glycophorin (erythroid) | |
| R6A | Band III (erythroid) | |
| Non-T acute lymphoblastic leukaemia and normal marrow lymphocyte progenitor associated[5] | | |
| J-5, VIL-1A | | gp100 |
| PI153/3 | | |
| BA-2 | | p24 |
| Leukocyte associated | | |
| 2D1 | | |
| OKT10 | | p40 |
| Ubiquitous, cell-proliferation associated transferrin receptor | | |
| OKT9, B3/25 | | gp90 (dimer)[6] |

[1]Antibodies screened in the ICRF Laboratories but developed originally in several different laboratories. This list falls far short of the total number of monoclonal "antihaemopoietic" reagents now described. See, for example, Knapp (1981). Taken from Greaves et al., 1981a (modified).
[2]May detect HLA linked DC-1 product.
[3]See Terhorst et al. (1980, 1981); Van Agthoven et al. (1981) for biochemical details.
[4]Verbi et al. (1982).
[5]These antigens may be expressed on some nonhaemopoietic cells also (Metzgar et al., 1981; Kemshead et al., 1981).
[6]Sutherland et al. (1981); Trowbridge and Omary (1981).

and OKT10 are included but not regarded as maturation stage specific since they are clearly associated with cell activation and proliferation (see below) (Terhorst et al., 1981; Greaves et al., 1981c; Sutherland et al., 1981).

Table 2 gives the corresponding maturation "positions" of 95 leukaemic T cell phenotypes analysed (Greaves et al., 1981f). T-ALL and T-NHL in childhood "correspond" to early or late thymocyte populations while T cell leukaemias/lymphomas of older (>35 yrs.) patients diagnosed as cutaneous lymphoma/leukaemia, CLL, PLL, or lymphoblastic lymphoma, all had "mature" T phenotypes. As in the normal thymus, no thymic leukaemias were HLA-DR$^+$. Some mature T leukaemias (3/30) were HLA-DR$^+$ as are normal activated T cells (Evans et al., 1978; Greaves et al., 1979c). Nineteen out of sixty-five thymic lymphoma/leukaemias were positive for the gp100 common ALL antigen (monoclonal J-5); all mature T leukaemias were negative. All T leukaemias were unreactive with anti-immunoglobulins. Nineteen T-ALL were tested for the erythroid specific marker glycophorin (monoclonal antibody LICR.LON.R10) and all were negative. Most thymic leukaemias were positive with OKT10 and many with OKT9. These antigens should not, however, be regarded as "differentiation" antigens since OKT10 reacts with myeloid

TABLE 2. Phenotypic subgroups in T cell malignancy (95 cases)[1]

| Dominant phenotype | | Markers: OKT series 1 | 3 | 4 | 6 | 8 | 9 | 10 | 11 | Others "T"[2] | $E_R$ | TdT | Clinical diagnosis (n) | | | | | |
|---|---|---|---|---|---|---|---|---|---|---|---|---|---|---|---|---|---|---|
| T precursors | | | | | | | | | | | | | | | | | | |
| Early/prethymic | 1 | + | – | – | – | – | | + | – | + | – | + | (2) | | | | | |
| | 2 | +w | – | – | – | – | + – | + | + | + | + | + | (10) | ALL (53) | | | (1) | |
| | 3 | +w | ± | +/– | + – | +/– | | + | + | + | + | + | (38) | | | | (4) | NHL (12) |
| Cortical thymic | 4 | +w | –/– | + | + | + | – | + | + | + | + | + | (3) | | | | (7) | |
| Mature T | | | | | | | | | | | | | *M. fungoides* Sézary | | CLL | PLL | Lympho-blastic lymphoma | |
| Inducer/helper T | | + | + | + | – | – | – | – | + | + | + | – | 9 | | 3 | 6 | 4 | |
| | | | | | | | | | | | | | | | | 2[3] | | 30 |
| Suppressor/cytotoxic T | | + | + | – | – | + | – | – | + | + | + | – | 1 | | 4 | 0 | 1 | |

[1]Taken from Greaves et al. (1981f) and updated to September 1981.
[2]"T", horse or rabbit antithymus serum.
[3]Two cases of T-PLL were reactive with both OKT4 and OKT8.

cells and activated mature T cells (Terhorst et al., 1981; Greaves et al., 1981c) and OKT9 reacts with the proliferation associated transferrin receptor (Sutherland et al., 1981). Reactivity with OKT9 and OKT10 reflects in part at least the physiological state and proliferative activity of the cell population investigated. It is likely that these features are themselves regulated in part by a differentiation-linked programme.

The simplest interpretation of these data is that T leukaemia phenotypes closely mimic those of normal T cells at various stages of differentiation and reflect the "position" of maturation arrest (cf. Fig. 1) imposed on the dominant subclone. Significantly, some thymic ALL cell lines can be induced by phorbol ester to acquire a mature T cell phenotype (Nagasawa and Mak, 1980; Greaves et al., 1981c; Delia et al., 1982).

The marker phenotypes observed in T leukaemia are not, however, perfect replicas. A number of "abnormalities" can be detected, including HLA-ABC expression (Bradstock et al., 1980a; Greaves, 1980; Greaves et al., 1981f); these are perhaps best regarded as asynchronies of gene expression since they maintain fidelity to the T lineage but distort the normal integrated sequence or pattern of expression presumed for normal maturation of T cells.

The finding that T cell leukaemias and lymphomas express the common ALL antigen/gp100 (Roberts et al., 1978; Greaves et al., 1981f; Minowada et al., 1978; Greaves, 1981a; Bernard et al., 1981) is of some importance since it has so far not been clearly detected on normal thymocytes or blood T cells. Using monoclonal anti-gp100 (J-5) we do find this antigen, however, on some foetal thymocytes. Whether this structure exists in a more cryptic form (Negoro et al., 1981) on most thymocytes is an important issue to resolve. This could, however, be another example of asynchrony of expression in leukaemia.

### Phenotypic heterogeneity of non-T ALL

Figure 2 summarises the phenotypes we have documented in non-T ALL (Greaves et al., 1981a,b,c; Kersey et al., 1981). These are arranged in a possible antigenic sequence. Note that one third of common ALL synthesize $\mu$ heavy chain (Vogler et al., 1978; Greaves et al., 1979a) and that minor cell populations within individual cALL may show a single light chain type (i.e., $\kappa$ or $\lambda$). In addition, the majority of cALL that do not synthesize detectable $\mu$ chain nevertheless have rearranged $\mu$ chain genes as revealed by molecular hybridization studies (Korsmeyer et al., 1981; Molgaard, Gould, and Greaves, unpublished observations). The majority of non-T ALL probably represent B cell precursors. Moreover, the presence of light chain restricted B cells, albeit in low numbers, in some of those patients suggests that the clonogenic cells in these patients are probably B lineage committed. However, this does not prove nor necessarily indicate that the original target cell population for leukaemic transformation was a similar B cell progenitor (see below). B-ALL is a rare subclass of ALL; these cells have a mature B cell phenotype and probably represent disseminated lymphoma (McGrath and Ziegler, 1979; Greaves, 1981b).

As documented previously (Janossy et al., 1979; Greaves et al., 1980a; Greaves et al., 1981e), we were able to detect in normal or nonleukaemic bone marrow a TdT positive "lymphoid" cell whose monoclonal antibody defined phenotype was essentially identical to that of common ALL. The common ALL associated antigen gp100, as well as the antigens detected by the monoclonal anti-ALL BA-2 (= p24) and monoclonal antineuroblastoma, PI153/3, are present on this normal cell. Furthermore, a similar gp100 molecule can be isolated from normal bone marrow (Newman et al., 1981). There is no evidence therefore that any of these reagents are leukaemia specific.

As with T cell leukaemias, some possible asynchronies of marker expression were observed in leukaemias with "early B cell" phenotypes. Thus ~90% of pre-B ($\mu^+$) cALL are gp100$^+$ and TdT$^+$, whereas only a very small fraction of normal marrow pre-B cells are detectably positive for the cALL antigen (gp100) (Greaves et al., 1980a) and even fewer (~1–2%) for TdT (Janossy et al., 1979). It could of course be argued that normal pre-B cells are quite heterogeneous and that leukaemias correspond quite accurately to a small fraction of the least mature pre-B cells, which could be gp100$^+$/TdT$^+$.

### "Lymphoid" phenotypes appearing in Ph$^1$ positive chronic granulocytic leukaemia (CGL)

One of the most striking observations documented on lymphoid markers in leukaemia is the presence of ALL associated markers in blast crisis of CGL (Janossy et al., 1977, 1978). These observations have recently been re-

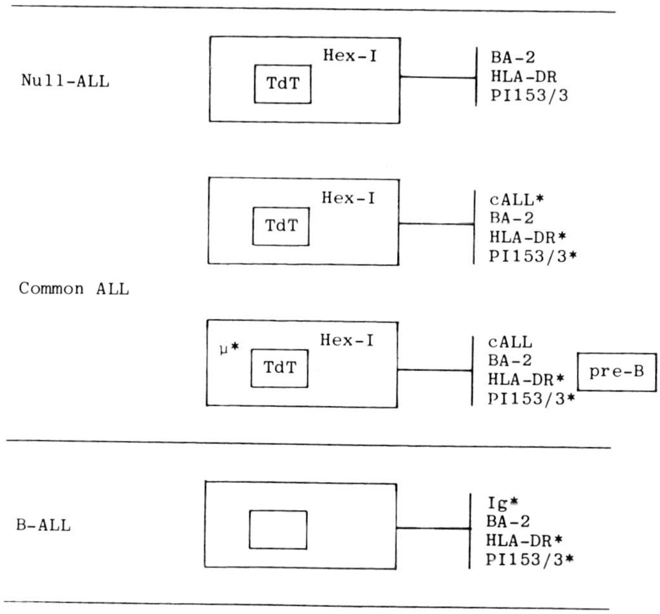


Fig. 2. Phenotypic sequence in non-T ALL. *Invariably expressed. Other markers expressed commonly but not invariably. TdT, terminal deoxynucleotidyl transferase; Hex-I, hexosaminidase I (intermediate) isoenzyme; cALL, Common ALL antigen (gp100) detected by rabbit antisera or monoclonal antibodies J-5, VIL-1A; HLA-DR, detected by monoclonal antimonomorphic DR reagents DA2, OKIa-1.

viewed (Greaves, 1981d); Table 3 summarises the relative dominance of different cell types in a large series of blast crisis we have been able to examine over several years. Approximately one third are ALL-like or "lymphoid". Indeed, by monoclonal antibody typing (cf. Table 1) they are indistinguishable from $Ph^1$ negative ALL; this similarity includes the expression of free $\mu$ chains (Greaves et al., 1979b). It is significant that no thymic (T-ALL-like) phenotypes have been detected. Note that in 2 cases (Table 3) monoclonal light chains were detected. This is an interesting observation since it suggests the presence of clonogenic cells committed to the B axis.

Although blast crises may frequently contain mixed populations of cells (i.e., the extent of clonal dominance varies), we have no evidence for mixed lineage phenotypes on individual cells, e.g., the cALL antigen, when present, is on the $TdT^+$ cell, the $\mu^+$ cells have "lymphoid" cell surface markers and $TdT^+$ cells lack nonlymphoid lineage restricted markers (e.g., glycophorin determinant detected by monoclonal LICR.LON.R10; platelet glycoprotein I detected by monoclonal AN51).

CGL is an excellent example of the importance of interpreting cellular phenotypes in cancer in relation to normal differentiation. In the absence of clonal markers (the $Ph^1$ chromosome and glucose 6 phosphatedehydrogenase) and any understanding of cell lineage relationships, the progression of malignancy observed in CGL might be readily interpreted as exemplifying both dedifferentiation (in "myeloblastic" blast crisis) and gross aberrant gene expression (in "lymphoblastic" blast crisis). While there could be some abnormalities of cellular markers (in addition to chromosomal changes), in CGL blast crises the dominant phenotypes observed are most rationally interpreted as an indication of the multipotential stem cell origin of CGL (Boggs, 1974; Janossy et al., 1976; Fialkow et al., 1978) and the subsequent selection of clonogenic cells that a) may be committed to a single lineage and b) show an apparent maturation arrest (reviewed in Greaves and Janossy, 1978; Greaves 1981d).

TABLE 3. Ph[1] positive CGL blast crisis[1]

| | | |
|---|---|---|
| Total | 352 | |
| Dominant lymphoid (L) | 101 | |
| cALL | 87 | 7 $\mu^+$<br>36 $\mu^-$<br>2 $\mu^+$ $\kappa^+$ TdT$^-$ |
| Null ALL | 14 | |
| B-ALL | 0 | |
| T-ALL | 0 | |
| Dominant nonlymphoid (M) | 210 | |
| Mixed L + M | 14 | |

[1]M, myeloid. Some myeloid blast crises are erythroid or megakaryoblastic but most are myeloblastic (granulocytic precursors)

The involvement of the B cell lineage in the CGL "clone" is directly demonstrated also by the enzyme and chromosome studies of Fialkow and colleagues (Fialkow et al., 1978; Martin et al., 1980; see also Fialkow et al. in this volume) and by recent observations of Bernheim et al. (1981). Where T cells fit into the picture is still unclear.

### Phenotypic shift in lymphoid malignancy

Alterations in phenotype during leukaemia progression or relapse may provide some evidence for distortion of "normal" genetic programmes as well as revealing the lineage affiliations of various clonogenic populations. Alterations in the latter population may indirectly indicate the possible target cell population involved (cf. Fig. 1). Thus shifts between "lymphoid" and "myeloid" blast crisis (Janossy et al., 1978; Greaves, 1981d; Catovsky, 1979) are possible because of the cellular origins of this leukaemia.

Andersson et al. (1980) have suggested that some common ALL may relapse with erythroid phenotypes; this might be taken to indicate either an aberrant phenotype or a multipotential stem cell origin for ALL. Over the past 6 years we have had the opportunity to retest a large number of leukaemias in relapse (Greaves et al., 1980b; Greaves 1981a,b). Figure 3 summarizes the essential points of these observations. None of 66 common ALL tested in relapse was glycophorin positive. No T-ALL has in relapse expressed a B-ALL or cALL phenotype. Similarly no Ph[1] negative cALL (out of >150 tested in up to 5 consecutive relapses) has relapsed with T cell features, myeloid, erythroid, or platelet markers. Some phenotypic alterations are observed in relapsed ALL. Thus within non-T ALL, the gp100/cALL antigen may be lost or acquired as compared with the phenotype at diagnosis (Greaves et al., 1980b; Greaves, 1981a,b). Within T-ALL, changes in the dominant phenotype may occur in relapse including the apparent loss of E rosette/OKT11A binding and TdT, which may reflect either a genuine loss of differentiation-linked markers or a more stringent maturation arrest imposed on leukaemic T cell precursors. Other phenotypic features are very stable; thus very few (<5%) of cALL in relapse "lose" expression of TdT or HLA-DR. No "mature" lymphoid leukaemias have in our series (or any other known to the author) "acquired" lymphoid precursor or ALL phenotypes. Cases of ALL presenting with the Philadelphia chromosome may, however, become predominantly myeloid (Catovsky, 1979) and are probably therefore part of the "CGL" family of leukaemias (see Catovsky, 1979; Greaves, 1981d). Some mixed lymphoid plus myeloid (Ph[1] negative) leukaemias are recorded (Mertlesman et al., 1978; Bradstock et al., 1981) and lymphoid leukaemias may occasionally switch to dominant myeloblastic leukaemia. These cases are, however, exceptional.

While these data fall short of a clonal analysis, they provide, by comparison with CGL, some insight into the likely "target" cells for lymphoid malignancy. As discussed elsewhere (Greaves et al., 1981b), the results suggest a) that the maturation associated phenotype is usually retained with disease progression (with some possible "deletions" imposed) and b) that 3 "target" cell populations (cf. Fig. 1) exist for lymphoid malignancy, namely, a pluripotential stem cell (Ph[1] positive ALL, Ph[1]-positive CGL in lymphoid blast crisis), lymphoid committed stem cells/progenitors (thymic or prethymic T for T-ALL and T-NHL, B cell progenitors for most non-T ALL), and mature immunocompetent T cells or B cells (for CLL, PLL, cutaneous lymphoma, and most non-Hodgkin lymphoma). The position of myeloma in this "target" cell map remains uncertain. Since lymphocytes with the immunoglobulin (Ig) idiotype of the myeloid protein can be found (Mellstedt et al., 1976) the tumour probably does not originate in plasma cells. Furthermore, myeloma Ig idiotypic determinants have been demonstrated in bone marrow cells that have features of pre-B cells (Kubagawa et al., 1979). This is a surprising observation and along with the report of myeloma Ig idiotype on patients' T cells (Preud'homme et al., 1977; Lea et al., 1979) has prompted the idea that myeloma may originate in lymphocyte progenitors, i.e., pre-B cells, a common T plus B progenitor, or even multipotential stem cells. Unfortunately, there are technical uncertain-

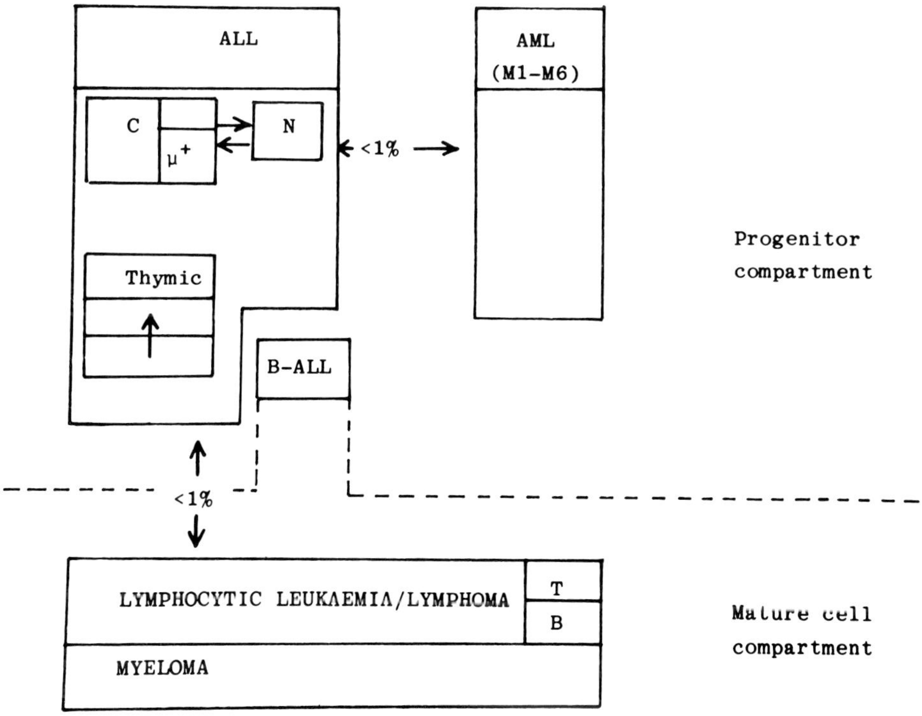


Fig. 3. Phenotypic 'shift' in leukaemia. C, common (acute lymphoblastic leukaemia); N, "null" (acute lymphoblastic leukaemia). Arrows indicate "direction" of phenotypic change and percentage observed (i.e., in less than 1% of patients followed). Ph[1] positive leukaemias excluded from survey. Changes do occur within the T-ALL compartment and between cALL and "null"-ALL.

ties associated with these observations that compromise the credibility of the interpretations offered. Myeloma may, in most cases, be derived from immunocompetent, or mature, B lymphocytes. Hairy cell leukaemia is also at present difficult to accommodate in any lineage map since these cells are claimed to express both T and B cell properties (see Guglielmi et al., 1980).

## DISCUSSION

These observations, based primarily on the application of monoclonal antibodies specific for cell surface associated determinants in association with other selective markers, e.g., TdT, immunoglobulins, suggest that leukaemic cells show a conservation or fidelity of lineage and cell type associated phenotype. The detailed or composite phenotypes detected are not perfect replicas; they show a number of "asynchronies" that may be more pronounced in relapse and in cell lines established from relapse cases (e.g., T-ALL; Greaves et al., 1981f). In these respects lymphoid leukaemias parallel the "minimally-deviated" hepatoma phenotypes described by Potter (1978). The phenotypic character, as anticipated (see Fig. 1) can be seen as being predominantly determined by the target cell population, the dominant clonogenic population (at one particular time), and the dominant "positive" of apparent maturation arrest. The ontogenic "distance" between the postulated "target" cell and the bulk of leukaemic cells in maturation arrest reflects the stringency of maturation prohibition, and it is clear that this varies considerably both between different leukaemias and at different times in the same individual (especially in CGL). Identification of the likely "target" cell and clonogenic populations in leukaemia has important implications for therapy (Till and McCulloch, 1980; Greaves, 1981b, c), including conventional strategies, i.e., chemotherapy, and also the projected use of monoclonal antibodies. In addition, the rational dissection of leukaemic cell heterogeneity can, to some extent, explain prognostic variability (Greaves et al, 1981d) and provide a sounder biological basis for disease classification.

Whether the asynchronies of marker expression are a reflection of aberrant gene control or an independent and asynchronous effect of "maturation arrest" of unlinked genes is unknown. More significantly, there is no evidence from these studies for gene expression which is illegitimate with respect to cell lineage affiliation. An obvious reservation is that the application of different markers or more markers might reveal some consistent abnormalities of a qualitative nature (see paper in this volume by McCulloch et al.); the broad fidelity of phenotype and absence of any common, leukaemia specific antigen (Greaves, 1979c) remains, however, a striking observation.

These observations and interpretations have some relevance to ideas on the nature of genetic changes in leukaemia. They clearly challenge those concepts that dictate the dedifferentiation (or retrogressive differentiation) or onco-foetal expressions are common in cancer and an important part of the underlying mechanism. While not disproving, they provide contrary evidence for the view that cell differentiation per se is qualitatively abnormal or that aberrant gene expression is common (in the sense of unexpected or illegitimate gene products).

Rather, it was suggested that a qualitatively normal phenotype of progenitor cells is both compatible with and sufficient for leukaemia, providing phenotypic features that are normally transitory are "fixed" (Greaves and Janossy, 1978; Greaves, 1979a,b). This hypothesis carries the corollary that all "structural" information required for acute leukaemia is programmed in normal haemopoietic progenitors. A similar view with respect to solid tumour cells was expressed by Pierce et al. (1978).

This perspective places emphasis on the regulation of phenotypic progression seen in normal cell maturation. Neither proliferative advantage or maturation arrest per se are sufficient to account for leukaemia phenotypes and therefore the simplest view is that these two independent but linked processes become uncoupled (Greaves, 1979b; Sachs, 1980).

Obviously leukaemic cells must have some genetic or inheritable abnormalities of gene expression. Nonrandom chromosome changes are at present the clearest example of leukaemia associated phenotypic changes (Rowley, 1978). Some of the genetic regions involved may therefore include genes, which in particular cell types exert a critical regulating role. Products of these chromosomal loci, or others as yet undetected by gross rearrangements, may play a direct causal role in leukaemogenesis yet endow the leukaemic clone with no unique extrachromosomal "product" to serve as a qualitative marker. Translocations or other more subtle genetic changes could disrupt the regulatory activity of these essential normal gene products via, for example, excessive or continual promotion or by inaccessibility of inhibitory signals.

These views are indirectly supported by the partial reversibility of maturation arrest in leukaemia (e.g., Harrison, 1977; Graf and Beug, 1978; Sachs, 1978; Gallagher et al., 1979) and accord with the recent evidence from molecular virology that indicates that leukaemogenic viruses may effectively uncouple maturation and proliferation (Graf and Beug, 1978; see also Beug et al. in this volume). Moreover they may do so by either inserting sequences (v-onc) homologous to normal, cell type specific genes (which may regulate proliferation and differentiation) or insert small sequences that regulate or promote these same normal (i.e., c-onc) genes (Neel et al., 1981; Payne et al., 1981; Wyke, 1981; see Hayward et al. in this volume). Alteration in quantitative levels of certain normal gene products (which may or may not be cell surface associated) or on-off control of these products (or both, i.e., level as a function of time) would create the behavioural pattern seen in leukaemic clones and be compatible with a qualitatively normal cellular phenotype as detected in a "static" or nonkinetic analysis of cellular antigens and other markers.

There is no reason why these general interpretations should not hold true for solid tumours, although here the issue of phenotypic correlates of metastasis is a crucial consideration (Poste, 1977). It is likely, however, that most carcinomas derive from tissue stem cells and retain many of the normal phenotypic features of those cells and their immature descendents, including perhaps migratory capacity. Many "solid" tumours undergo malignant progression with "apparent" dedifferentiation (Foulds, 1969; Pierce et al., 1978). As discussed elsewhere (Greaves, 1981d) there may be some remarkable similarities here with leukaemia requiring some reinterpretation of data. Malignant astrocytoma in particular provides a striking parallel with blast crisis of CGL (Greaves, 1981d).

## LITERATURE CITED

Andersson, L.C., Wegelius, R., Borgström, G.H., and Gahmberg, C.G. (1980) Change in cellular phenotype from lymphoid to erythroid in a case of ALL. Scand. J. Haematol., *24*:115–121.

Bernard A., Boumsell, L., Reinherz, E.L., Nadler, L.M., Ritz, J., Coppin, H., Richard, Y., Valensi, F., Dausset, J., Flandrin, G., Lemerle, J., and Schlossman, S.F., (1981) Cell surface characterization of malignant T cells from lymphoblastic lymphoma using monoclonal antibodies: Evidence for phenotypic differences between malignant T cells from patients with acute lymphoblastic leukemia and lymphoblastic lymphoma. Blood, *57:*1105–1110.

Bernheim, A., Berger, R., Preud'homme, J.L., Labaume, S., Bussel, A., and Barot-Ciorbaru, R. (1981) Philadelphia chromosome positive blood B lymphocytes in chronic myelocytic leukemia. Leuk. Res., *5:*331–340.

Boggs, D.R., (1974) Hematopoietic stem cell theory in relation to possible lymphoblastic conversion of chronic myeloid leukemia. Blood, *44:*449–453.

Bollum, F. (1979) Terminal deoxynucleotidyl transferase as a hemopoietic cell marker. Blood, *54:*1203–1215.

Bradstock, K.F., Hoffbrand, A.V., Ganeshaguru, K., Llewellin, P., Patterson, K., Wonke, B., Prentice, H.G., Bennett, M., Pizzolo, G., Bollum, F.J., and Janossy, G. (1981) Terminal deoxynucleotidyl transferase expression in acute nonlymphoid leukaemia: An analysis by immunofluorescence. Br. J. Haematol. *47:*133–143.

Bradstock, K.F., Janossy, G., Bollum, F.J., and Milstein, C. (1980a) Anomalous gene expression in human thymic acute lymphoblastic leukaemia (Thy-ALL). Nature, *284:*455–457.

Bradstock, K.F., Janossy, G., Pizzolo, G., Hoffbrand, A.V., McMichael, A., Pilch, J.R., Milstein, C., Beverley, P., and Bollum, F.J. (1980b) Subpopulations of normal and leukaemic human thymocytes: an analysis with the use of monoclonal antibodies. J. Natl. Cancer Inst., *65:*33–42.

Brown, G., Capellaro, D., and Greaves, M.F. (1975) Leukaemia associated antigens in man. J. Natl. Cancer Inst., *55:*1281–1289.

Catovsky, D. (1979) Ph$^1$-positive acute leukaemia and chronic granulocytic leukaemia: One or two diseases? Br. J. Haematol., *42:*493–498.

Coggin, J.H. (1978) Retrogressive differentiation in cancer. In: Immunological Aspects of Cancer. J. Castro, ed. MTP Press, Lancaster, pp. 89–100.

Delia, D., Greaves, M.F., Newman, R.A., and Sutherland, D.R. (1982) Modulation of T leukaemic cell phenotype with phorbol ester. Int. J. Cancer *29:*23–31.

Evans, R.L., Faldetta, T.J., Humphrey, R.E., Pratt, D.M., Yunis, E.J., and Schlossman, S.F. (1978) Peripheral human T cells sensitized in mixed leukocyte culture synthesize and express Ia-like antigens. J. Exp. Med., *148:*1440–1445.

Fialkow, P. (1976) Clonal origin of human tumors. Biochim. Biophys. Acta, *458:*283–321.

Fialkow, P.J., Denman, A.M., Jacobson, R.J., and Lowenthal, M.N. (1978) Chronic myelocytic leukemia: Origin of some lymphocytes from leukemic stem cells. J. Clin. Invest., *62:*815–823.

Foulds, L. (1969) Neoplastic Diseases. Academic Press, New York and London.

Gallagher, R., Collins, S., Trujillo, J., McCredie, K., Ahearn, M., Tsai, S., Metzgar, R., Aulakh, G., Ting, R., Ruscetti, F., and Gallo, R. (1979) Characterization of the continuous, differentiating myeloid cell line (HL-60) from a patient with acute promyelocytic leukemia. Blood, *54:*713–733.

Graf, T., and Beug, H. (1978) Avian leukaemia viruses: Interaction with target cells in vivo and in vitro. Biochim. Biophys. Acta, *516:*269–300.

Greaves, M.F. (1979a) Cell surface characteristics of human leukaemic cells. In: Essays in Biochemistry Campbell, P.N., and Marshall, R.D., eds. Academic Press, New York, Vol. 15, pp. 78–124.

Greaves, M.F. (1979b) Tumour markers, phenotypes, and maturation arrest in malignancy: A cell selection hypothesis. In: Tumour Markers Boelsma, E., and Rümke, P., eds. Elsevier, Amsterdam, pp. 201–211.

Greaves, M.F. (1979c) Immunodiagnosis of leukaemia. In: Immunodiagnosis of Cancer (Herberman, R.H. and McIntire, K.R., eds. Marcel Dekker, New York, pp. 542–587.

Greaves, M.F. (1980) Analysis of human leukaemic cell populations using monoclonal antibodies and the fluorescence activated cell sorter. In: Advances in Comparative Leukemia Research 1979. Yohn, D.S., Lapin, B.A., and Blakeslee, J.R., eds. Elsevier, Amsterdam, pp. 235–242.

Greaves, M.F. (1981a) Monoclonal antibodies as probes for leukaemic heterogeneity and haemopoietic differentiation. In: Leukemia Markers Knapp, W., ed. Academic Press, New York, pp. 19–32.

Greaves, M.F. (1981b) Analysis of lymphoid phenotypes in acute leukemia: Their clinical and biological significance. Cancer Res., *41:*4752–4766.

Greaves, M.F. (1981c) Biology of acute lymphoblastic leukaemia. Leukaemia Research Fund (United Kingdom) publication.

Greaves, M.F. (1981d) "Target" cells, differentiation, and clonal evolution in chronic granulocytic leukaemia: A "model" for understanding the biology of malignancy. In: Chronic Granulocytic Leukaemia Shaw, M.T., ed. W.B. Saunders, New York, pp. 15–47.

Greaves, M.F., Brown, G., Rapson, N., and Lister, T.A. (1975) Antisera to acute lymphoblastic leukaemia cells. Clin. Immunol. Immunopathol., *4:*67–84.

Greaves, M.F., Delia, D., Janossy, G., Rapson, N., Chessells, J., Woods, M., and Prentice, G. (1980a) Acute lymphoblastic leukaemia associated antigen. IV. Expression on nonleukaemia "lymphoid" cells. Leukemia Res., *4:*15–32.

Greaves, M.F., Delia, D., Newman, R., and Vodinelich, L., (1981a) Analysis of leukaemic cells with monoclonal antibodies. In: Monoclonal Antibodies in Clinical Medicine McMichael, A., and Fabre, J., eds. Academic Press, London (in press).

Greaves, M.F., Delia, D., Robinson, J., Sutherland, R., and Newman, R. (1981b) Exploitation of monoclonal antibodies: A "Who's who" of haemopoietic malignancy. Blood Cells, *7:*257–280.

Greaves, M.F., Delia, D., Sutherland, R., Rao, J., Verbi, W., Kemshead, J., Robinson, J., Hariri, G., Goldstein, G., and Kung, P. (1981c) Expression of the OKT monoclonal antibody defined antigenic determinants in malignancy. Int. J. Immunopharmacol. *3:*283–300.

Greaves, M.F., and Janossy, G. (1976) Antisera to human T lymphocytes. In: Methods in Cell Mediated Immunity, 2nd Edition Bloom, B., and David, J., eds. Academic Press, New York, pp, 89–104.

Greaves, M.F., and Janossy, G. (1978) Patterns of gene expression and the cellular origins of human leukaemias. Biochim. Biophys. Acta, *516:*193–230.

Greaves, M.F., Janossy, G., Peto, J., and Kay, H. (1981d) Immunologically defined subclasses of acute lymphoblastic leukaemia in children: Their relationship to presentation features and prognosis. Br. J. Haematol., *48:*179–197.

Greaves, M.F., Paxton, A., Janossy, G., Pain, C., Johnson, S., and Lister, T.A. (1980b) Acute lymphoblastic leukaemia associated antigen. III. Alterations in expression during treatment and in relapse. Leuk. Res., *4:*1–14.

Greaves, M.F., Rao, J., Hariri, G., Verbi, W., Catovsky, D., Kung, P., and Goldstein, G. (1981f) Phenotypic heterogeneity and cellular origins of T cell malignancies. Leuk. Res., *5:*281–299.

Greaves, M.F., Robinson, J.B., Delia, D., Ritz, J., Schlossman, S., Sieff, C., Goldstein, G., Kung, P., Bollum, F., and Edwards, P. (1981e) Comparative antigenic phenotypes of normal and leukemic hemopoietic precursor cells analysed with a "library" of monoclonal antibodies. In: Modern Trends in Human Leukemia IV. Neth, R., Gallo, R.C.,

Graf, T., Mannweiler, K., and Winkler, K., eds. Springer-Verlag, Berlin, Heidelberg, pp. 296–304.

Greaves, M.F., Verbi, W., Festenstein, H., Papasteriadis, C., Jaraquemada, D., and Hayward, A. (1979c) "Ia-like" antigens on human T cells. Eur. J. Immunol., *9:*356–362.

Greaves, M.F., Verbi, W., Kemshead, J., and Kennett, R. (1980c) A monoclonal antibody identifying a cell surface antigen shared by common acute lymphoblastic leukemias and B lineage cells. Blood, *56:*1141–1144.

Greaves, M.F., Verbi, W., Reeves, B.R., Drysdale, H.C., Jones, L., Sacker, L.S., and Samaratunga, I. (1979b) "Pre-B" phenotypes in blast crisis of Ph[1] positive CML: Evidence for a pluripotential stem cell "target." Leuk. Res., *3:*181–191.

Greaves, M.F., Verbi, W., Vogler, L., Cooper, M., Ellis, R., Ganeshaguru, K., Hoffbrand, V., Janossy, G., and Bollum, F.J. (1979a) Antigenic and enzymatic phenotypes of the pre-B subclass of acute lymphoblastic leukaemia. Leuk. Res., *3:*353–362.

Guglielmi, P., Preud'homme, J.L., and Flandrin, G. (1980) Phenotypic changes of phytohaemagglutinin-stimulated hairy cells. Nature, *286:*166–168.

Harrison, P.R. (1977) The biology of the Friend cell. Int. Rev. Biochem., *15:*227–267.

Hynes, R.O., and Fox, C.F. (eds). (1981) Tumor Cell Surfaces and Malignancy. Alan R. Liss, Inc., New York.

Janossy, G., Bollum, F., Bradstock, K., Rapson, N. and Greaves, M.F. (1979) Terminal deoxynucleotidyl transferase positive human bone marrow cells exhibit the antigenic phenotype of common acute lymphoblastic leukaemia. J. Immunol., *123:*1525–1529.

Janossy, G., Greaves, M.F., Sutherland, R., Durrant, J., and Lewis, C. (1977) Comparative analysis of membrane phenotypes in acute lymphoid leukaemia and in lymphoid blast crisis of chronic myeloid leukaemia. Leuk. Res., *1:*289–300.

Janossy, G., Roberts, M., and Greaves, M.F. (1976) Target cell in chronic myeloid leukaemia and its relationship to acute lymphoid leukaemia. Lancet, *ii:*1058–1061.

Janossy, G., Woodruff, R.K., Paxton, A., Greaves, M.F., Capellaro, D., Kirk, B., Innes, E.M., Eden, O.B., Lewis, C., Catovsky, D., and Hoffbrand, A.V. (1978) Membrane marker and cell separation studies in Ph[1]-positive leukemia. Blood, *51:*861–877.

Kemshead, J.T., Fritschy, J., Asser, U., Sutherland, R., and Greaves, M.F. (1981) Monoclonal antibodies defining markers with apparent selectivity for particular haemopoietic cell types may also detect antigens on cells of neural crest origin. Hybridoma (in press).

Kersey, J.H., LeBien, T.W., Abramson, C.S., Newman, R., Sutherland, R., and Greaves, M. (1981) p24: A human leukemia-associated and lymphohemopoietic progenitor cell surface structure identified with monoclonal antibody. J. Exp. Med., *153:*726–731.

Knapp, W. (ed) (1981) Leukemia Markers. Academic Press, New York/London.

Korsmeyer, S.J., Hieter, P.A., Ravetch, J.V., Poplack, D.G., Leder, P., and Waldmann, T.A. (1981) Patterns of immunoglobulin gene arrangement in human lymphocytic leukemias. In: Leukemia Markers. Knapp, W., ed. Academic Press, New York/London, pp. 85–97.

Kubagawa, H., Vogler, L.B., Capra, J.D., Conrad, M.E., Lawton, A.R., and Cooper, M.D. (1979) Studies on the clonal origin of multiple myeloma: Use of individually specific (idiotype) antibodies to trace the oncogenic event to its earliest point of expression in B-cell differentiation. J. Exp. Med., *150:*792–807.

Lea, T., Følre, Ø.T., Michaelson, T.E., and Natvig, J.B. (1979) Shared idiotype on human peripheral blood B and T lymphocytes. J. Immunol., 122:2413–2417.

Marchesi, V.T. (ed) (1976) Membranes and Neoplasia. Progr. Clin. Biol. Res., *9.*

Martin, P.J., Najfeld, V., Hansen, J.A., Penfold, G.K., Jacobson, R.J., and Fialkow, P.J. (1980) Involvement of the B-lymphoid system in chronic myelogenous leukemia Nature, *287:*49–50.

McGrath, I.T., and Ziegler, J.L. (1979) Bone marrow involvement in Burkitt lymphoma and its relationship to acute B cell leukaemia. Leuk. Res., *4:*33–59.

Mellstedt, H., Petersson, D., and Holm, G. (1976) Monoclonal B lymphocytes in peripheral blood of patients with myeloma: Relation to activity of disease. Scand. J. Haematol., *16:*112–120.

Mertlesman, R., Koziner, B., Ralph, P., Filippa, D., McKenzie, S., Arlin, E.A., Yee, T.S., Moore, M.A.S., and Clarkson B.D. (1978) Evidence for distinct lymphocytic and monocytic populations in a patient with terminal transferase positive acute leukemia. Blood. *51:*1051–1056.

Metzgar, R.S., Borowitz, M.J., Jones, N.H., and Dowell, B.L. (1981) Distribution of common acute lymphoblastic leukemia antigen in nonhematopoietic tissues. J. Exp. Med., *154:*1249–1254.

Minowada, J., Janossy, G., Greaves, M.F., Tsubota, T., Srivastava, B.I.S., Morikawa, S., and Tatsumi, E. (1978) Expression of an antigen associated with acute lymphoblastic leukemia in human leukemia-lymphoma cell lines. J. Natl. Cancer Inst., *60:*1269–1277.

Nagasawa, K., and Mak, T.W. (1980) Phorbol esters induce differentiation in human malignant T lymphoblasts. Proc. Natl. Acad. Sci. U.S.A., *77:*2964–2968.

Neel, B.G., Hayward, W.S., Robinson, H.L., Fang, J., and Astrin, S.M. (1981) Avian leukosis virus-induced tumors have common proviral integration sites and synthesize discrete new RNAs: Oncogenesis by promoter insertion. Cell, *23:*323–334.

Negoro, S., Yoshizaki, K., Henderson, E.S., and Seon, B.K. (1981) New antigenic determinant common to human thymus leukaemia-associated antigens and common acute lymphoblastic leukaemia antigen. Clin. Exp. Immunol., *45:*487–495.

Newman, R.A., Sutherland, R., and Greaves, M.F. (1981) The biochemical characterization of a cell surface antigen associated with acute lymphoblastic leukemia and lymphocyte precursors. J. Immunol., *126:*2024–2030.

Nicolson, G.L. (1976) Transmembrane control of the receptors on normal and tumor cells. Biochim. Biophys. Acta, *458:*1–72.

Nowell, P.C. (1976) The clonal evolution of tumor cell populations. Science, *194:*23–28.

Payne, G.S., Courtneidge, S.A., Crittenden, L.B., Fadly, A.M., Bishop, J.M., and Varmus, H.E. (1981) Analysis of avian leukosis virus DNA and RNA in bursal tumors: Viral gene expression is not required for maintenance of the tumor state. Cell, *23:*311–322.

Pierce, G.B., Shikes, B., and Fink, L.M. (1978) Cancer: A Problem in Developmental Biology. Prentice Hall, New Jersey.

Poste, G. (1977) The cell surface and metastasis. In: Cancer Invasion and Metastasis: Biology Mechanisms and Therapy Day, S.B., ed. Raven Press, New York, pp. 19–47.

Potter, V.R. (1978) Phenotypic diversity in experimental hepatomas: The concept of partially blocked ontogeny. Br. J. Cancer, *38:*1–23.

Preud'homme, J.L., Klein, M., Labaume, S., and Seligmann, M. (1977) Idiotype-bearing and antigen binding receptors produced by blood T lymphocytes in a case of human myeloma. Eur. J. Immunol. *7:*840–846.

Reinherz, E.L., Kung, P.C., Goldstein, G., Levey, R.H., and Schlossman, S.F. (1980) Discrete stages of human intrathymic differentiation: Analysis of normal thymocytes and leukemic lymphoblasts of T lineage. Proc. Natl. Acad. Sci. U.S.A., *77:*1588–1592.

Roberts, M., Greaves, M.F., Janossy, G., Sutherland, R., and

Pain, C. (1978) Acute lymphoblastic leukaemia (ALL) associated antigen. I. Expression in different haematopoietic malignancies. Leuk. Res., *2:*105–114.

Rowley, J.D. (1978) Chromosomes in leukemia and lymphoma. Semin. Hamatol., *15:*301–319.

Sachs, L. (1978) Control of normal cell differentiation and the phenotypic reversion of malignancy in myeloid leukaemia. Nature, *274:*535–539.

Sachs, L. (1980) Constitutive uncoupling of pathways of gene expression that control growth and differentiation in myeloid leukemia: A model for the origin and progression of malignancy. Proc. Natl. Acad. Sci. U.S.A., *77:*6152–6156.

Sutherland, R., Delia, D., Schneider, C., Newman, R., Kemshead, J., and Greaves, M. (1981) Ubiquitous, cell surface glycoprotein on tumour cells is proliferation-associated receptor for transferrin. Proc. Natl. Acad. Sci. U.S.A., *78:*4515–4519.

Sutherland, R., Smart, J., Niaudet, P., and Greaves, M.F. (1978) Acute lymphoblastic leukaemia associated antigen. II. Isolation and partial characterisation. Leuk. Res., *2:*115–126.

Terhorst, C., Van Agthoven, A., Reinherz, E.L., and Schlossman, S. (1980) Biochemical analysis of human T lymphocyte differentiation antigens T4 and T5. Science, *209:*520–521.

Terhorst, C., Van Agthoven, A., LeClair, K., Snow, P., Reinherz, E., and Schlossman, S. (1981) Biochemical studies of the human thymocyte cell-surface antigens T6, T9, and T10. Cell, *23:*771–780.

Tidman, N., Janossy, G., Bodger, M., Granger, S., Kung, P.C., and Goldstein, G. (1981) Delineation of human thymocyte differentiation pathways utilizing double-staining techniques with monoclonal antibodies. Clin. Exp. Immunol., *45:*457–467.

Till, J.E., and McCulloch, E.A. (1980) Hemopoietic stem cell differentiation. Biochim. Biophys. Acta, *605:*431–459.

Trowbridge, I.S., and Omary, M.B. (1981) Human cell surface glycoprotein related to cell proliferation is the receptor for transferrin. Proc. Natl. Acad. Sci. U.S.A., *78:*3039–3043.

Van Agthoven, A., Terhorst, C., Reinherz, E., and Schlossman, S. (1981) Characterization of T cell surface glycoproteins T1 and T3 present on all human peripheral T lymphocytes and functionally mature thymocytes. Eur. J. Immunol., *11:*18–21.

Verbi, W., Greaves, M.F., Koubek, K., Janossy, G., Stein, H., Kung, P., and Goldstein, G. OKT11 and OKT11A: Monoclonal antibodies with pan T reactivity which block sheep erythrocyte "receptors" on T cells. Eur. J. Immunol. (in press).

Vogler, L.B., Crist, W.M., Bockman, D.E., Pearl, E.R., Lawton, A.R., and Cooper, M.D. (1978) Pre-B cell leukemia: A new phenotype of childhood lymphoblastic leukemia. N. Engl. J. Med., *298:*872–878.

Wallach, D.F.H. (1975) Membrane Molecular Biology of Neoplastic Cells. Elsevier, Oxford.

Wyke, J. (1981) Strategies of viral oncogenesis. Nature, *290:*629–630.

JOURNAL OF CELLULAR PHYSIOLOGY SUPPLEMENT 1:127–132 (1982)

# Acute Lymphocytic Leukemia: Biological Markers of Clinical Response

ALVIN M. MAUER,* W. PAUL BOWMAN, LOIS W. DOW, A. THOMAS LOOK, AND DOROTHY WILLIAMS

*St. Jude Children's Research Hospital, Memphis, Tennessee 38101*

Treatment for acute lymphocytic leukemia (ALL) has been successful in producing cures for some children with that disease. It has become increasingly important therefore to develop treatment regimens for curable ALL that are effective for disease control but carry minimal risks of late detrimental consequences. For other children current therapy is still inadequate and treatment failure occurs. For these children altered treatment strategy must be developed that includes intensification of some treatment components, since improved methods of disease control are the primary consideration. In this paper the clinical and laboratory methods by which risk groups can be defined at diagnosis have been described. We have increasingly effective means by which the expected treatment outcome with current therapy can be defined at diagnosis. This assignment to a specific risk group is important to provide children with ALL with the most appropriate therapy for their expected clinical response. These studies of clinical utility also pose biological questions concerning the relationship of such things as white blood count, leukemic cell karyotype, and initial proliferative activity to the likelihood of drug resistance emerging. During the coming decade there will no doubt be an increasing emphasis on a biological characterization of leukemic cell populations. Answers must be developed to such questions as the mechanism for leukemic cell growth regulation. Biological information developed in this model tumor cell system in humans will likely produce information of value in understanding human cancer in general.

Progress in our understanding of acute lymphocytic leukemia (ALL) can be roughly defined by the three decades since the first description of an effective chemotherapeutic agent by Farber et al. in 1948. The subsequent decade of the 1950s can be characterized as one in which several other effective agents were introduced and the important concept of combining agents to increase the proportion of patients responding was developed.

The decade of the 1960s was one in which the emphasis was on comparative clinical therapeutic trials. The strategy of treatment was outlined and the independent contributions of the remission induction phase and the continuation phase to outcome were demonstrated. Perhaps the single most important event of the 1960s was the recognition that the central nervous system acted as a chemotherapeutic sanctuary for leukemic lymphoblasts; therefore a phase of CNS prophylaxis was a necessary part of treatment strategy. With the addition of that phase of treatment by means of cranial irradiation and intrathecal methotrexate in the middle 1960s, the proportion of patients alive and without evidence of disease 5 years after diagnosis approached the 50% mark. Therapy was actually being stopped for children who had been in continuous complete remission for 2½–3 years by the end of that decade.

While a great deal had been learned about treatment strategy during the 1960s, little had been added to our knowledge of the biology of ALL. Tritium labeled thymidine had been in

*To whom reprint requests/correspondence should be addressed.

0021-9541/82/SUPPL. 1-0127$02.00 

troduced in 1959 and information was acquired with this cell label about the characteristics of proliferation of normal and leukemic blood cells. Information was acquired by studies of transplantable murine leukemias that contributed to concepts of chemotherapy. It remained for the decade of the 1970s, however, to see an accelerating acquisition of information about the biological features of ALL. It had been recognized that certain clinical features present at diagnosis such as the white blood count, presence of early CNS involvement, the finding of a mediastinal mass on chest roentgenogram, and age bore a relationship to the expected clinical course and treatment outcome. To these clinical risk features was added an important laboratory observation by Sen and Borella (1975). They described leukemic lymphoblasts in a group of clinically high-risk ALL patients that formed rosettes with sheep erythrocytes as was characteristic for normal thymic lymphocytes. This observation was the beginning of a biologically oriented classification system for ALL. The description by Greaves et al. (1975) of a marker defined by a heterologous antibody that was characteristic of a group of children who predominantly had standard-risk clinical features was another important step in the developing biological classification of ALL. Rapidly therafter many new markers were defined aided especially by the monoclonal antibody technique. The cell marker systems have been recently summarized by Bowman et al. (1980). Of great biological interest has been the finding that the cell marker systems essentially describe the process of differentiation of normal lymphoid cells albeit in an aberrant fashion in the leukemic cell populations.

It will be the purpose of this paper to introduce some new biological observations that have clinical importance. The current therapeutic problems will be described an an attempt will be made to predict the interrelationship of biologically oriented studies with clinical therapeutic strategies. It would seem that the real opportunity of the coming decade of the 1980s would be to wed biology and therapy in a mutually advantageous manner.

## CURRENT THERAPEUTIC PROBLEMS

Progress in the therapy of ALL has been rapid and gratifying during the past three decades. It is now realistic to speak of the curability of ALL in some children. As defined by George et al. (1979), at this time it becomes increasingly important to design treatment strategies for curable ALL that will provide effective disease control and reduce the risks of the consequences of disease and treatment. Among the important problems are the risks of central nervous system damage by CNS prophylaxis as described by Price and Jamieson (1975) and the finding of intellectual impairment in some long-term survivors of treatment by Eiser and Lansdown (1977).

For some children, however, the major problem remains the lack of initial responsiveness to treatment or the emergence of a drug-resistant population of leukemic lymphoblasts during or following cessation of continuation therapy. For these children, the goal must be primary control of disease by altered treatment strategy, new drug regimens or intensification of current treatment modalities.

It has become increasingly important, therefore, to further refine our concept of risk factors so that patients can be assigned to one or the other of these basic two groups of patients at standard or high risk of treatment failure for a treatment program designed specifically for their needs. The search has been on for means by which risk assessment can be defined with increasing precision to avoid overtreatment of standard-risk patients and undertreatment for high-risk patients.

## RISK FACTOR CONCEPTS

Clinical features associated with specific prognostic importance have been summarized by Simone (1975) and Miller (1975). The single most important risk factor is the white blood count at diagnosis. The greater the count the less likely the opportunity for long-term disease control. Other factors associated with poor prognosis are early CNS involvement, the presence of a mediastinal mass, an age at diagnosis of less than 2 years or greater than 10 years, and bulky lymphadenopathy and hepatosplenomegaly.

When the biological cell markers became available it was found that risk features tended to cluster in specific subgroups of ALL and one could generally assign expected risk for treatment failure as shown in Table 1. However, exceptions exist and some common ALL patients have clinical high-risk features and some T-cell ALL patients have clinical features usually associated with standard risk for treatment failure. Current studies, such as those described by Bowman and Muer (in press), are designed to assess the independent significance of the biological markers with respect to risk of treatment failure. Obviously, of great-

TABLE 1. Biologic classification of ALL

| Leukemic blast cell type | Approximate percent of patients | Unusual cell characteristics | Risk assessment |
|---|---|---|---|
| Common ALL | 65 | Common ALL antigen, Ia-like antigen, cytoplasmic corticosteroid binding, nuclear TdT activity | Standard |
| Pre-B-cell ALL[1] | | Above characteristics with cytoplasmic IgM | |
| T-cell ALL | 20 | T-cell antigen, 50% have $E^+$ blasts, reduced cytoplasmic corticosteroid binding, nuclear TdT activity, acid phosphatase positive | High |
| B-cell ALL | <5 | Surface Ig positive, Ia-like antigen | High |
| Undifferentiated[2] | 10 | May have Ia-like antigen | Variable |

[1]Approximately 20% of ALL patients have blast cells with cytoplasmic IgM demonstrable by immunofluorescence. Studies to date indicate no clinical differences from patients with common ALL not having this feature (Crist and co-workers, Blood, *54* (Suppl):183a, 1979).
[2]This group currently cannot be identified by defined cell marker systems. It is probably heterogenous and the clinical outcome is variable.

est importance is the development of information relating to the role of these biological features in the mechanisms by which treatment failure occurs. The identification of patients at high-risk allows for the focus of studies in this group to develop this type of mechanistic information.

## BIOLOGICAL CONSIDERATIONS OF TREATMENT FAILURE

The problem for a small group of patients with ALL is the presence of an initially resistant population of blast cells. Overall this patient group is small being about 5% or less of ALL patients. For a newly defined subgroup as will be described below, primary treatment failure may occur in as many as 30% of newly diagnosed patients.

For most treatment failure patients, the problem is an initial, prompt and complete response followed subsequently in a variable period of time by reemergence of drug-resistant leukemic blast cells and clinical relapse. Although a remission can be reinduced in most of these treatment failures, relapse occurring during or even following cessation of therapy is the worst of all prognostic features and almost all of these patients will eventually die from their disease.

There have been many studies attempting to define the biochemical mechanisms by which drug resistance occurs. These mechanisms vary with the specific chemotherapeutic agent and its mechanism of action. There also seem to be some cell characteristics that impart resistance of a general nature to a variety of drugs. One of the important biological considerations is whether drug resistance is acquired as a factor of modulation of the residual leukemic population during remission or if there is a preexisting resistant subpopulation which is selected by chemotherapy for subsequent growth and reemergence in relapse. As will be shown below, the initial growth rate of the leukemic cell population is related to the timing and likelihood of relapse.

## RECENT OBSERVATIONS OF RISK FACTORS

I would briefly like to describe four studies from this institution that have provided further information about risk assessment, but more importantly raise biological questions about their relationship to clinical outcome.

### White blood count

As indicated above, one of the most closely associated factors with prognosis has been the white blood count at diagnosis. At this institution we have used a white blood count of $100{,}000/mm^3$ as the dividing line between standard and high-risk patients. One of us (W.P.B.) has analyzed the results of the current treatment program for standard-risk patients by further dividing this group into patients with counts less than or greater than $25{,}000/mm^3$. As of 1 July 1981, 135 patients had been entered into the treatment program for standard-risk patients. One hundred one of these patients had counts less than $25{,}000/mm^3$ and to date none of those patients have relapsed. However, there are 34 patients with counts greater than $25{,}000/mm^3$ and 9 of those have already relapsed. Thus to date, all relapses on this treatment program have occurred in the patient group with the higher white blood count. This clinical finding provides an even more sensitive index for further categorization of patients by risk from the standpoint of treatment planning.

It also provides an interesting challenge to determine the biological features of ALL that associate the likelihood of leukemic blast cells spilling out of the bone marrow into the blood with a greater risk of acquisition of drug resistance and early relapse. The clinical feature of white blood count has utility for treatment planning, but even more important, it provides a biological clue to disease behavior.

### Proliferative activity

Dow et al. (1981) has analyzed the relationship of initial blood and bone marrow proliferative activity of leukemic cells to prognosis in patients with circulating blasts. Proliferative activity has been assessed by determining the tritiated thymidine labeling index of blood and bone marrow leukemic blast cells by means of autoradiography. The initial range of bone marrow labeling indices has been wide from 0.8 to 32.7 with a median value of 6.2%. For the blood the range has also been wide from 0.3 to 20.0 and having a median value of 3.2%.

A significant difference in treatment outcome was found when the group was divided into two parts based on a higher or lower level of proliferative activity in the blood and bone marrow blast cells. In the bone marrow treatment failure occurred in 42% of patients with LI less than 6%, but in 74% of patients with LI greater than 6%. For the blood studies, treatment failure occurred in 41% of patients with LI less than 4% and 82% of patients with LI greater than 4%. Both of these observations were statistically different. Furthermore, when relapse did occur in those patients with greater proliferative activity in blood and bone marrow the onset was earlier than in patients with lesser proliferative activity. Moreover, this measure of proliferative activity added statistically significant information to the leukocyte count described previously.

One might expect a more rapid onset of relapse in patients having blast cell populations of greater proliferative activity. The puzzle in these observations is why the greater proliferative activity is associated with a greater propensity for acquiring drug resistance. Does the greater proliferative activity of a preexisting resistant subpopulation allow for greater opportunity for emergence? Is it possible that greater proliferative activity is associated with a greater tendency to modulation of blast cell characteristics associated with drug resistance?

Once again, an observation has been made of clinical utility in defining patients at greater or lesser risk of relapse; however, these observations also provide a provocative clue for further studies of the biological characteristics of acute leukemia as they relate to treatment response.

### Chromosomal characteristics

Williams et al. (1981) have analyzed the initial blast cell karyotypes of 136 patients with ALL. She had found that five groups of the patients can be defined according to the characteristics and modal number distribution of these karyotypes. The first group of 41 patients had hyperdiploid karyotypes with more than 50 chromosomes. A second group of 18 patients also had hyperdiploid karyotypes with 47–50 chromosomes. A third group of 46 patients had normal numbers and, by current banding techniques, normally appearing karyotypes. A small group of 3 patients had hypodiploid karyotypes and a final group of 28 patients had normal numbers of chromosomes, but abnormal morphology and were classified as pseudodiploid.

The patients with greater than 50 chromosomes have had only two treatment failures to date. The clinical behavior of the 67 patients who had hypodiploid, normal or 47–50 chromosomes were similar and formed an intermediate response group with 21 treatment failures to date. A most important group was found to be those patients with pseudodiploid karyotypes. Only 70% of these 28 patients achieved initial remission and already 19 (68%) have experienced treatment failure. Again, relapse, when it has occurred, is earlier than in either of the other two groups.

Once again, an observation of clinical utility has emerged from these studies. However, it is intriguing to realize that whatever transforming process results in a cell line that has a pseudodiploid karyotype also causes that cell population to have increasing initial resistance to treatment and a greater likelihood of earlier relapse. Equally intriguing, of course, is the observation that the burden of an excess number of chromosomes marks a cell population predictable in its initial response and unlikely to have a subsequent acquisition of drug resistance. Clearly, there is a biological clue in these observations of fundamental importance in our understanding of cancer.

### Aneuploidy and surface markers

One of us (A.T.L.) has analyzed for leukemic blast cell DNA content by the method of flow microfluorometry and correlated this information with the results of cell surface marker studies. As an extension of the observations on karyotype, it has been found that cells having hyperdiploid DNA content are more likely to have an expression of surface common ALL antigen. However, those cells having T-lymphocyte markers rarely have abnormal DNA content detectable by flow cytometry. Thus, hyperdiploidy is most frequently found in a form of ALL usually associated with good treatment response. However, one form of leukemia, the T-cell variety, which generally has a poor treatment outcome, rarely has large deviations from normal with respect to DNA content. These observations certainly suggest that there is an interrelationship among these biological features associated with treatment response.

## INFLUENCE OF GROWTH CHARACTERISTICS

Certainly one dominant theme of interrelationship with clinical characteristics is the growth potential of the leukemic blast cells at diagnosis. It has been a consistent observation that the growth potential of the blast cells at relapse is greater than at the time of diagnosis. The changing growth potential might be related to modulation of the leukemia cell population by environmental factors according to the Gompertzian scheme. It also may be related to the selection of a preexisting, more rapidly growing, subpopulation of leukemia cells.

The growth rate of the leukemia cell population depends upon several factors, including the currently unmeasurable rate of cell loss by death. The proliferation rate is determined by the fraction of the cell population in the mitotic cycle. Two concepts of leukemia cell growth have been proposed. In one there is a clonogenic subpopulation of leukemia cells responsible for self-renewal and also population expansion. This scheme is in Figure 1. Although cells produced by the clonogenic fraction may go on to further divisions, they may not be capable of self-renewal and thus not involved in population maintenance.

A second method of conceiving of the proliferative characteristics of the leukemia cell population is shown in Figure 2. In this scheme the leukemia cell population might be thought of as a family of cell cycle times ranging from some cells in a continuous cycle of cell division while other cells have progressive lengthening of interphase times and, in fact, may have temporary characteristics of resting or $G_0$ cells. In this model, even the resting cells retain the capacity for self-renewal.

Obviously, these two hypothetical models are clearly different with respect to their impact on treatment strategy planning. It would be important to know which model is the more nearly correct and how concepts of growth regulation could be developed to help in a further understanding of why some leukemia cell populations have greater growth potential than others and how this growth potential affects treatment outcome.

## FUTURE RESEARCH DIRECTIONS

There is a growing body of information relating biological characteristics of leukemic blast cells to clinical outcome. One of the immediate needs is to attempt to correlate these various biological characteristics in order to determine if there are common factors which might help account for their influence on clinical features. Although each of the features mentioned above have an independent contribution to treatment outcome, there may very well be important relationships.

Certainly important future studies should deal with the alterations of the leukemic cell population that occur during therapy and account for the emergence of a drug-resistant population of cells. The biological information

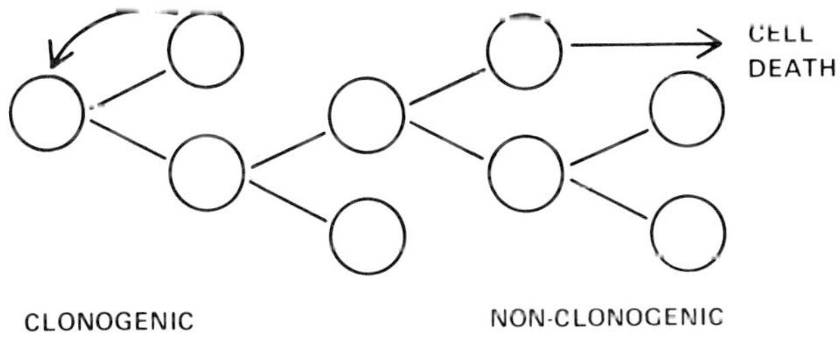


Fig. 1. A model of tumor cell proliferation in which a clonogenic subpopulation of cells is responsible for self-renewal and also population expansion.

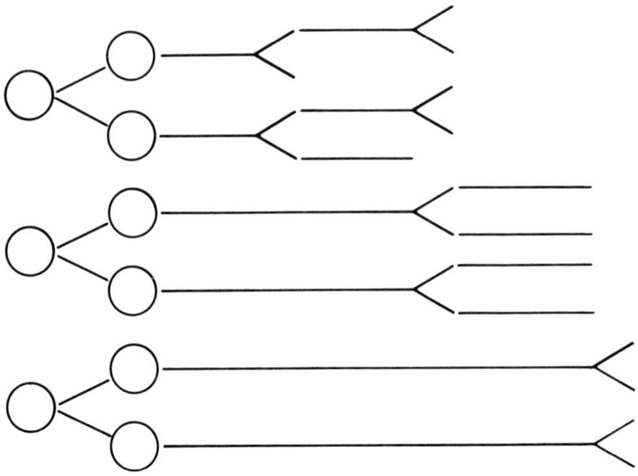

Fig. 2. A model of tumor cell proliferation in which a family of cell cycle times is present although some cells may be in a continuous cycle of cell division and others having long interphase delays. All cells retain the capacity for self-renewal.

described above may help define cells with an increased likelihood of developing drug resistance and allow for studies of cellular alterations during the course of disease associated with the acquisition of drug resistance. Two critical questions that need attention are the mechanism by which drug resistance develops and the characteristics of the leukemic cell proliferative pattern. The answers to both of these questions are critical in developing more appropriate treatment strategies during this coming decade.

## ACKNOWLEDGMENTS

This research was supported in part by the National Cancer Institute, Cancer Center Support Core grant CA21765, the Leukemia Program Project grant CA20180, from the National Institutes of Health, Division of Research Resources, the Biomedical Research Support Grant RR05584, and ALSAC.

## LITERATURE CITED

Bowman, W.P., Melvin, S., and Mauer, A.M. (1980) Cell markers in lymphomas and leukemias. In: Advances in Internal Medicine. G.H. Stollerman, ed. Yearbook Medical, Chicago, Vol. 25, pp. 391–425.

Bowman, W.P. and Mauer, A.M. The role of cell markers in the management of leukemia. In: Progress in Hematology, XII. Elmer B. Brown, ed. (in press).

Dow, L.W., Chang, L.J.A., Tsiatis, A.M., Melvin, S.L., and Bowman, W.P. (1981) Comparison of blood and marrow pretreatment labeling indices (LI) in acute lymphoblastic leukemia (ALL). Proc. Am. Assoc. Cancer Res., *21*:480 (C-577 abst).

Eiser, C., and Lansdown, R. (1977) Retrospective study of intellectual development in children treated for acute lymphoblastic leukaemia. Arch. Dis. Child., *52*:525–529.

Farber, S., Diamond, L., Mercer, R., Silvester, R.F., and Wolff, J.A. (1948) Temporary remission in acute leukemia in children produced by folic acid antagonist, 4-aminopteroyl glutamic acid (Aminopterin). N. Engl. J. Med., *238*:787–793.

George, S.L., Aur, R.J.A., Mauer, A.M., and Simone, J.V. (1979) A reappraisal of the results of stopping therapy in childhood leukaemia. N. Engl. J. Med., *300*:269–273.

Greaves, M.F., Brown, G., Rapson, N.T., et al. (1975) Antisera to acute lymphoblastic leukemia cells. Clin. Immunol. Immunopathol., *4*:67–84.

Miller, D.R. (1975) Prognostic factors in childhood leukemia. J. Pediatr., *87*:672–676.

Price, R.A., and Jamieson, P.A. (1975) The central nervous system in childhood leukemia. II. Subacute leukoencephalopathy. Cancer, *35*:306–318.

Sen, L., and Borella, L. (1975) Clinical importance of lymphoblasts with T markers in childhood acute leukemia. N. Engl. J. Med., *292*:828–832.

Simone, J.V. (1975) Prognostic factors in childhood acute lymphocytic leukemia. In: Proceedings of the Workshop of the International Institute for Scientific Cooperation on Prognostic Factors in Human Acute Leukemia, Reisenberg, Germany, October 1973. T.M. Fliedner and S. Perry, eds. Pergamon, New York.

Williams, D.L., Tsiatis, A., Brodeur, G.M., Look, A.T., Melvin, S.L., Bowman, W.P., Dahl, G.V., Kalwinsky, D.K., and Rivera, G. (1981) Prognostic importance of chromosome ploidy in 136 untreated children with acute lymphoblastic leukemia. N. Engl. J. Med. (in press).

JOURNAL OF CELLULAR PHYSIOLOGY SUPPLEMENT 1:133–135 (1982)

# Some Thoughts on the Nature of Erythropoietin-Responsive Cells

EUGENE GOLDWASSER*
*Department of Biochemistry, University of Chicago, Chicago, Illinois 60637*

The organizers of this meeting, in celebration of the 20th anniversay of the operational definition of the pluripotent hemopoietic stem cell (CFU-S), have generously offered those of us who are chairing sessions, but not on the program, the opportunity of contributing a paper to the proceedings on the work of our laboratories. I would like to offer this short essay on one aspect of our research, in tribute to the "founding fathers," even though they will probably not agree with our interpretation.

I assume, in this discussion, that there exist substances analogous to erythropoietin (epo) that have similar roles as primary inducers of granulopoiesis (granulopoietin, gpo) and of megakaryopoiesis (megakaryopoietin, mkpo). Additionally, for the purpose of this paper, the following definitions are used: pluripotent hemopoietic stem cells (PSC) are conceptually the same as spleen colony-forming cells (CFU-S), operationally defined as self-renewing cells that are the precursors of, at least, three lineages of blood cells (erythrocytic, granulocytic, and megakaryocytic), as detected by the appearance of colonies in, or on, the spleens of lethally irradiated hosts after the injection of cells. These colonies must contain PSC as well as one or more of the differentiated blood cell types. Frequently short cuts are taken, almost of necessity, because of the tediousness of experimentally defining each spleen colony. I think we can agree that, when the experiment involves simply scoring the number of spleen colonies, some reserved judgment about the interpretation is justified.

An alternative definition of PSC may be made by consideration of the mixed colonies in vitro first described by Johnson and Metcalf (1977). Since such a colony contains cells of more than one differentiated type and contains cells that can yield similar colonies, the progenitor cell of such a colony may be defined as a PSC.

Other experimentally defined cells are: granulocyte/macrophage precursor cells (CFU-C), megakaryocyte precursors (CFU-M), erythroid burst progenitors (BFU-E), erythroid colony precursors (CFU-E) and epo-responsive cells (ERC). This last "cell" is really a compartment that includes all cells capable of being affected by epo and this paper will present some ideas concerning what cells are ERC and what the relationships between ERC and the other hemopoietic precursor cells may be. For the purpose of this discussion, I will restrict my attention to cells of the adult bone marrow.

In the case of the erythroid system, we are all aware that although we may use terms like BFU-E and CFU-E (with an indication of the day of scoring), there is, in fact, a continuum of cells from the most primitive to the end-stage cell. We also know that epo responsiveness is restricted to a particular "window" within this continuum, but the precise limits of this window have not been defined. A similar situation most probably occurs in the granulocytic series; there must be a subset of cells (GRC), which are targets for gpo, within the developing granulocyte-macrophage lineage. There is probably more than one inductive factor (gpo) that accounts for branching. The megakaryocytic line may be a fork in the line yielding erythroid cells from PSC or may branch independently. The data available are too scanty to permit more than speculation.

The prevailing thought (Lajtha, 1979) concerning the interrelationships between all of these cell types, defined by their end-product cells (red cells, granulocytes, macrophages, and platelets) is that each pathway of differentiation arises by an undefined process (which has not even prompted much published speculation) ultimately yielding unipotent, committed stem cells (USC). These cells are considered to be the targets of inductive substances (epo, gpo, and mkpo) and thus have only two major options: either to go down the pathway of specific differentiation or not to.

*To whom correspondence/reprint request should be addressed.

This model tells us little about the mechanism by which blood cell differentiation actually occurs, because USC are already differentiated from PSC. It can be called a noninstructional model in that the program defining the ultimate fate of the cell is considered to be inherent in the USC. The inducer substances, according to this model, act to increase the number and/or rate of cell divisions yielding a cohort of end product cells from the programmed USC.

Some time ago we set out to answer the question: What cells are acted on by epo to yield red cells? In vivo experiments had shown clearly that epo was directly involved in red cell production and these observations suggested a simplistic model in which the instructions to become erythrocytes are the results of the interaction of epo with a primitive subset of ERC.

Because of a conviction that whole animals are too complex to permit detailed study of the cellular and molecular biology of erythropoiesis, we concentrated on studies of primary bone marrow cell cultures. Using cells grown in liquid cultures we could ask whether epo addition had an effect on CFU-S. Our results (Van Zant and Goldwasser, 1977) did not agree with the established opinion that CFU-S were unaffected by epo, since we found not only an increase due to epo in erythroid spleen colonies (in the normal host) but, unexpectedly, a decrease in nonerythroid spleen colonies (in the plethoric host) due to epo. These data, first derived in a study comparing normal with polycythemic irradiated hosts, to distinguish between erythroid and nonerythroid colonies, have recently been confirmed by morphological studies of the colonies (N. Wolf, G. Van Zant, and E. Goldwasser, unpublished observations). The suggestion, from these experiments, that ERC might include PSC does not fit with the USC hypothesis and led to further tests of whether epo acted only on committed cells. If epo, and the other inducers, act only on irreversibly committed cells, then they ought to act independently of each other, and the effect of epo on erythroid processes should not be influenced by what may happen in the granulocytic and megakaryocytic compartments.

Evidence for the identity of a granulopoietin with one, or more, of the granulocyte/macrophage colony-stimulating factors (CSF) is not complete, but it is reasonable to tentatively accept CSF as being a gpo. With that assumption, the independence of the differentiating pathways can be tested. We have accumulated extensive data (Van Zant and Goldwasser, 1977b, 1979 a and b) showing that the action of epo is influenced by the presence of CSF and vice versa. High concentrations of CSF cause decreased epo-dependent bursts and burst-associated hemoglobin synthesis, and high concentrations of epo cause decreased CSF-dependent granulocyte/macrophage colony development. In addition, low concentrations of CSF caused a significant increase in epo-dependent erythroid differentiation (VanZant and Goldwasser, 1979a; Metcalf et al, 1980), additional evidence that CSF and epo may act on the same cell. These effects of CSF on epo action are restricted to the earlier phases of erythroid differentiation, since CSF does not affect epo-dependent CFU-E or hemoglobin synthesis associated with those later cells. It is possible, of course, that these findings may indicate some processes more complex than the interaction of CSF and epo with the same cell but in the absence of compelling evidence for that possibility it seems safe to rely on Occam's razor and infer that there is at least one common locus of action for these two substances.

The suppressive effect of CSF on epo action had, however, an aspect that made the simple explanation we offered somewhat equivocal. We found that the "competition" was discernible only at marrow cell concentrations greater than $2 \times 10^5$ ml. At $10^5$ cells/ml, a concentration usually used for BFU-E experiments, there was essentially no effect of CSF on epo action, except at extraordinarily high concentration. This puzzling dependence on cell concentration is not yet fully understood but there are some observations that tend to indicate that it is not some subtle artefact. We think that, in contrast to the simple scoring of bursts, physiologically significant responses of hemopoietic cells may be dependent on cellular interactions that would be promoted at high cell concentration. As a point of reference, it may be important to remember that marrow cells normally function within the bone cavity where their concentration is about $2 \times 10^9$/ml (Crafts and Meineke, 1959), i.e., in a close-packed array. Recently we found that the suppressive effect on CSF on BFU-E can be demonstrated when the cells are at a final concentration of $10^5$/ml if the cells are first treated with epo and CSF for a short time as a pellet, then diluted for culturing in semisolid medium (G. VanZant and E. Goldwasser, unpublished observations). In a typical experiment we added 5 U/ml of epo or 5 U/ml of epo and 280 U/ml of CSF to 10 ml of a cell suspension containing $10^8$ mouse marrow cells at 37°C. The cells were then pelleted at 220g for 10 min and incubated under the original medium for 1 hr at 37°C. After that time the medium was removed, the cells washed with 10 ml of fresh cold medium without epo or CSF, resuspended and plated in semisolid medium at a concentration of $10^5$/ml with 5 U/ml of epo. Bursts were scored 8 days later with the help of benzidine staining. The controls (epo only in the pellet) had 14 $\pm$ 3 bursts per $10^5$ cells, whereas the addition of CSF at the pellet step caused a 57% suppression (6 $\pm$ 2). This finding suggests that study of regulatory functions among hemopoietic cells may require techniques that permit us to investigate their close range interactions. Perhaps a new concept of how cells of the marrow communicate with each other may emerge.

While this type of research is being contemplated, I would like, once again, to suggest an alternative to the idea that epo (and the other inducers by analogy) act solely on irreversibly committed unipotent cells. In this model, which we have termed the "transient receptor" model (Goldwasser, 1970), the inductive factors interact with PSC as the first locus of action. This interaction is with specific membrane receptors for a particular inducer. When the cell, perhaps as a function of where it is in the cell cycle, has a critical number of specific receptors for an inducer and when the concentration of that inducer is at, or beyond, a critical level, then the cell can be triggered to go down a specific differentiated pathway. If neither receptors nor inducer is at the requisite level then the cell does not follow that pathway, but as other specific receptors and inducers change in concentration may be induced down a different pathway. One, as yet untestable, attribute of this model, is that specific membrane receptors will have only a relatively short period of expression on the PSC.

As originally formulated, neither this nor any other model was capable of explaining a surprising finding (Vainchenker et al, 1979; Axelrad and McLeod, personal communication): that pure epo caused the formation of megakaryocytic colonies in vitro. After these experiments were reported, we found that the "pure" epo was contaminated with a chemically insignificant, but assayable, quantity of endotoxin. If we can assume that it was not the endotoxin that caused the megakaryocyte colonies formation, we are faced with a formidable problem: How can epo cause nonerythroid differentiation?

One possible way out of this dilemma is to make still one more ad hoc addition to the transient receptor model: inducers also have a mitogenic effect on PSC. The first action of a factor that might eventually cause a reduction in the PSC population may be as a signal to some cells to replicate so that subsequent induction of differentiation does not cause a depletion of the pool. If that were the case, then the effect of epo would be first to cause a fraction of the PSC to divide, then to induce some cells to take the erythroid pathway. If the culture medium, which contained fetal calf serum, had a small amount of mkpo in it, the number of potential target cells for mkpo might be increased by epo action so that megakaryocytic colonies could develop. The effect of CSF in increasing epo-induced burst formation mentioned above may also represent this type of action. It may be profitable to look for an effect of epo on granulocyte/macrophage colonies similar to that seen for megakaryocytic colonies.

Almost certainly these ideas are far too oversimplified, as most models are. They are also, most likely, inaccurate and incomplete descriptions of the complex processes going in the marrow. Nevertheless they do provide a point of departure for the design of experiments and may add to that combination of concept and experiment that will permit a deeper understanding of this fascinating branch of biology.

## ACKNOWLEDGMENTS

The research described in this paper was supported, in part, by grant CA 18375 from the National Cancer Institute.

## LITERATURE CITED

Crafts, R.C. and Meineke, H.A. (1959) Ann. N.Y. Acad. Sci., *77*:501.

Goldwasser, E. (1970) Fed. Proc., *34*:2285.

Johnson, G.R., and D. Metcalf, (1977) Proc. Natl. Acad. Sci. U.S.A., *74*:3879

Lajtha, L.G. (1979) Differentiation, *14*:23.

Metcalf, D., Johnson, G.R. and Burgess, A.W. (1980) Blood, *55*:138.

Vainchenker, W., Bouget, J., Guichard, J., and Breton-Gorius, J. (1979) Blood, *54*:940.

VanZant, G. and Goldwasser, E. J. Cell Physiol., *90*:241.

VanZant, G. and Goldwasser, E. (1977b) Experimental Hematology Today, S.J. Baum and G.D. Lodnoy eds., Springer Verlag, New York, p. 63.

JOURNAL OF CELLULAR PHYSIOLOGY SUPPLEMENT 1:137–143 (1982)

# The Control of Globin and Other Eukaryotic Genes

O. SMITHIES*
*Genetics Laboratory, University of Wisconsin—Madison, Madison, Wisconsin 53706*

The hypothesis is advanced that eukaryotes control different parts of their genomes by the selective use of origins of replication. Replication of a gene from an origin upstream of the gene is postulated to open it for transcription under appropriate circumstances; replication of a gene from an origin downstream of the gene closes it. The applicability of this hypothesis to the control of the human β-globin cluster of genes is considered.

At an evening session of a symposium such as this, a detailed review of past work that has been published or soon will be published seems somewhat inappropriate. So I have decided not to tell you about the work my laboratory has been involved in over the last five years in cloning and sequencing the human fetal globin genes, $^{G}\gamma$ and $^{A}\gamma$. You will, if you are interested, find a description of this work and references to the appropriate earlier papers in Shen et al. (1981). Instead I propose to try out on you a hypothesis that we are now using in my laboratory for designing future experiments. I make no claims that the hypothesis is correct, but I hope you may find it useful, as we do, in thinking about new approaches to the basic question: How are eukaryotic genes in general and globin genes in particular controlled at different stages of development?

The hypothesis is conceptually simple. I propose that, in eukaryotes, control over the potential expression of a given region of the genome is determined by the particular origin of DNA replication that was used to replicate the region in the immediately preceding cell division. A gene within a region is "open" for transcription under appropriate circumstances if the region was replicated from an origin lying upstream of the gene in the transcriptional sense. If the region was replicated from an origin downstream of the gene, then the gene is "closed" to transcription under all circumstances. Figure 1 illustrates this hypothesis in its simplest form and shows a single gene as open or closed, depending on which of two nearby origins of replication was used.

I shall refer to origins and genes as being homodirectional if they are oriented so that replication and transcription occur in the same direction; heterodirectional is the reverse. My hypothesis can then be restated as: genes are opened by homodirectional origins on their upstream 5′ sides and closed by heterodirectional origins on their downstream 3′ sides. I define the "open" state as being necessary but not always sufficient for transcription. The term does not preclude the possibility that open regions may only be transcribed after additional events such as hormone stimulation. The "closed" state forbids transcription.

My hypothesis assumes that the open and closed states of a gene are determined by the direction in which it was replicated. This could be achieved simply and unequivocally if the structure of chromatin is asymmetric with respect to the direction of its assembly during replication (as indicated by the plus signs in Fig. 1), and if this asymmetry controls the strand of DNA to be used for RNA templating. It is self-evident that a gene will be closed if its transcription is attempted in the wrong direction using the wrong DNA strand as template.

Chromatin could be asymmetric with respect to the direction of its assembly in many mutually compatible ways, which could be reproduced in future cell divisions that utilized the same origins of replication. Thus chromatin might be asymmetric at the DNA level; for example, methylation of specific bases on only one of the two newly synthesized DNA strands at a replication fork would lead to a replicatable asymmetry. Chromatin might be asymmetric at the level of nucleosome assembly; since there is only one molecule of histone H1 per nucleosome, nucleosomes cannot be completely symmetric structures. It is therefore

*To whom correspondence/reprint requests should be addressed.

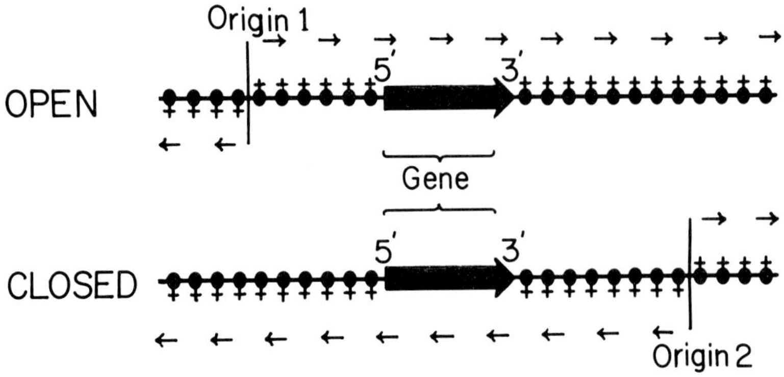


Fig. 1. A diagrammmatic representation of the proposed mechanism for controlling different regions in eukaryotic genomes. The horizontal lines represent DNA molecules with their associated nucleosomes. Two origins of DNA replication are considered. If origin 1 were used for the preceding replication in the direction indicated by the small arrows, then any homodirectional gene within its range (such as the gene indicated by the solid arrow with its 5′ and 3′ ends shown) is postulated to be open for potential transcription. If origin 2 on the 3′ side of the gene were used for its replication, then the (now) heterodirectional gene is postulated to be closed. The plus signs above or below the nucleosomes symbolize an asymmetry in the chromatin, at either the DNA or nucleosome level (see text), which is assumed to reflect the direction of the preceding replication. This asymmetry could control which DNA strand is to be used for RNA templating and so could determine the open versus closed state of a gene within the region.

likely that they can be assembled in only one orientation with respect to the direction in which their DNA was replicated. Chromatin might be asymmetric with respect to histone modifications; for example, if only one of any of the paired core histones was acetylated, a replicatable asymmetry could again result. Clearly there are many ways in which chromatin could retain and propagate a memory of the direction of its replication and could transmit this information to the transcriptional machinery.

The hypothesis that differential gene control in eukaryotes is primarily mediated by changes in origins of replication requires that complex eukaryotes use different sets of origins of replication at different stages of development and in different tissues. It follows from this that there must be some control over the usage of origins of replication, and that in a sense my hypothesis replaces one uncertainty by another. Nevertheless the hypothesis that the differential use of multiple origins can modify transcriptional potentials is sufficiently unifying to be worth testing and to warrant an investigation into variations in orgin usage. A specific prediction of the hypothesis appears to be potentially, although not easily, testable; namely, that all active genes must have been replicated from an upstream origin and that genes which have been replicated from downstream origins cannot be active. The assumption that signals of open versus closed states are transmitted to the transcriptional and related machinery by some asymmetric structure in chromatin that reflects the direction of its assembly should also be testable experimentally. As already indicated, there are many ways in which an asymmetric chromatin structure could result from the direction of its replication.

The hypothesis which I am discussing here is not novel in its suggestion that some relationship exists between the use of different origins of replication, chromatin structure, and the control of gene expression; its novelty lies in proposing a precise and experimentally testable relationship between these factors. Three papers were particularly important in developing this relationship. In 1976, Wilkins published a stimulating discussion of the hypothesis that programming of DNA replication plays a direct role in assigning distinct transcriptional potentials to different cells during development. Specifically he proposed that the conversion of a set of replicative maps to a set of transcription potentials is accomplished by a substance that binds preferentially to one replicative state of DNA and can affect its subsequent transcription. The second paper, published in 1977 by Weintraub et al., discusses in detail the generation of variegated chromosomes during development as a consequence of multiple changes in the configurations of nucleosomes accompanying the use of different origins of replication. The third paper is that published by Seidman et al. in 1979. These authors describe experimental observations on cultured chicken cells that are consistent with the location of active origins of replication on the 5′ side of active transcriptional units: this is a prerequisite of my hypothesis. The authors first demonstrated that parental histones preferentially segregate with the leading, continuously synthesized side of replicating forks in SV40. They then showed that in the chicken cells parental cellular histones also segregate asymmetrically and are preferentially associated with the DNA strands that serve as templates for RNA. They concluded that if the pattern of nucleosome segregation seen in SV40 can be generalized to the chromatin of chicken cells, the origins of DNA replication must be located on the 5′ side of the relevant transcriptional units in these cells. The authors also pointed out that any changes in the arrangements of gene coding sequences relative to origins of replication, either by DNA rearrangements or by changes in the origins of replication, would alter the sense of the coding strand of DNA with respect to the leading or lagging side of the replication forks.

Now I want to shift direction somewhat and try to indicate how this simple idea (genes are closed when replicated from a downstream origin and are potentially open when replicated from an upstream origin) may help us understand the control of globin genes. I shall use the human β-globin gene cluster as an example. The arrangement of the cluster is shown in Figure 2. It covers over 60 kbp (kilobase pairs) of DNA, and all of it has been cloned (see Efstratiadis et al., 1980, for references). The five productive genes, ε, $^{G}\gamma$, $^{A}\gamma$, δ and β are all oriented in the same transcriptional direction (left to right in the figure). They are expressed differentially during development (see Wood, et al., 1977, for a review).

The earliest studied embryos have nucleated erythrocytes derived from the yolk sac. At this stage the ε, $^{G}\gamma$, and $^{A}\gamma$ genes are active, but the adult δ and β genes are not. A possible pattern of replication (A) is shown in Figure 2 with origins 1 and 4 being in use.

During midfetal life non-nucleated red cells derived from the liver are present. These cells are larger than adult-type red cells and have various antigen and isozyme characteristics which make them recognizably fetal-type cells. At this stage the $^{G}\gamma$ and $^{A}\gamma$ genes are active and the ε gene is inactive. Up to 25% of cells contain adult hemoglobin, although only in an occasional cell is it abundant (Boyer and Dover, 1979). The presence of even a small amount of adult hemoglobin implies, however, that the β-globin gene is open in some cell lineages, although it appears to be closed in the majority of lineages. Accordingly I suggest that there are two possible patterns of origin usage available to cells during fetal life, (B) and (C) in Figure 2, but that they are not equally employed.

Around birth nearly all cells contain fetal hemoglobin and about 35% contain adult hemoglobin. The pattern is thus similar to that in the fetus but with a greater preponderance of cells in which adult hemoglobin is abundant. Not until around one month of age are significant numbers of cells produced that contain only adult hemoglobins.

By one year of age human babies are hematological adults. Their red cells are then all of the mature type, which are smaller than fetal red cells and have different amounts of cell surface antigens and isozymes. The majority of these adult cells contain no fetal hemoglobins, and I assume that they use replication pattern (D) in which origin 3 is active. This origin of replication is the one whose activation I assume shuts off the fetal genes. It is in a very real sense the fetal-to-death globin switch, and I shall discuss it in more detail later.

In hematological adults a minority of red cells still contain fetal hemoglobin. These so-called F-cells are of the adult type in size, an-

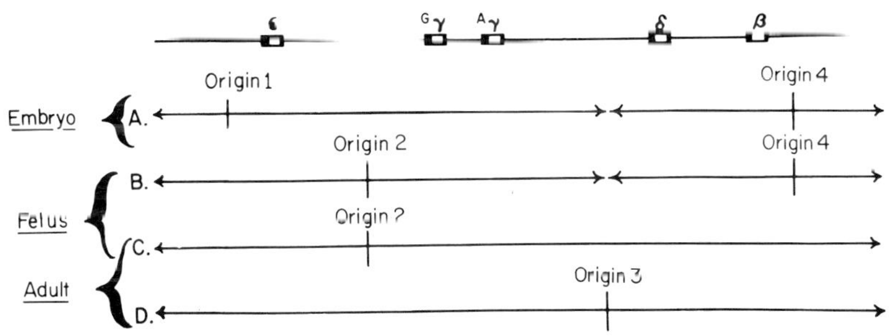


Fig. 2. Assumed patterns of DNA replication in the human β-globin gene cluster. The top line shows a map at the DNA level of the human β-globin gene cluster. Four assumed patterns of DNA replication, A–D, in which several origins of replication, 1–4, are differentially employed (see text) are shown below. The relative usages of each pattern of replication in the embryo, fetus, and adult are represented by the thicknesses of the brackets.

tigens and isozymes. Accordingly I assume that some adult cell lineages are still using the fetal pattern of replication (C) in the globin region.

Studies of in vitro cultures of red cell precursors from fetal liver, newborn peripheral blood, and adult blood (see for example Stamatoyannopoulos et al., 1981; Papayannopoulou et al., 1981) give some clues to the relationship between cells that contain mainly HbF, those which contain both HbF and HbA, and those that contain only HbA. These studies have already been reviewed this evening by Dr. Thalia Papayannopoulou, who has been a great proponent of their usefulness, and so I will not discuss them in detail. But I will make the point that a single burst-forming precursor cell can give subcolonies, all of which may be uniform, or a burst-forming cell may give subcolonies that differ. For example, some precursors from adult peripheral blood give bursts in which all the derived subcolonies are positive for both HbA and HbF; these burst-forming cells could be precursors of the minority class of adult red cells, that is, the F-cells. More instructive, from the point of view of understanding the nature of controlling events, are those burst-forming cells that give two different sorts of subcolony. It is very common, for example, to find single bursts in which some of the subcolonies are positive for both HbF and HbA, while some are only positive for HbA. Examples of sectoring within subcolonies are also found. It therefore appears that cells containing HbA only and those containing both HbF and HbA can be derived from a common precursor, that the distinction between the two categories of colony is due to a change in a daughter of the precursor cell, and that the change is replicated in subsequent progeny cells. I am suggesting, of course, that the clonal change is a change in the usage of the critical origin of replication, origin 3 in Figure 2.

Unfortunately there is only limited information about the timing of this type of clonal change relative to other events in the cell lineage leading to mature red cells. Consider, for example, a precursor cell with both the fetal and adult globin genes open giving rise to a daughter cell with the fetal genes closed but with the adult genes still open. If transcription commences after the fetal genes are closed, then the resulting red cells will all contain only HbA. If some of the precursor cells never close the fetal genes, then their progeny red cells will all be F-cells containing substantial amounts of both HbF and HbA. If, however, the fetal gene shutdown occurs subsequent to the start of transcription, then the progeny red cells will still be F-cells but the relative proportion of HbF to HbA will be decreased in these cells. Clearly there are many possibilities.

A number of authors (see, for example, Boyer and Dover, 1979) have stressed that the percentage of F-cells in peripheral blood is not directly correlated with the percentage of HbF in each cell. This suggests that the fraction of precursor cells in which the fetal genes are closed is controlled independently of the time in the maturation of the erythron when this shutdown occurs. It means practically that the blood level of HbF in adults is affected by two independent variables: the percentage of F-cells and the percentage of HbF in a given F-cell. A third variable, the survival in the circulation of cells containing HbF relative to those lacking it, will also materially affect the adult level of HbF. This is particularly important in cases where any of the globin genes are defective, since an imbalance of α-globin chains relative to β-type chains can decrease red cell survival.

A further complication is the effect of stress on the timing of maturation events in the erythron. For example, during an acute and substantial expansion of bone marrow function the level of F-cells increases and the F-cells may start to be associated with antigens and isozymes normally only seen in fetal type cells. It is as if the terminal differentiation to globin gene transcription and red cell formation has been initiated at a much earlier stage in the maturation of the total erythron than is normally the case.

I have pointed out these various complications in order to alert you to the problems facing attempts to make simple statements in this field. With these caveats in mind I now want to see how the effects of deletions in the β-globin gene cluster can be interpreted in terms of the hypothesis I have been developing.

Eleven different deletions have been described in the β-globin gene cluster. They are summarized in Figure 3. All but one of them (case 10) involve the adult β-globin gene in such a way that normal adult HbA could not conceivably be produced. In six of the ten cases with damaged or absent adult β-globin genes (cases 1–6), the fetal globin genes are both intact. Yet the expression of the fetal genes differs substantially in these six cases. Fritsch et al., (1979) following a proposal first made by Huisman et al. (1974), suggested that there

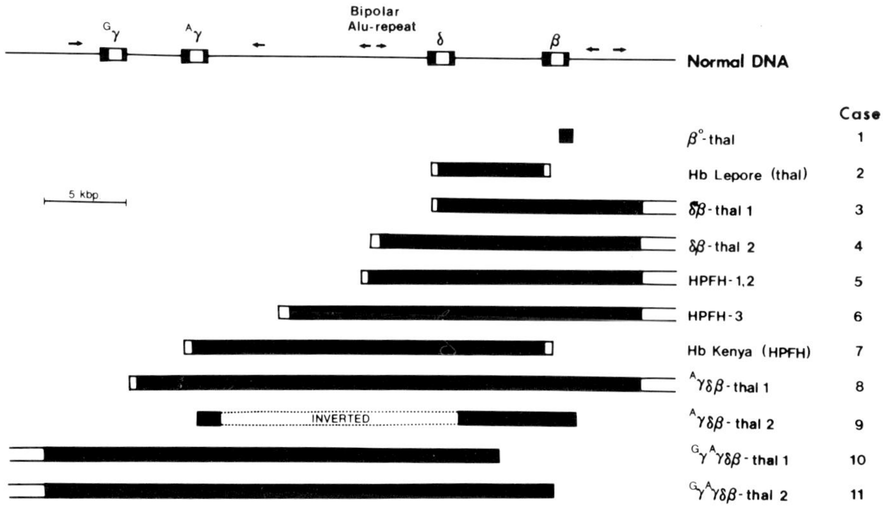


Fig. 3. Deletions affecting the normal expression of β-type globin genes. The extent of eleven deletions are indicated by the bars below the DNA map of the β-globin gene cluster. Uncertainty of the endpoints of some of the deletions are shown by unfilled parts of the bars. The position of the bipolar Alu-repeat (see text) considered important in controlling the fetal to adult changeover is indicated. Cases 1–4 are thalassemias; cases 5–7 are HPFH, and cases 8 and 9 are thalassemias (but see text); cases 10 and 11 are thalassemias. References to cases are 1) Flavell et al., 1979; Orkin et al. 1979a; 2) Flavell et al. 1978; 3) Fritsch et al., 1979; Bernards et al., 1979; 4) Ottolenghi et al., 1981; 5) Fritsch et al., 1979; Tuan et al., 1980; Jagadeeswaran, P., Tuan, D., Forget, B., and Weissman, S., personal communications; 6) Tuan et al., 1980; 7) (This deletion has not been characterized by DNA mapping); 8) Fritsch et al., 1979; Orkin et al., 1979b; 9) Jones et al., 1981; 10) Van der Ploeg et al., 1980; and 11) Orkin et al., 1981.

might be a critical region between the $^{A}\gamma$ and δ globin genes that affects the normal developmental shutdown of the fetal globin genes (an interpretation shared by Ottolenghi et al. (1981), who have added some additional cases). If this region is deleted (cases 5 and 6), then normal shutdown of the fetal genes does not occur and all red cells in both homozygotes and heterozygotes contain substantial quantities of HbF. The amount of HbF present is such that homozygotes for this hereditary persistence of fetal hemoglobin (HPFH) are virtually symptomless and are able to lead full and normal lives; the hematological picture shows no indication of stress (e.g., reticulocyte levels are normal). In heterozygotes the level of HbF is still high, with a mean level of 26% for the 54 $(^{G}\gamma^{A}\gamma)^{+}$ HPFH heterozygotes compiled by Wood et al., (1979).

If the critical region is not deleted (cases 1–4), then the fetal globin genes appear to be much more completely shutdown. The effects of the deletions are somewhat graded and appear to be less severe, in that there is more fetal globin synthesis the closer the deletion gets to the critical region. For example, 60 heterozygotes for $(^{G}\gamma^{A}\gamma)^{+}$ $(\delta\beta)^{0}$ thalassemia had a mean level of HbF of 10.5%, compared to the HPFH heterozygotes that had 36%. Nonetheless there is a clear difference between individuals who have the critical region deleted and those who do not. The former are well. The latter are clinically thalassemic and show signs of hematologic stress. For example, the reticulocyte levels of 11 $(^{G}\gamma^{A}\gamma)^{+}$ $(\delta\beta)^{0}$ thalassemia homozygotes were about three times normal, and these individuals are the least affected of the thalassemics.

How well can the critical region be defined? What is its nature? Could it be the optional origin of replication 3 that I have implicated?

The end points of the deletions in cases 4 and 5 are within a few hundred bases of each other; yet case 4 is classified as thalassemia and case 5 is HPFH. In this region, Fritsch et al. (1981) have demonstrated the occurrence of two opposite-oriented representatives of the Alu-family of repeated DNA (Houck et al., 1979). I shall therefore call this region the bipolar Alu-repeat. Its position is indicated in Figure 3. In case 5 the right hand (3′) element of the bipolar Alu-repeat is completely deleted, as is about half of the left-hand (5′) element (P. Jagadeeswaran, D. Tuan, B. Forget, and S. Weissman, personal communication). In case 4, the left-hand element of the repeat is intact but the status of the right-hand element is not yet known, although it may be intact (Ottolenghi et al., 1981). It appears likely therefore that this bipolar Alu-repeat is a critical element whose deletion seriously deranges the shutdown of the fetal globin genes.

At the present time there is no definitive evidence that this bipolar Alu-repeat could be the optional origin of replication 3 postulated in my hypothesis. Indeed attempts to find out experimentally if it is or can be used as an origin of DNA replication form an important aim of my current research. However it is of considerable interest that Jelinek et al. (1980) pointed out that the Alu-family of repeated DNA shows features in common with the origins of replication of several mammalian viruses; these authors suggested that Alu-repeats might be origins.

I do not want to leave you with the impression that the simple presence or absence of the bipolar Alu-repeat between the $^{A}\gamma$ and δ-globin genes leads to an absolutely clear cut distinction between fetal globin genes being off and fetal globin genes being on. It does not. As I indicated, there is a somewhat graded increase in the expression of the fetal globin genes as the deletion endpoint approaches the bipolar Alu-repeat, (see Wood et al., 1979 for a detailed discussion of the differences and similarities between HPFH and the $(\delta\beta)^0$ thalassemias). Possibly the intermediate effects on fetal globin expression of deletions that spare the bipolar Alu-repeat but include nearby sequences on its right-hand side are indications that the bipolar Alu-repeat is necessary but not sufficient to enable the region to function in a completely normal fashion as an optional origin of DNA replication.

Case 7 is straightforwardly interpreted on my hypothesis. The bipolar Alu-repeat is deleted so that the $^{G}\gamma$ gene is left open, as is also the hybrid γ-β-globin gene, which directs the synthesis of the β-type chain of Hb Kenya. Essentially, symptomless HPFH results.

Cases 8 and 9 are ambiguous in interpretation, since only a single fetal globin gene remains intact. It is quite doubtful that one fetal globin gene could lead to normal hematological values, even if maximally active.

Case 11 would be expected to be a thalassemia by any hypothesis, since all the β-type structural genes are damaged or deleted.

Case 10 remains and is of considerable interest. The individual concerned is heterozygous and shows the symptoms of a heterozygote for thalassemia. Yet the β-globin structural gene is apparently intact on the affected chromosome, together with about 2.5 kbp of DNA on its 5′ side. Since this length of DNA is sufficient to include all sequences known to be important in transcription the case is difficult to explain conventionally. I interpret it simply as being the consequence of deleting optional origin 3. One of two possibilities then remains: either the gene is replicated from a different upstream origin that was juxtaposed by the deletion, or it is replicated from some origin on the downstream side of the gene that normally does not replicate the β-gene. In the latter case my hypothesis would say that the gene could never be functional. In the former case it could be functional. Clearly we need to start looking at these deletions from the point of view of origins of replication.

In conclusion, I want to reemphasize that I have presented this hypothesis to you not because I am confident that it is correct, but because it suggests new ways of thinking about old problems, and because it has suggested to me many new experiments that are worthwhile in their own right. Who of us here would not like to know which origins of replication are used in different cell types at different stages of development, whether these origin usages are subject to programmed genetic control and whether their differential usage determines cell types. And who of us would not like to know if chromatin is asymmetric with respect to its direction of synthesis, and whether this asymmetry controls which strand of DNA can be used for RNA templating.

## ACKNOWLEDGMENTS

This work was supported by NIH grants AM20120 and GM20069.

## LITERATURE CITED

Bernards, R., J. M. Kooter, R. A. Flavell (1979) Physical mapping of the globin gene deletion in $(\delta\beta)^0$-thalassemia. Gene, *6*:265–280.

Boyer, H. B., and G. J. Dover (1979) The in vivo biology of F cells in man. In Stamatoyannopoulos, G., Nienhuis, A. W. eds.: Cellular and Molecular Regulation of Hemoglobin Switching. New York: Grune and Stratton, pp. 47–71.

Efstratiadis, A., J. W. Posakony, T. Maniatis, R. M. Lawn, C. O'Connell, R. A. Spritz, J. K. DeRiel, B. G. Forget, S. M. Weissman, J. L. Slightom, A. E. Blechl, O. Smithies, F. E. Baralle, C. C. Shoulders, and N. J. Proudfoot (1980) The structure and evolution of the human β-globin gene family. Cell, *21*:653–668.

Flavell, R. A., R. Bernards, J. M. Kooter, E. deBoer, P. F. R. Little, G. Annison, and R. Williamson (1979) The structure of the human β-globin gene in β-thalassemia. Nucleic Acids Res., *6*:2749–2760.

Flavell, R. A., J. M. Kooter, E. deBoer, P. F. R. Little, and R. Williamson (1978) Analysis of the β-δ-globin gene loci in normal and Hb lepore DNA: Direct determination of gene linkage and intergene distance. Cell, *15*:25–41.

Fritsch, E. F., J. C. K. Shen, R. M. Lawn, and T. Maniatis (1981) The organization of repetitive sequences in mammalian globin gene clusters. Cold Spring Harbor Symp. Quant. Biol. *XLV*:761–775.

Fritsch, E. F., R. M. Lawn, and T. Maniatis (1979) Characterisation of deletions which affect the expression of fetal globin genes in man. Nature, *279*:598–603.

Houck, C. M., F. P. Reinhart, and C. W. Schmid (1979). A ubiquitous family of repeated DNA sequences in the human genome. J. Mol. Biol., *132*:289–306.

Huisman, T. H. J., W. A. Schroeder, G. D. Efremov, H. Duma, B. Mladenovski, C. Brodie, E. A. Rachmilewitz, N. Bouver, A. Miller, A. Brodie, J. R. Shelton, J. B. Shelton, and G. Apell (1974) The present status of the heterogeneity of fetal hemoglobin in β-thalassemia: An attempt to unify some observations in thalassemia and related conditions. Ann. N.Y. Acad. Sci., *232*:107–124.

Jelinek, W. R., T. P. Toomey, L. Leinwand, C. H. Duncan, P. A. Biro, P. V. Choudary, S. M. Weissman, C. M. Rubin, C. M. Houck, P. L. Deininger, and C. W. Schmid (1980) Ubiquitous, interspersed repeated sequence in mammalian genomes. Proc. Natl. Acad. Sci., *77*:1398–1402.

Jones, R. W., J. M. Old, R. J. Trent, J. B. Clegg, and D. J. Weatherall (1981) Major rearrangement in the human β-globin gene cluster. Nature, *291*:39–44.

Orkin, S. H., J. M. Old, D. J. Weatherall, and D. G. Nathan, (1979a) Partial deletion of β-globin gene DNA in certain patients with $\beta^0$-thalassemia. Proc. Natl. Acad. Sci U.S.A., *76*:2400–2404.

Orkin, S. H., B. P. Alter, and C. Altay (1979b) Deletion of the Aγ-globin gene in Gγ-δβ-thalassemia. J. Clin. Invest., *64*:866–869.

Orkin, S. H., S. C. Goff, and D. G. Nathan (1981) Heterogeneity of DNA deletion in γδβ-thalassemia. J. Clin. Invest., *67*:878–884.

Ottolenghi, S., B. Giglioni, R. Taramelli, P. Comi, U. Mazza, G. Saglio, C. Camaschella, R. Izzo, A. Cao, R. Galanello, E. Gimferrer, M. Baiget, and A. M. Gianni (1981) Molecular comparison of δβ thalassemic and HPFH DNA: Evidence for a regulatory area? Proc. Natl. Acad. Sci. U.S.A. (in press).

Papayannopoulou, Th., B. Nakamato, S. Kurachi, D. Kurnit, and B. Stamatoyannopoulos (1981) Cell biology of hemoglobin switching II. Studies on the regulation of fetal hemoglobin synthesis in human adults. In Hemoglobins in Development and Differentiation. New York: Alan R. Liss, Inc. pp. 307–320.

Seidman, M. M. , and A. J. Levine (1979) The Asymmetric segregation of parental nucleosomes during chromosome replication. Cell, 18:439–449.

Shen, S., J. L. Slightom, and O. Smithies (1981) A history of the human fetal globin gene duplication. Cell, *26*:191–203 .

Stamatoyannopoulos, G., T. Papayannopoulou, M. Brice, S. Kurachi, B. Nakamoto, G. Lim, and M. Fargular (1981) Cell biology of hemoglobin switching I. The switch from fetal to adult hemoglobin formation during ontogeny. In Hemoglobins in Development and Differentiation. New York: Alan R. Liss, Inc. pp. 287–305.

Tuan, D., M. J. Murnane, J. K. deRiel, and B. G. Forget (1980) Heterogeneity in the molecular basis of hereditary persistence of fetal hemoglobin. Nature, *285*:335–337.

Van der Ploeg, L. H. T., A. Konings, M. Oort, D. Roos, L. Bernini, and R. A. Flavell (1980) γ-β-Thalassaemia studies showing that deletion of the γ- and δ-genes influences β-globin gene expression in man. Nature, *283*:637–642.

Weintraub, H., S. J. Fling, I. M. Leffak, M. Groudine, and R. M. Grainger (1977) The generation and propagation of variegated chromosome structures. Cold Spring Harbor Symp. Quant. Biol., *42*:401–407.

Wilkins, A. S. (1976) Replicative patterning and determination. Differentiation, *5*:5–19.

Wood, W. G., J. B. Clegg, and D. J. Weatherall (1977) Developmental biology of human hemoglobins. Prog. Hematol., *13*:43–90.

Wood, W. G., J. B. Clegg, and D. J. Weatherall (1979) Hereditary persistence of fetal haemoglobin (HPFH) and δβ thalassaemia. Br. J. Haematol., *43*:509–520.

JOURNAL OF CELLULAR PHYSIOLOGY SUPPLEMENT 1:145–149 (1982)

# Human Hemoglobin Switching: Insights From Studies of Erythroid Cultures

THALIA PAPAYANNOPOULOU* AND GEORGE STAMATOYANNOPOULOS
*Department of Medicine (Divisions of Medical Genetics and Hematology), University of Washington, Seattle, Washington 98195*

Clonal erythroid cultures have been used in the investigation of the cellular mechanisms underlying the switch from fetal to adult Hb formation during ontogeny and the expression of fetal hemoglobin in adult life. It has been shown that adult BFUes have the potential to form F cells and A cells in culture. The cellular segregation in the expression of Hb F within a clone leads to the formation of characteristic bursts displaying a bimorphic $F^+/F^-$ pattern of Hb F expression. Analysis of distribution of F-expressing and not F-expressing subclones and sectors among bursts of various sizes is compatible with the assumption that progenitor cells exercise options on whether they will or they will not form progeny expressing the Hb F phenotype. Distributions of Hb F expressing subclones among bursts fit the expectations of a stochastic model; probabilities for F expression in the adult erythroid cells may be influenced by the environment in which the cells differentiate and mature. Experimental data from cultures of progenitors from earlier stages of human development indicate that the change in globin expression during ontogeny is controlled at the level of progenitor cells. Observations of Hb F levels produced by single erythroid bursts of the neonate make it unlikely that hemoglobin switching during ontogeny is accomplished through replacement of stem cell lines. The findings in erythroid cultures are compatible with the hypothesis that hemoglobin switching during ontogeny is accomplished through intrinsic or interactive changes in the programs of differentiation of progenitors.

During human development the red cell phenotype gradually changes from predominantly "fetal" to predominantly "adult" by the first six months of postnatal life. This "switching" includes changes in globin types, as well as in other red cell characteristics, such as the expression of iI determinants and of carbonic anhydrase isozymes (Boyer et al., 1981). Switching of the fetal phenotype to the adult phenotype never becomes complete; a minority of red cells expressing either fetal hemoglobin, the F cells (Boyer et al, 1981; Wood et al, 1975) or the i-antigenic determinants (Papayannopoulou et al., 1980), or low levels of carbonic anhydrase (Boyer et al., 1981) continues to be present throughout adult life. This changing of phenotypes in the erythroid system during development provides an opportunity to investigate the cellular events that underly the programs of expression of specialized proteins in terminally differentiated cells. In the last several years we have utilized clonal erythroid cultures to study the cellular regulation of Hb switching. The purpose of this paper is to provide a summary of this work and the major conclusions we have reached.

## OBSERVATIONS IN ADULT BFUe CULTURES

The most prominent characteristic of fetal hemoglobin synthesis in the adult life is its cellular restriction. The F cells, best studied with immunochemical methods, contain both adult and fetal hemoglobin, the latter composing approximately 20% of the hemoglobin of the red cells (Boyer et al., 1981; Wood et al., 1975). In addition, in the adult individual, fetal hemoglobin synthesis can be activated to significant levels during the course of various hematologic or nonhematologic disorders (reviewed by Wood et al., 1977). Under these circumstances, the activated Hb F synthesis is expressed as an increase in the number of F-

*To whom reprint requests/correspondence should be addressed.

cells (Papayannopoulou et al., 1981). The cellular phenotypes of Hb F synthesis in the adult, pose several questions: Why the cellular bimorphism in fetal hemoglobin production? Do the F cells and A cells (i.e., the cells with no demonstrable fetal Hb) have a common stem cell origin? Are these cells submitted to the same regulation? Does the presence of F cells show that sometimes during the process of differentiation, decisions are made on whether a cell will express the fetal Hb program? If this is so, at what level of hemopoietic cell differentiation are these decisions being made?

The clonal erythroid cultures provide an excellent system to probe these questions for several reasons. The clones formed are of single cell origin (Prchal et al, 1976; Papayannopoulou et al., 1977). If there are differences in "programs" or in "commitment" among progenitors, these are expected to manifest in their progeny formed in culture. The progenitors may also undergo changes during their proliferative and differentiative lifespan, in vitro, before the formation of their terminal progeny; if these changes influence "programs" or "commitment," they will produce phenotype differences within clones. Finally, the culture system can be easily manipulated by changing environmental conditions; if the environment can influence the "programs" of the progenitors, changes in expression could be detected following manipulations of the culture environments.

Since the most informative observations were expected to derive from studying the progeny of single progenitors, in our studies we focused on the evaluation of the expression of fetal Hb in single erythroid clones, by labeling these with fluorescent anti-Hb antibodies. With this approach we had the opportunity to test expression of Hb F in populations of clones present in a given culture, as well as in the cells present within each individual clone. The results from these studies have been presented in detail before (Papayannopoulou, et al., 1976, 1977, 1979). We found that in the adult BFUe cultures, some clones express Hb F, whereas in others Hb F is undetectable. This heterogeneity in Hb F expression among clones is not the result of differences in total accumulation of hemoglobin within the cells of each clone, since with double labeling with both anti-Hb F and anti-Hb A antibodies, F-positive and F-negative clones appear to be equally mature and to have accumulated similar amounts of Hb A. Most interestingly, the studies using fluorescent antibodies revealed that heterogeneity in Hb F expression is present within clones. Erythroid bursts may contain portions (either in the form of subcolonies or as sectors not physically separable as subcolonies), which express Hb F and portions that do not express Hb F (bimorphic $F^+/F^-$ bursts). That the heterogeneity of Hb F expression within single bursts is not owing to initiation of these bimorphic $F^+/F^-$ bursts by two progenitors was demonstrated with coculture experiments of homozygous Hb S and homozygous Hb C peripheral blood BFUes (Papayannopoulou, et al, 1977). By using monospecific anti-Hb S and anti-Hb C antibodies conjugated to two different fluorochromes we were able to show that, in such cocultures, the bursts expressed either Hb S or Hb C and they were not mosaics consisting of S and C cells. In such bursts of homozygous Hb S or homozygous Hb C origin, subclones with Hb F expression and subclones without Hb F expression were present. This phenomenon of bimorphic $F^+/F^-$ bursts is characteristic of all adult BFUe cultures under all cultures conditions we have tried and in all supportive media used (i.e., plasma clot, methycellulose, or agar).

### The significance of bimorphic $F^+/F^-$ bursts

The detection of heterogeneity within a single clone originating from a single BFUe was informative from several points of view: a) it showed that from a single erythroid progenitor cell, progeny with Hb F expression or without detectable Hb F expression can be produced. This finding, extrapolated in vivo, suggests that F cells and A cells are produced by a single erythroid progenitor (Papayannopoulou et al., 1977; Stamatoyannopoulos and Papayannopoulou, 1979) and not by two different populations of stem cells, as previously hypothesized (Weatherall et al., 1976). b) Since the cells forming colonies or subcolonies are products of the differentiation events that the BFUe undergoes in vitro, decisions about the formation of F cells appear to be taken not at the pluripotent stem cell level but during the process of differentiation of early erythroid progenitors. c) The clear-cut segregation of Hb F expression among subclones or sectors of a single burst suggests that, at least in vitro, the cell that initiates subclones or sectors exercises options on whether to express or not express Hb F.

Cultures of BFUes from several normal adult individuals have shown that the bimorphic $F^+/F^-$ bursts are always present and in most experiments they are the most frequent type of bursts in a culture. The high frequency of in vitro clones expressing fetal hemoglobin suggests that the majority (if not all) of the normal erythroid progenitors have the potential to express Hb F. Production of F cells or F normoblasts appears therefore to be an inherent characteristic of adult erythroid progenitors. Since these progenitors form large numbers of Hb F expressing progeny in culture, but not in vivo, a mechanism must exist, whereby their potential to form F cells is limited under conditions of normal in vivo differentiation.

### Analysis of phenotypes of bursts

To interpret the findings in culture we formulated a working hypothesis the basic concept of which is that the early adult erythroid progenitor cell contains a "primitive" program of globin synthesis but this program changes during normal differentiation to one allowing the "adult" pattern of globin synthesis. This latter program is expressed when terminal cells are formed in vivo under normal conditions of cellularity and differentiation of adult marrow cells. When there is an abnormality along the pathway of differentiation, such as when progenitors become committed to Hb synthesis at stages earlier than would normally have done so, Hb F is expressed in the end stage cells (Stamatoyannopoulos and Papayannopoulou, 1979).

We examined whether expression of Hb F in the erythroid clones in vitro is linked to the divisions each BFUe undergoes in order to form the subcolonies composing a burst. In ten experiments, several hundred bursts were doubly labeled with anti-Hb A and anti-Hb F, and in each burst, the number of subcolonies, the number of cells in each subcolony (by surface counting in plasma clot preparations), and the types of subcolonies, i.e., whether they did or did not express Hb F, were scored (Stamatoyannopoulos et al., 1981 a,b). Per experiment the frequencies of bimorphic $F^+/F^-$ bursts, of monomorphic $F^+$, and of monomorphic $F^-$ bursts were determined. We have previously observed that the $F^+/F^-$ bursts are of larger sizes (number of cells per burst) and contain more subcolonies compared to the monomorphic $F^+$ or $F^-$ bursts (Stamatoyannopoulos and Papayannopoulou, 1979). The frequency distributions of $F^+/F^-$ or $F^+$ or $F^-$ bursts among bursts of 2,4, . . . ,n subcolonies was derived. We observed that among bursts with a small number of colonies per burst, the monomorphic $F^+$ or $F^-$ bursts predominate (Stamatoyannopoulos et al. 1981 a,b). Among the largest bursts the mixed $F^+/F^-$ bursts are predominant. These frequency distributions can be interpreted by two possibilities: The BFUes forming the mixed $F^+/F^-$ bursts are parental to those forming the monomorphic $F^+$ or the $F^-$ bursts; alternatively, expression of Hb F in the subcolonies could be random.

To test the possibility of random expression, the findings were compared to the expectations of a simple stochastic model. The model assumes that options about Hb F expression are exercized at the level of cells forming subclones or sectors and they occur with a probability (P) that is constant per experiment. By assuming that the total frequency of F positive subclones in each experiment provides a measure of P and the frequency of F negative subclones is equivalent with the value 1-P (the probability for non-F expression), we estimated per experiment the frequencies of $F^+/F^-$ bursts among bursts of 2,4, . . . ,n subclones. The predicted distributions were very similar to the experimental observations (Stamatoyannopoulos et al., 1981a,b). Also, using stochastic assumptions we predicted, per experiment, the ratio of $F^+$:$F^-$ subclones within $F^+/F^-$ bursts of 2, 4, . . . , n size. The predicted distributions were strikingly similar to the experimental observations. These findings strongly suggest that the phenomenon whereby the progenitor cells choose between the $F^+$ or $F^-$ expression in their progeny is random. These choices may be influenced by the environment in which the progenitor cells differentiate (Papayannopoulou and Zanzani, 1981; Stamatoyannopoulos et al., 1981 a,b).

## PROGENITORS FROM DIFFERENT ONTOGENETIC STAGES

The availability of clonal erythroid cultures provided the opportunity to ask questions concerning the programs of globin synthesis in clones formed from hemopoietic cells of fetuses at various stages of development or from various hemopoietic tissues (fetal liver, bone marrow, and peripheral blood). Consistently we and other investigators (Rowley et al., 1979; Stamatoyannopoulos et al., 1979; Hassan et al, 1979a,b) have observed that bursts formed by fetal BFUes contain erythroblasts that produce mainly $\gamma$ globins. When neonatal BFUes are used for culture, the levels of $\gamma$-chain synthesis decrease to those expected for the neonatal period (Kidoguchi et al., 1978; Stamatoyannopoulos et al., 1979; Hassan et al., 1979a). These findings suggest that the BFUes contain ontogenetic stage specific programs of globin synthesis and support the concept that hemoglobin switching is controlled at the level of progenitors.

Initial measurements of globin synthesis in fetal and neonatal cultures were carried out in pooled populations of bursts. Do all the BFUes in the fetus have the same program or is there heterogeneity in their globin programs? Studies of individual bursts have shown only a small degree of heterogeneity in γ-globin synthesis among fetal bursts; also, all fetal bursts produce small levels of adult hemoglobin. Neonatal bursts show a higher degree of heterogeneity in Hb F synthesis (Comi et al., 1980; Stamatoyannopoulos et al., 1981b). The studies of individual neonatal erythroid bursts are especially informative. One of the previously suggested possibilities for cellular regulation of Hb switching assumes replacements of stem cell lines: a line of fetal stem cells is replaced by a line of adult stem cells (Weatherall et al., 1976). If this was correct, one would expect to detect two types of bursts in the neonatal BFUe cultures: either bursts with globin expression similar to that observed in fetal bursts or bursts with Hb F levels similar to those found in adult bursts. This bimodality in globin synthetic values has not been found in the studies of single neonatal bursts we have done. In neonates at an early stage of switching the total distribution of $\gamma/\gamma + \beta$ ratios in bursts is shifted toward higher levels; in neonates at later stages of switching the distribution of $\gamma/\gamma + \beta$ ratios in bursts is shifted toward lower, adult values (Stamatoyannopoulos et al., 1981b). The findings suggest that during switching there are changes in the "programs" contained in the progenitor cells rather than a change in stem cell populations.

Studies of Hb F synthesis in selected bursts, such as macroscopic bursts or mixed (erythroid and nonerythroid) bursts, of fetal or adult origin have indicated that all forms of assayable precursors in vitro are precommitted to ontogenetic stage specific programs (Stamatoyannopoulos et al., 1981b). Measurement of globin expression on bursts which were transferred from their original plates, at a stage of the first 3–4 divisions, and allowed to mature in separate individual plates make it unlikely that commitment to an ontogenetic stage specific program is accomplished in vitro (Stamatoyannopoulos et al., 1981b). The results of the studies in culture thus suggest that the ontogenetic stage specific globin programs are most likely controlled at the level of stem cells or in their immediate progeny. The mechanisms underlying the change in the programs of progenitors during the period of Hb F to Hb A switching remain speculative.

It might be proposed that the change in programs relates to changes in the kinetics of hematopoiesis. In its simplest formulation this hypothesis would be that "fast in vivo stem cell cycling" is associated with activation of a program allowing transcription of γ-globin genes, whereas "slow stem cell kinetics or prolonged resting of precursors" is leading to activation of a program allowing transcription of the β globins. Indirect evidence from transplantation experiments in sheep argues against this possibility. When sheep fetal liver cells, at a time when these cells are producing predominantly fetal Hb, are transplanted to adult lethally irradiated recipient animals, the fetal cells upon engraftment in the irradiated animal switch earlier than expected to adult globin (Zanjani et al., 1979). If the regulation of switching was solely kinetic, one would expect in this set up of fetal to adult transplantation that continuation of the expression of the fetal program by the fetal stem cells will occur. However, a significant activation of Hb F occurs in primates when their hemopoiesis is accelerated (DeSimone et al., 1978). Alternately, switching in the program of progenitor cells might be totally intrinsically controlled. One can envision that the stem cells contain a developmental clock that determines when or after how many divisions the cells will change their developmental program. Furthermore, the cells may change their program because of changes in instructive environments. During ontogeny, the hemopoietic environment may change and interactions between the stem cells and their new environments may be responsible for programming of the stem cells to the formation of progeny, which expresses the ontogenetic stage specific program of globin synthesis. We hypothesize that through either of these mechanisms, the change in the program of stem cells is such as to influence the way by which the committed cells differentiate and mature. These changes in patterns of committed cell differentiation influence, in turn, the pattern of globin expression.

The application of erythroid cultures has been instrumental for the accumulation of experimental data that led to the development of the current concept that Hb switching is a developmental phenomenon regulated in progenitor cells. It is however unlikely that further studies of phenomenology using in vivo derived committed erythroid progenitors will allow dissection of the events that, during ontogeny, make the progenitor cells change their programs. New in vitro experimental strategies may be required to study directly the cellular mechanisms of Hb switching. Further progress in the understanding of hemoglobin switching appears to be tightly linked to the development of methods allowing direct studies of progenitors and their responses to environmental components controlling differentiation and commitment.

## ACKNOWLEDGMENTS

The studies summarized in this paper have been supported by NIH grant HL 20899.

## LITERATURE CITED

Boyer, S.H., G.J. Dover, K.D. Smith, and A. Scott (1981) Some interpretations of in vivo studies of globin gene switching in man and primates. In: Hemoglobins in Development and Differentiation. G. Stamatoyannopoulos, A.W. Nienhuis, eds. Alan R. Liss, Inc., New York, pp. 225–243.

Comi, P., B. Giglioni, S. Ottolenghi, A.M. Gianni, E. Polli, P. Barba, A. Covelli, G. Migliaccio, M. Condorelli, and C. Peschle (1980) Globin chain synthesis in single erythroid bursts from cord blood: Studies on $\gamma \rightarrow \beta$ and $G\gamma \rightarrow A\gamma$ switches. Proc. Natl. Acad. Sci. U.S.A., *77*:362–365.

De Simone, J., S.I. Biel, and P. Heller (1978) Stimulation of fetal hemoglobin synthesis in baboons by hemolysis and hypoxia. Proc. Natl. Acad. Sci. U.S.A., *75*:2937–2940.

Hassan, M.W., A. Ibrahim, R.F. Rieder, and L.L. Cederquist (1979a). Synthesis of Hb A and Hb F in erythroid colonies cultured from human fetal liver and umbilical cord blood. Blood, *54*:140–1151.

Hassan, M.W., J.D. Lutton, R.D. Levene, R.F. Rieder, and L.L. Cederquist (1979b) In vitro culture of erythroid colonies from human fetal liver and umbilical cord blood. Br. J. Haemat., *41*:477–484.

Kidoguchi, K., M. Ogawa, J.D. Karam, J.M. Wilson, and M.S. Fitch (1978) Synthesis of fetal and adult heomglobins in culture by human umbilical cord blood erythropoietic precursors. Hemoglobin, *2*:503–512.

Papayannopoulou, Th., M. Brice, and G. Stamatoyannopoulos (1976) Stimulation of fetal hemoglobin synthesis in bone marrow cultures from adult individuals. Proc. Natl. Acad. Sci. U.S.A., *73*:2033–2037.

Papayannopoulou, Th., M. Brice, and G. Stamatoyannopoulos (1977) Hemoglobin F-synthesis in vitro: Evidence for control at the level of primitive erythroid stem cells. Proc. Natl. Acad. Sci. U.S.A., *74*:2923–2927.

Papayannopoulou, Th., M. Brice, and B. Nakamoto (1979) Patterns of fetal Hb synthesis from fetal neonatal and adult erythroid progenitors in culture studied by immunofluorescence. In: Fetal Liver Transplantation. I.G. Lucarelli, R.P. Gale, eds. Excerpta Medica, Amsterdam, pp. 85–94.

Papayannopoulou, Th., P. Chen, A. Maniatis, and G. Stamatoyannopoulos (1980) Simultaneous assessment of i-antigenic expression and fetal hemoglobin in single red cells by immunofluorescence. Blood, *55*:221–232.

Papayannopoulou, Th., and E. Zanjani (1981) Humoral factors in preswitching fetal sheep influence the growth and γ-globin synthesis in human erythroid cells in culture. Clin. Res., *29*:521A.

Papayannopoulou, Th., L. Halpap, S.H. Chen, M. Fukuda, R. Hoffman, L. Dow, and S. Hill (1981) Fetal red cell markers and their relationships in patients with hematologic malignancies. In: Hemoglobins in Development and Differentiation. G. Stamatoyannopoulos and A.W. Nienhuis, eds. Alan R. Liss, Inc., New York, pp. 443–448.

Prchal, J.R., J.W. Adamson, L. Steinmann, and P.J. Fialkow (1976) Human erythroid colony formation in vitro: Evidence for clonal origin. J. Cell Phys., *89*:489–492.

Rowley, P.T., B.M. Ohlsson-Wilheim, B.A. Farley, and B. Kosciolek (1979) Hemoglobin synthesis in cultures of hepatic erythroid cells from the human fetus. Proc. Natl. Acad. Sci. U.S.A., *76*:1477–1480.

Stamatoyannopoulos, G., and Th. Papayannopoulou (1979) Fetal hemoglobin and the erythroid stem cell differentiation process. In: Cellular and Molecular Regulation of Hemoglobin Switching. G. Stamatoyannopoulos and A.W. Nienhuis, eds. Grunne-Stratton, New York, pp. 323–341.

Stamatoyannopoulos, G., B.B. Rosenblum, Th. Papayannopoulou, M. Brice, B. Nakamoto, and T.H. Shepard (1979) Hb F and Hb A production in erythroid cultures from human fetuses and neonates. Blood, *54*:440–450.

Stamatoyannopoulos, G., D.M. Kurnit, and Th. Papayannopoulou (1981a) Stochastic expression of fetal hemoglobin in adult erythroid cells. Proc. Natl. Acad. Sci. U.S.A., *78*:7005–7009.

Stamatoyannopoulos, G., Th. Papayannopoulou, M. Brice, S. Kurachi, B. Nakamoto, G. Lim, and M. Farquhar (1981b) Cell biology of hemoglobin switching. I. The switch from fetal to adult hemoglobin formation during ontogeny. In: Hemoglobins in Development and Differentiation. G. Stamatoyannopoulos and A.W. Nienhuis, eds. Alan R. Liss, Inc., New York, pp. 287–307.

Weatherall, D.J., J.B. Clegg, and W.G. Wood (1976) A model for the persistence or reactivation of fetal hemoglobin production. Lancet, *2*:660–663.

Wood, W.G., G. Stamatoyannopoulos, G. Lim, and P.E. Nute (1975) F-cells in the adult: Normal values and levels in individuals with hereditary and acquired elevations of Hb-F. Blood, *46*:671–682.

Wood, W.G., J.B. Clegg, and D.J. Weatherall (1977) Developmental biology of human hemoglobins. In: Progress in Hematology. E.B. Brown, ed. Grune & Stratton, New York, *10*:43–90.

Zanjani, E.D., P.B. McGlave, A. Bhakthavathsalan, and G. Stamatoyannopoulos (1979) Sheep fetal hematopoietic cells produce adult hemoglobins when transplanted in the adult animal. Nature, *280*:495–496.

JOURNAL OF CELLULAR PHYSIOLOGY SUPPLEMENT 1:151–164 (1982)

# Control of Growth and Normal Differentiation in Leukemic Cells: Regulation of the Developmental Program and Restoration of the Normal Phenotype in Myeloid Leukemia

LEO SACHS*
*Department of Genetics, Weizmann Institute of Science, Rehovot 76100, Israel*

*"The described cultures thus seem to offer a useful system for a quantitative kinetic approach to hematopoietic cell formation and for experimental studies on the mechanism and regulation of hematopoietic cell differentiation."*

—Ginsburg and Sachs, 1963

The origin and evolution of malignancy involves a sequence of genetic changes. Evidence has, however, been obtained with various types of tumors that malignant cells have not lost the genes that control normal growth and differentiation. An in vitro cloning and cell culture system has been developed to determine and dissect the controls that regulate normal myeloid cell growth, differentiation, and malignancy and to suggest a new approach to the therapy of myeloid leukemia based on the induction of normal differentiation in malignant cells. Experiments on induction of normal cell differentiation and restoration of the normal phenotype in myeloid leukemia by the physiological inducer of differentiation, the appropriate molecular form of the macrophage and granulocyte-inducing protein MGI, have also been used to formulate a general model for regulation of developmental programs and the origin and evolution of the malignancy.

The change of normal into malignant cells involves a sequence of changes including specific chromosome changes (see Sachs, 1974; Azumi and Sachs, 1977; Rowley, 1977; Klein, 1979). Evidence has, however, been obtained with various types of tumors including sarcomas (Sachs, 1974), myeloid leukemia (Sachs, 1974, 1978) and teratocarcinomas (Mintz and Illmensee, 1975; Dewey et al., 1977) that malignant cells have not lost the genes that regulate the normal control of growth and differentiation. This was first shown in sarcomas by the finding that it was possible to reverse the malignant to a nonmalignant phenotype with a high frequency in cloned sarcoma cells (Rabinowitz and Sachs, 1968, 1969, 1970a, 1970b; Hitosumachi et al., 1971; Sachs, 1974). The comparison of sarcomas with myeloid leukemias then showed that reversion of the malignant phenotype can be achieved by 2 mechanisms. In one mechanism, found with sarcomas, this reversion was obtained by chromosome segregation resulting in a change in the balance of specific chromosomes. This reversion of malignancy by chromosome segregation, with a return to the normal gene balance required for expression of the nonmalignant phenotype, occurred without hybridization between different types of cells (Rabinowitz and Sachs, 1968, 1969, 1970a, 1970b; Hitosumachi et al., 1971; Yamamoto et al., 1973; Sachs, 1974; Bloch-Shtacher and Sachs, 1977). Reversion of malignancy associated with chromosome changes has also been found after hybridization between different types of cells (Harris et al., 1969; Klein et al., 1973; Wiener et al., 1974; Ringertz and Savage, 1976).

## RESTORATION OF THE NORMAL PHENOTYPE IN MYELOID LEUKEMIA BY INDUCTION OF NORMAL DIFFERENTIATION BY THE PHYSIOLOGICAL INDUCER OF DIFFERENTIATION

In addition to this reversion of malignancy by chromosome segregation, another mechanism of reversion to a nonmalignant phenotype was also obtained in certain clones with a high frequency, but in contrast to sarcomas this reversion was not associated with chromosome segregation. Phenotypic reversion of malig-

*To whom correspondence/reprint requests should be addressed.

nancy in these leukemic cells was obtained by induction of the normal sequence of cell differentiation by the physiological inducer of differentiation (Paran et al., 1970; Fibach et al., 1972, 1973; Fibach and Sachs, 1974, 1975; Sachs, 1974, 1978, 1980; Lotem and Sachs, 1974, 1977a, 1977b, 1978; Hayashi et al., 1974). These myeloid leukemic cells also have specific chromosome changes compared to normal cells (Sachs, 1974; Hayashi et al., 1974; Azumi and Sachs, 1977; Lotem and Sachs, 1977a). The chromosome changes in these leukemic cells thus seem to involve changes in genes other than those involved in the induction of normal differentiation. Induction of normal differentiation in teratocarcinoma cells (Mintz and Illimensee, 1975; Dewey et al., 1977) may also be due to this mechanism. Studies on the phenotypic reversion of malignancy by induction of normal cell differentiation can result in a new approach to therapy based on the induction of normal differentiation in malignant cells (Paran et al., 1970; Fibach et al., 1972; Sachs, 1974, 1978, 1981; Lotem and Sachs, 1978a, 1981). They can also provide an insight into the mechanism of differentiation and the origin and evolution of malignancy. I shall summarize our results on the controls that regulate growth (proliferation) and differentiation in myeloid leukemic cells and indicate how these may serve as a model to explain the origin and further evolution (progression) (Foulds, 1954) of malignancy.

## MACROPHAGE AND GRANULOCYTE INDUCING (MGI) PROTEINS AND THE REGULATION OF GROWTH AND DIFFERENTIATION IN NORMAL AND LEUKEMIC MYELOID CELLS

The culture of normal hematopoietic cells in vitro has made it possible to study the controls that regulate the proliferation and differentiation of normal and leukemic myeloid cells (see Sachs, 1974, 1978, 1980). We first showed (Ginsburg and Sachs, 1963; Pluznik and Sachs, 1965), as was then confirmed by others (Bradley and Metcalf, 1966), that normal myeloid precursor cells cultured with a feeder layer of other cell types can form colonies of granulocytes and macrophages in vitro. We also found that the formation of these colonies is due to secretion, by cells of the feeder layer, of a specific inducer that induces both the formation of colonies and the differentiation of cells in these colonies to macrophages or granulocytes (Pluznik and Sachs, 1965, 1966; Ichikawa et

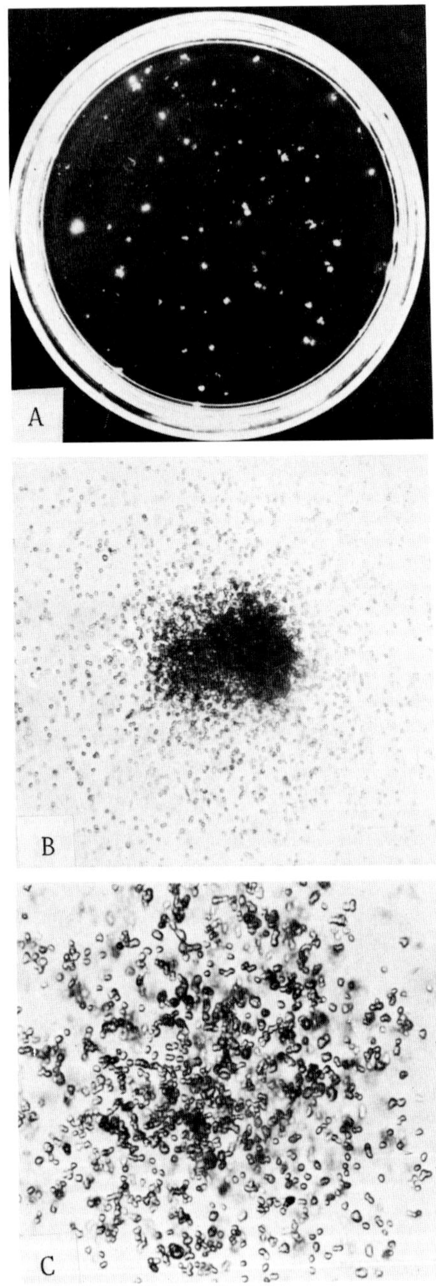


Fig. 1. Colonies of macrophages and granulocytes from normal hemotopoietic cells induced by MGI. Petri dish with colonies (A) (Pluznik and Sachs, 1965), granulocyte colony (B), and macrophage colony (C) (Ichikawa et al., 1966).

al., 1966) (Fig. 1). After we first detected its presence in culture supernatants (Pluznik and Sachs, 1966; Ichikwawa et al., 1966), this protein inducer has been referred to by a number of names including mashran gm (Ichikawa et al., 1967), colony-stimulating factor (CSF) (Metcalf, 1969), macrophage and granulocyte inducer (MGI) (Landau and Sachs, 1971), and colony-stimulating activity (CSA) (Austin et al., 1971). I shall use the name MGI (Table 1). This protein inducer can be produced and secreted by various normal and malignant cells in culture and in vivo (Ginsburg and Sachs, 1963; Pluznik and Sachs, 1965, 1966; Ichikawa et al., 1966, 1967; Bradley and Metcalf, 1966; Paran et al., 1968; Metcalf, 1969; Landau and Sachs, 1971; Austin et al., 1971; Mintz and Sachs, 1973; Weiss and Sachs, 1978; McCulloch, 1979; Douer and Sachs, 1979). MGI is a family of proteins that exist in a number of molecular forms that have different biological activities (Fibach and Sachs, 1976; Maeda et al., 1977; Hozumi et al., 1979; Lotem et al., 1980; Sachs, 1980; Falk and Sachs, 1980; Burgess and Metcalf, 1980; Liebermann et al., 1981b). The functions of these different forms of MGI will be discussed later.

Experiments with normal myeloblasts, the normal counterpart of myeloid leukemic cells, have shown that these normal cells require the normal macrophage and granulocyte inducing protein MGI for cell viability, proliferation, and differentiation to mature macrophages or granulocytes (Fibach and Sachs, 1976; Sachs, 1978). There are, however, clones of myeloid leukemic cells that can be induced to differentiate normally to mature macrophages or granulocytes by the physiological inducer MGI (Fig. 2), but that no longer require MGI for cell viability and proliferation. These clones, which in mice have been found in 4 different inbred strains (Ichikawa, 1969; Fibach et al., 1972, 1973; Sachs, 1974; Ichikawa et al., 1976; Lotem and Sachs, 1977a; Burgess and Metcalf, 1980), have also been found in humans (Paran et al., 1970; Lotem and Sachs, 1979), and are referred to as $MGI^+D^+$ ($MGI^+$ to indicate that they can be induced to differentiate by MGI; $D^+$ for differentiation to mature cells). This loss in the requirement of MGI for cell viability and proliferation then allows these leukemic cells, unlike the normal cells, to multiply in the absence of MGI. This gives the leukemic cells a growth advantage over the normal cells in vivo when there is a limiting amount of MGI. It has been suggested (Sachs, 1978) that this loss, either partial or complete, is the origin of myeloid leukemia. There are other clones of myeloid leukemic cells that can also proliferate without MGI, but that are either partially ($MGI^+D^-$) or almost completely ($MGI^-D^-$) blocked in their ability to be induced to differentiate by MGI (Fibach et al., 1973; Sachs, 1974, 1978; Lotem and Sachs, 1974, 1977a, 1979; Ichikawa et al., 1976; Hoffman-Liebermann and Sachs, 1978; Symonds and Sachs, 1979; Liebermann et al., 1980) (Fig. 3). These clones, which have specific chromosome changes (Hayashi et al., 1974; Azumi and Sachs, 1977; Lotem and Sachs, 1977a), have been considered as further stages in the development of myeloid leukemia (Sachs, 1978). $MGI^-D^-$ cells can revert to $MGI^+D^+$ cells by segregation of the appropriate chromosomes.

TABLE 1. History of cloning and clonal differentiation of normal hematopoietic cells in culture

| | |
|---|---|
| Cloning and differentiation in liquid medium | (Ginsburg and Sachs, 1963) |
| In semisolid medium: agar | (Pluznik and Sachs, 1965) |
| Inducer of cloning and differentiation secreted by cells | (Pluznik and Sachs, 1965) |
| Inducer in conditioned medium from cells | (Pluznik and Sachs, 1966; Ichikawa et al., 1966) |
| Different inducer for macrophages and granulocytes | (Ichikawa et al., 1966) |
| Cloning and differentiation in methylcellulose | (Ichikawa et al., 1966) |
| Conformation of cloning and differentiation in agar | (Bradley and Metcalf, 1966) |
| Cloning and differentiation of human cells | (Paran et al., 1970; Pike and Robinson, 1970) |
| Inducer of differentiation that does not induce cloning | (Fibach and Sachs, 1976) |
| **Terminology used for proteins that induce cloning and differentiation of normal macrophages and granulocytes** | |
| Mashran gm | (Ichikawa et al., 1967) |
| Colony-stimulating factor (CSF) | (Metcalf, 1969) |
| Colony-stimulating activity (CSA) | (Austin et al., 1971) |
| Macrophage and granulocyte inducer (MGI) | (Landau and Sachs, 1971) |
| MGI-1 for cloning; MGI-2 for differentiation | (Sachs, 1980; Lotem et al., 1980) |

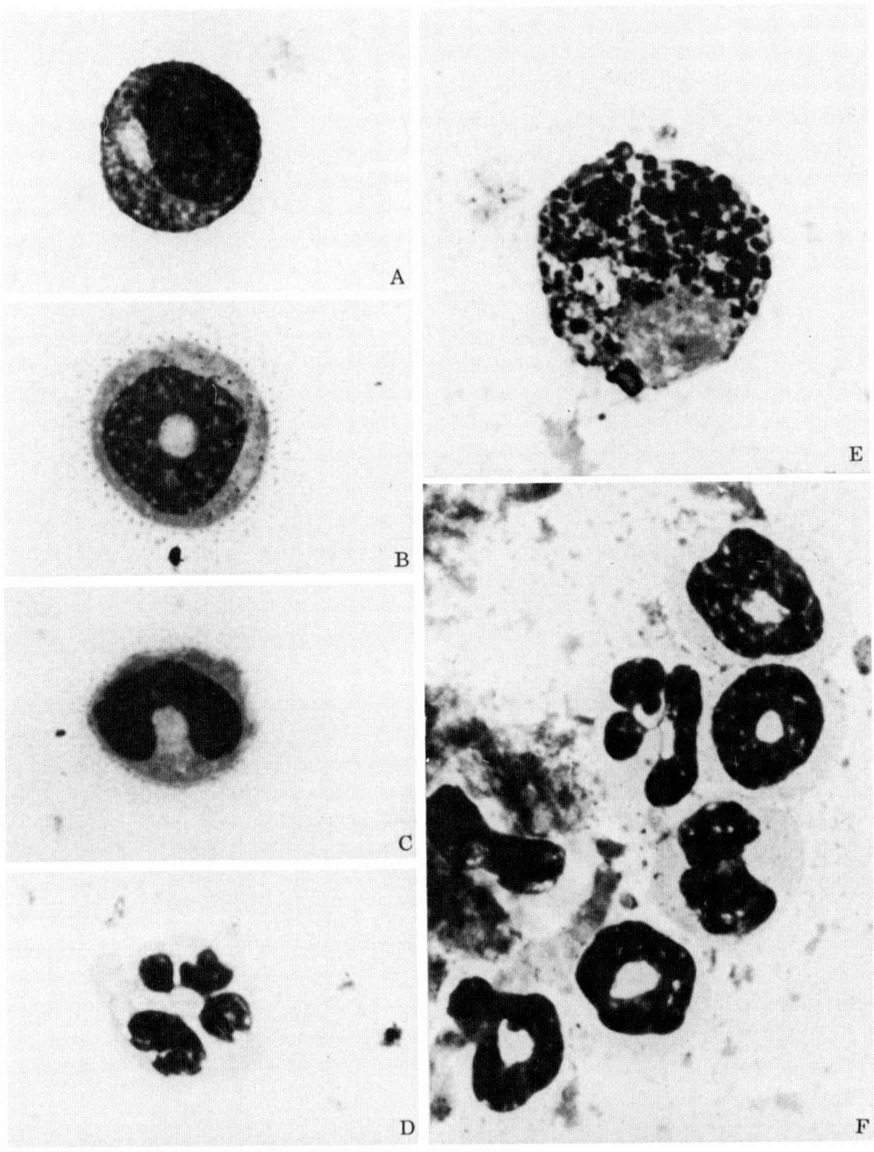

Fig. 2. Differentiation of $MGI^+D^+$ myeloid leukemic cells to mature macrophages and granulocytes by MGI (MGI-2). Leukemic cell (A), stages in differentiation to mature granulocytes (B–D), macrophage (E), and group of granulocytes in different stages of differentiation (F) (Sachs, 1978).

These chromosome changes then restore the appropriate gene balance required for induction of normal differentiation by MGI (Azumi and Sachs, 1977).

## UNCOUPLING OF CONTROLS FOR GROWTH AND DIFFERENTIATION

Proliferation in the normal cells is regulated at 2 control points. The first control is that which requires MGI to produce more cells that can then differentiate, and the second control is the stopping of cell multiplication that occurs in the program of terminal differentiation in the mature cells. In the presence of MGI, proliferation of the normal cells is followed by initiation of differentiation and termination of the differentiation program, and it is assumed that these 3 processes are coupled in the normal cells. In the $MGI^+D^+$ leukemic cells, the first proliferation control is uncoupled from the

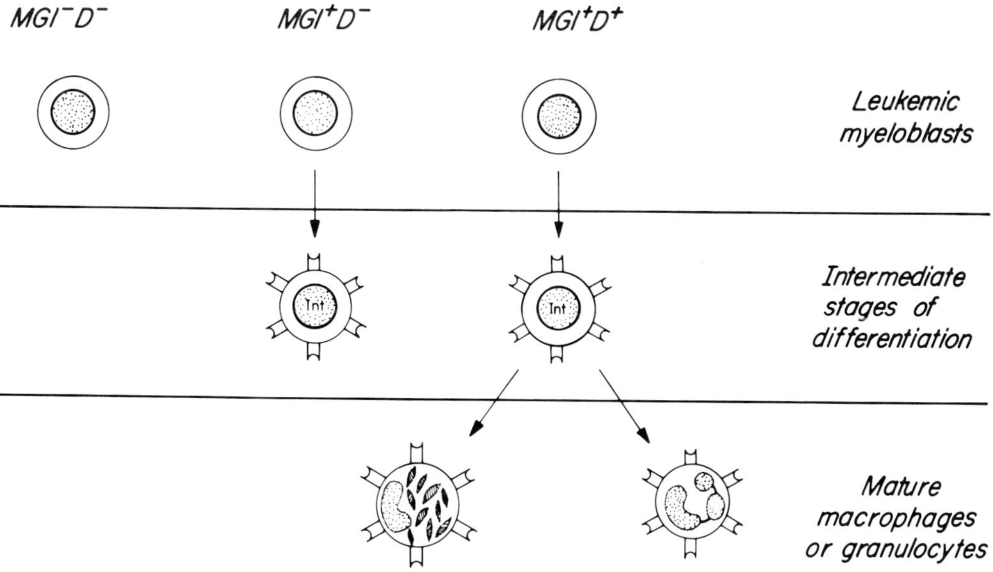


Fig. 3. Classification of different types of myeloid leukemic cell clones according to their ability to be induced to differentiate by MGI-2.

TABLE 2. Uncoupling of controls for proliferation and differentiation in myeloid leukemia

| Cell type | MGI induction of: Initial proliferation[1] (A) | Partial differentiation (B) | Terminal differentiation (C) | Coupling or uncoupling of different controls |
|---|---|---|---|---|
| Normal myeloblasts | + | + | + | Coupling of A, B, and C |
| Leukemic MGI⁺D⁺ | − | + | + | Uncoupling of A from B and C |
| Leukemic MGI⁺D⁻ | − | + | − | Uncoupling of B from A and C |
| Leukemic MGI⁻D⁻ | − | − | − | Uncoupling of A and blocking of B and C |

[1]Initial proliferation before differentiation requires addition of MGI to normal myeloblasts, whereas the 3 types of leukemic cells can proliferate without adding MGI (Sachs, 1980).

process that initiates differentiation, but the coupling between the initiation of differentiation and the stopping of proliferation in the mature cells is still maintained. In the $MGI^+$ $D^-$ leukemic cells, both the first proliferation control and the terminal proliferation control have been uncoupled from the initiation of differentiation. In the final stage, $MGI^-D^-$, uncoupling of the first proliferation control is associated with almost no inducibility for differentiation by MGI (Table 2). The origin and further evolution of myeloid leukemia is thus associated with the uncoupling of controls (see also Simantov and Sachs, 1978; Simantov et al., 1980), presumably due to the chromosome changes found in the leukemic cells that can produce changes in gene balance (Sachs, 1974, 1978; Azumi and Sachs, 1977) or the integration of appropriate viruses at specific sites in the cellular genome (Liebermann and Sachs, 1979). The results presented below can explain the mechanism of this uncoupling.

## CONSTITUTIVE GENE EXPRESSION, THE NORMAL DEVELOPMENTAL PROGRAM, AND THE ORIGIN AND EVOLUTION OF MYELOID LEUKEMIA

Since the normal but not the leukemic myeloblasts require MGI for cell viability and proliferation (Sachs, 1978), the molecular changes required for cell viability and proliferation that have to be induced in the normal cells are constitutive in the leukemic cells. This suggests that the origin of myeloid leukemia is due to a change from an induced to a constitutive expression of genes that control cell viability and proliferation. Studies on changes in the synthesis of specific proteins in normal myeloblasts, $MGI^+D^+$, $MGI^+D^-$, and $MGI^-D^-$ leukemic clones at different times after treatment with MGI, using two-dimensional gel electrophoresis (Liebermann et al., 1980; Hoffman-Liebermann et al., 1981a; Cohen and Sachs, 1981) have provided evidence at the level of protein synthesis that support this explanation and also indicate the relationship between constitutive gene expression and uncoupling of other controls. The leukemic cells were found to be constitutive for changes in the synthesis of specific proteins that were only induced in the normal cells after treatment with MGI. These constitutive changes, which included the constitutive presence of some proteins and the constitutive absence of other proteins, can be divided into 2 classes. One class consisted of changes in 14 specific proteins that were induced by MGI in the normal cells and were constitutive in all the 8 leukemic clones studies derived from 4 different tumors (Table 3, Fig. 4). The other class consisted of changes in other specific proteins that were induced by MGI in normal and $MGI^+D^+$ leukemic cells, but were constitutive in $MGI^+D^-$ and $MGI^-D^-$ leukemic cells. In this second class there were more constitutive changes in $MGI^-D^-$ than in $MGI^+D^-$ cells (Table 4, Fig. 4) (Liebermann et al., 1980).

The protein changes during the normal proliferation and differentiation of myeloblasts seem to be induced by MGI as a series of parallel multiple pathways of gene expression (Liebermann et al., 1980). It can be assumed that the normal developmental program that appears to couple proliferation and differentiation in normal cells requires a synchronous initiation and progression of these multiple parallel pathways. The presence of constitutive gene expression for some pathways can be expected to produce asynchrony in the coordination required for the normal developmental program. Depending on the pathways involved, this asynchrony could then result in

TABLE 3. Consecutive expression of protein changes in all the leukemic clones

| State of protein in leukemic cells compared to normal myeloblasts | Change in synthesis induced by MGI in normal myeloblasts | Number of proteins | Time when change in synthesis initiated in normal myeloblasts (number of proteins) | | | | | |
|---|---|---|---|---|---|---|---|---|
| | | | 1H[1] | 3H | 9H | 1D[1] | 2D | 4D |
| Absent | Decrease | 5 | — | 3 | 2 | — | — | — |
| Present | De novo | 9 | — | 2 | — | 2 | 3 | 2 |

[1]H, hours; D, days (Liebermann et al., 1980).

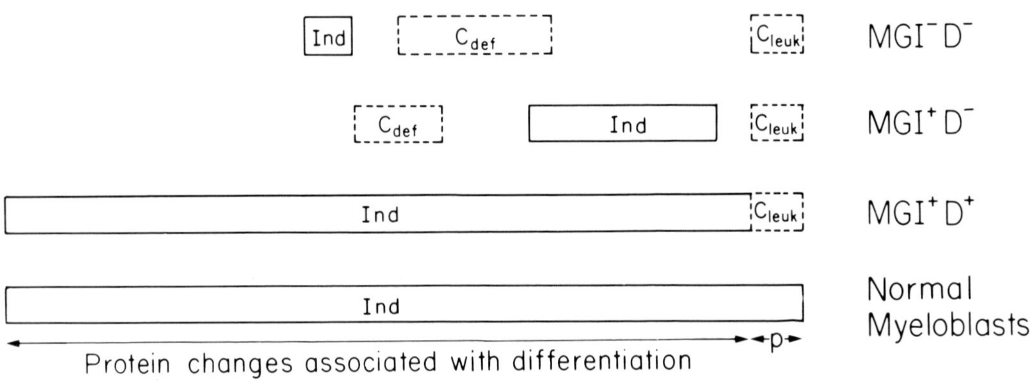


Fig. 4. Schematic summary of protein changes associated with proliferation and differentiation in normal myeloblasts and different types of myeloid leukemic cells. $C_{leuk}$, constitutive expression of changes in leukemic compared to normal myeloblasts. $C_{def}$, constitutive expression of changes in the differentiation defective $MGI^+D^-$ or $MGI^-D^-$ clones compared to $MGI^+D^+$ clones. Ind, MGI-induced changes. p, protein changes associated with proliferation to produce more cells before differentiation (Liebermann et al., 1980).

TABLE 4. Constitutive expression of protein changes in differentiation-defective leukemic clones

| Type of defective clone | State of protein in defective clones compared to MGI$^+$D$^+$ cells | Change in synthesis induced by MGI in MGI$^+$D$^+$ cells | Number of proteins | Time when change in synthesis initiated in MGI$^+$D$^+$ cells (number of proteins) | | | | | | |
|---|---|---|---|---|---|---|---|---|---|---|
| | | | | 1H[1] | 3H | 9H | 1D[1] | 2D | 4D | 6D |
| MGI$^+$D$^-$ | Absent | Decrease | 3 | 2 | — | — | 1 | — | — | — |
| | Lower synthesis | Decrease | 7 | 5 | — | 1 | 1 | — | — | — |
| | Higher synthesis | Increase | 4 | — | 3 | — | — | 1 | — | — |
| | Present | De novo | 13 | — | 3 | — | — | 7 | 2 | 1 |
| MGI$^-$D$^-$ | Absent | Decrease | 19 | 12 | — | 2 | 3 | 2 | — | — |
| | Lower synthesis | Decrease | 14 | 6 | — | 1 | 2 | 5 | — | — |
| | Higher synthesis | Increase | 2 | — | — | 2 | — | — | — | — |
| | Present | De novo | 8 | — | 3 | 1 | — | 3 | 1 | — |

[1]H, hours; D, days (Liebermann et al., 1980)

an uncoupling of the controls for proliferation and differentiation and produce different blocks in the ability to be induced for and to terminate the differentiation process. The suggestion derived from these results (Sachs, 1978, 1980; Liebermann et al., 1980) is, therefore, that myeloid leukemia originates by a change that produces certain constitutive pathways of gene expression and that this uncouples the normal requirement for MGI for proliferation and differentiation. This can be followed by constitutive expression of other pathways resulting in the uncoupling of other controls and thus interfering with the normal program of terminal differentiation (Fig. 5).

## ROLE OF DIFFERENT MOLECULAR FORMS OF MGI

The MGI independent proliferation of myeloid leukemic cells appears to proceed through stages (Collins et al., 1977), probably starting from a small decrease to finally a complete loss of the requirement for MGI for cell viability and proliferation. There are different molecular forms of MGI. One form, MGI-1, induces in normal myeloblasts the cell viability and proliferation required for colony formation by these normal cells. There are 2 forms of MGI-1: MGI-1M that induces macrophage colonies and MGI-1G that induces granulocyte colonies (Lotem et al., 1980). MGI-1 does not appear to affect the behavior of the myeloid leukemic cells. Another form, MGI-2, induces differentiation in MGI$^+$ myeloid leukemic cells that no longer require MGI-1 for cell viability and proliferation (Sachs, 1980; Lotem et al., 1980; Falk and Sachs, 1980; Lipton and Sachs, 1981; Liebermann et al., 1981b). MGI-2 also induces differentiation in normal myeloblasts without inducing colony formation. In addition, MGI-1 induced MGI-2 in the normal but not in the leukemic myeloblasts, and it has been suggested that induction of the differentiation inducer by the proliferation inducer could ensure the coupling between proliferation and differentiation in the normal cells (Sachs, 1980; Liebermann et al., 1981b). There may be induction of different forms of MGI-2 for differentiation to macrophages or granulocytes.

Our results also indicate that cells in suspension can express either the macrophage or

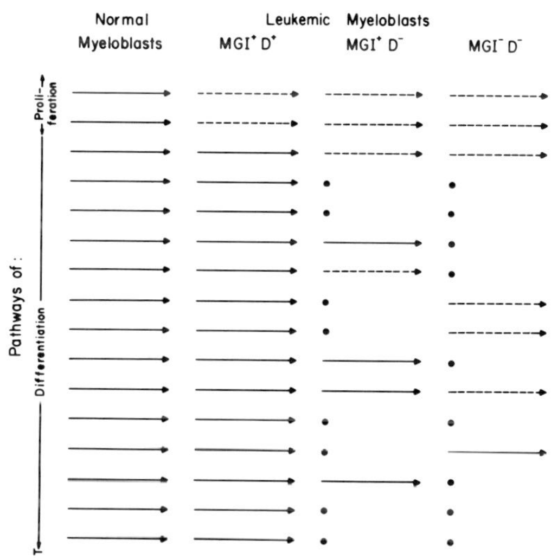


Fig. 5. Schematic relationship between MGI induced and constitutive changes in gene expression and the control of cell proliferation and differentiation in normal and different types of leukemic myeloblasts. ——→, MGI-induced changes; ----→, constitutive changes; •, blocked for induction of changes by MGI; T, terminal stage of differentiation (Sachs, 1980).

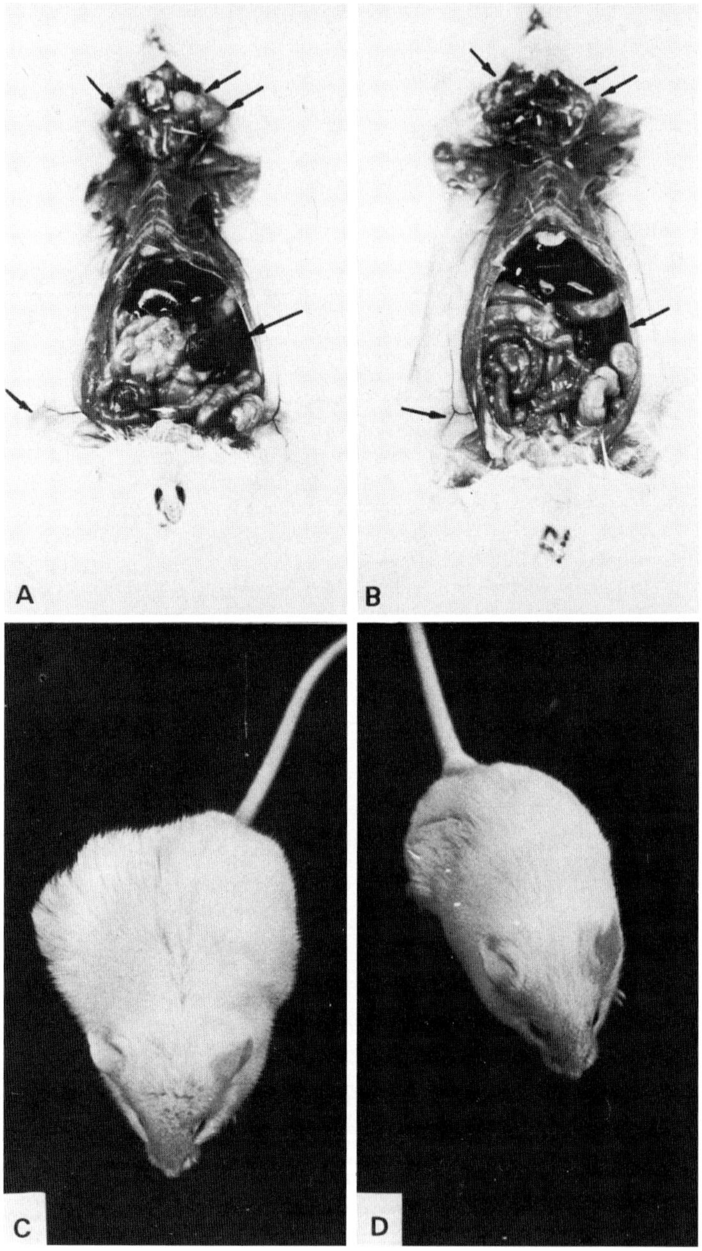


Fig. 6. Internal organ involvement and subcutaneous growth of leukemic cells in mice that were untreated or treated with MGI. Mice 38 days after intravenous inoculation of $MGI^{+}D^{+}$ leukemic cells that were either untreated (A) or had been treated with MGI (B). Note enlargement of cervical and inguinal lymph nodes and spleen in the untreated (A) compared to the MGI-treated mouse (B). Mice 25 days after subcutaneous inoculation with $MGI^{+}D^{+}$ leukemic cells. Untreated mouse, (C); MGI-treated mouse (D). The inhibition of leukemia development was mediated by MGI-2 (Lotem and Sachs, 1981).

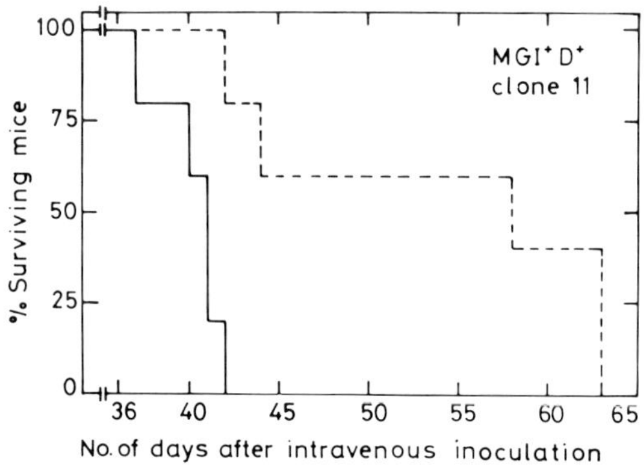


Fig. 7. Effect of treatment with MGI on the survival of mice intravenously inoculated with MGI$^+$D$^+$ leukemic cells. Mice were inoculated intravenously with MGI$^+$D$^+$ leukemic cells and they were then injected with normal culture medium (solid line) or MGI (dashed line). This inhibition of leukemia development was mediated by MGI-2 (Lotem and Sachs, 1981).

granulocyte program depending on the inducer, and that changes in cell shape associated with cell attachment can regulate specific proteins and restrict the developmental program to macrophages. The in vivo environment of cells in relation to the possibilities for cell adhesion may thus play a major role in determining the developmental program of myeloid cells (Liebermann et al., 1981a). The regulation of MGI-1 and MGI-2 appears to be under the control of different genes (Falk and Sachs, 1980) and the family of MGI proteins is presumably a multigene family. We have also shown that MGI-2, which induces differentiation of MGI$^+$ leukemic cells in vitro, can inhibit the development of myeloid leukemia in vivo (Figs. 6 and 7) (Lotem and Sachs, 1981). The relationship between the chemical structures by MGI-1 and MGI-2, and whether they originate from a common precursor (Lotem et al., 1980), remains to be elucidated. There may also be hybrid molecules of the different forms of MGI including hybrid molecules with MGI-1 and MGI-2 activity. The study of the stages that finally result in a loss of the requirement for MGI-1 for cell viability and proliferation would allow a further dissection of the constitutive pathways of gene expression associated with the initial uncoupling of the requirement for MGI-1 for proliferation from the MGI-2 requirement for differentiation. In addition to stages in the loss of the requirement for MGI-1 for cell viability and proliferation, there are presumably also stages in the origin of other leukemic cells that constitutively produce their own MGI-1 (Paran et al., 1968).

## INDUCTION OF DIFFERENTIATION BY TREATMENT WITH OTHER COMPOUNDS

Myeloid leukemic cells show a clonal difference in inducibility for differentiation by the physiological inducer MGI-2. They also show a clonal difference in inducibility for differentiation by steroid hormones such as dexamethasone (Fig. 8), by nonphysiological inducers such as actinomycin D, dimethylsulfoxide, or tumor-promoting phorbol esters (Sachs, 1974, 1977, 1978; Lotem and Sachs, 1974, 1975, 1979; Krystosek and Sachs, 1976, 1977; Honma et al., 1977; Maeda and Sachs, 1978; Oshima et al., 1979), and by compounds including interferon, cordycepin, and 2-deoxy-D-glucose that by themselves did not induce differentiation but enhanced the effect of some inducing compounds (Krystosek and Sachs, 1976; Lotem and Sachs, 1978b, 1979; Hoffman-Liebermann et al., 1981a). The inducing compounds include x-irradiation and compounds used in cancer chemotherapy (Table 5). Some of these compounds can also induce the production of MGI in the cells that then differentiate (Weiss and Sachs, 1978; Lotem and Sachs, 1979; Falk and Sachs, 1980; Liebermann et al., 1981a). Erythroleukemic cells that do not respond to the physiological inducer of erythroid differentiation also show clonal differences in their response to nonphysiological inducers (Marks and Rifkind, 1978; Friend, 1978). Studies with MGI$^+$D$^+$ myeloid leukemia cell clones resistant to induction of differentiation by the steroid hormone dexamethasone have shown that defective induction of protein changes by dexamethasone in clones with normal steroid binding to cytoplasmic receptors and normal transport and nuclear binding of the receptor-steroid complex can also be associated with constitutive expression of specific protein changes in the mutant cells. The constitutive protein changes that prevented normal induction of differentiation by dexamethasone were different from the constitutive changes that prevented normal induction of differentiation by MGI-2 (Cohen and Sachs, 1981). The clonal differences in response to nonphysiological inducers and compounds such as interferon, like the difference in response to the physiological inducer MGI and the steroid hormone dexamethasone, may also be due to constitutive

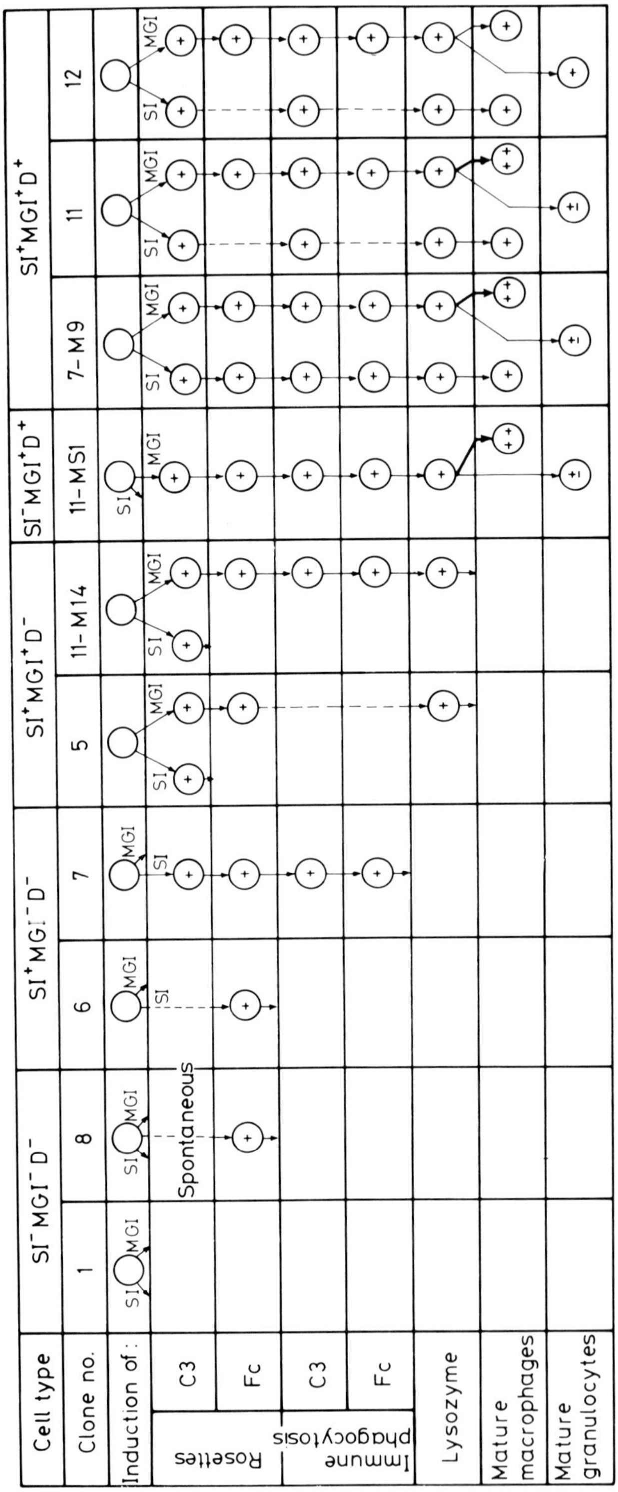


Fig. 8. Inducibility for differentiation-associated properties in different clones of myeloid leukemic cells by the steroid inducer (SI) dexamethasone and by MGI (MGI-2) (Sachs, 1978).

TABLE 5. Inducers and noninducers for normal cell markers in MGI⁺D⁺ myeloid leukemic cells[1]

| Type of compound | Inducers | Noninducers |
|---|---|---|
| Peptide hormones | MGI | Erythropoietin, nerve growth factor, ubiquitin, thymopoietin, interferon |
| Steroids | Dexamethasone, prednisolone, hydrocortisone, oestradiol | Progesterone, testosterone, epitestosterone, androstenedione, cortisone |
| Lectins | Concanavalin A, phytohaemagglutinin, pokeweed mitogen | |
| Polycylic hydrocarbons | Benzo(a)pyrene, dimethylbenz(a)-anthracene | Benz(*a*)anthracene, dibenz(*a,c*)anthracene, dibenz(*a,h*)anthracene, phenathrene |
| Other compounds | Lipopolysaccharide, lipid A, mitomycin C, dimethyl sulphoxide cytosine arabinoside hydroxyurea, thymidine, 5-iododeoxyuridine, 5-bromodeoxyuridine, 5-fluorodeoxyuridine, nitrosoguanidine, actinomycin D, adriamycin, daunomycin X-irradiation 12-0-tetradecanoylphorbol-13-acetate | Colchicine, vinblastine, Na butyrate, cycloheximide, dibutyryl cyclic AMP, dibutyryl cyclic GMP, cordycepin, deoxyglucose, ouabain. ionophore 23187 |

[1]The different inducers were not all active on the same clone and did not all induce the same markers (Sachs, 1978).

TABLE 6. Induction of differentiation in a MGI⁻D⁻ clone by combined treatment with the tumor-promoting phorbol ester TPA and MGI-2

| | Treatment with | | Rosettes (percent) | | Phagocytosis of rosettes (percent) | | | | Cell type (percent) | |
|---|---|---|---|---|---|---|---|---|---|---|
| Clone | TPA | MGI-2 | Fc | C3 | Fc | C3 | Lysozyme | Blast | Intermediate | Macrophage |
| MGI⁻D⁻ | – | – | 0.7 | 0.1 | 0 | 0 | 0 | 100 | 0 | 0 |
| clone 1 | – | + | 0.6 | 0.2 | 0 | 0 | 0 | 100 | 0 | 0 |
| | + | – | 0.6 | 0.1 | 0 | 0 | 0 | 100 | 0 | 0 |
| | + | +[1] | 10.8 | 2.9 | 1.1 | 0.4 | 3.6 | 94 | 6 | 0 |

[1]Replacement of MGI-2 by dexamethasone, lipopolysaccharide, or epidermal or fibroblast growth factor did not result in induction of differentiation (Lotem and Sachs, 1979).

expression of different specific pathways of gene expression in the nonresponding cells.

There are also MGI⁻D⁻ clones that can be induced to differentiate by treatment with steroid hormones such as dexamethasone (Lotem and Sachs, 1974, 1977a) (Fig. 8), by some nonphysiological compounds such as the tumor-promoting phorbol ester TPA (Lotem and Sachs, 1979), or by combined treatment with different compounds (Krystosek and Sachs, 1976; Lotem and Sachs, 1979; Symonds and Sachs, 1982) (Table 6). Different compounds can induce different changes in gene expression, and the induction of differentiation in these clones by combined treatment with different compounds can be due to complementation of changes in gene expression by regulating mRNA production and mRNA translation (Hoffman-Liebermann et al., 1981b). We have been able to induce all our MGI⁻D⁻ clones for some differentiation-associated properties by the appropriate combination of compounds (Krystosek and Sachs, 1976; Lotem and Sachs, 1979; Symonds and Sachs, 1972).

## DEVELOPMENTAL PROGRAMS AND MALIGNANCY IN DIFFERENT CELL TYPES

The model for the origin and evolution of myeloid leukemia may be applicable to malignant tumors derived from other types of normal cells, whose viability, proliferation, and differentiation is induced by other physiological inducers. Different tumor cells can also express certain functions that are induced in normal development and yet the cells are malignant. These include myeloma proteins (Potter, 1972; Natvig and Kunkel, 1973), fetal proteins, a variety of cell growth factors and other horones, and specialized products (Paran et al., 1968; Ichikawa et al., 1969; Weinhouse, 1972; Norgaard-Pederson and Axelsen, 1978; Potter, 1978; De Larco and Todaro, 1978; Lotem and Sachs, 1979; Lehmann, 1979; Siden et al., 1979; Unkeless et al., 1979). Such tumors could include cases where some of the tumor cells behave like $MGI^+$ cells and are induced for expression of these products by the appropriate physiological inducer. It can, however, also be suggested that these products, including myeloma and fetal proteins, which are induced by the appropriate physiological inducer during normal development, are constitutive in the malignant cells and that this prevents the cells proceeding to terminal differentiation. The synthesis of these products may indeed be an indication of constitutive changes in these malignant cells and this can be experimentally tested. The presence of a fetal protein thus does not necessarily mean that the malignant cells have the same controls, even for this protein, as the normal embryonic cells. The constitutive expression of some genes could also produce an abnormality in the developmental program that results in the formation of abnormal proteins that are not found in normal cells. The model of specific constitutive pathways of gene expression that uncouple the controls for proliferation and differentiation in myeloid leukemia, may thus be a general model to explain the origin and evolution of malignancy and the expression of fetal and some other specialized functions of normal development in various types of malignant cells.

## LITERATURE CITED


Austin, P.E., E.A. McCulloch, and J.E. Till (1971) Characterization of the factor in L cell conditioned medium capable of stimulating colony formation by mouse marrow cells in culture. J. Cell Physiol., *77:*121–134.

Azumi, J., and L. Sachs (1977) Chromosome mapping of the genes that control differentiation and malignancy in myeloid leukemic cells. Proc. Natl. Acad. Sci. U.S.A., *74:*253–257.

Bloch-Shtacher, N., and L. Sachs (1977) Identification of a chromosome that controls malignancy in Chinese hamster cells. J. Cell Physiol., *93:*205–212.

Bradley, T.R., and D. Metcalf (1966) The growth of mouse bone marrow cells in vitro. Aust. J. Exp. Biol. Med. Sci., *44:*287–300.

Burgess, A.W., and D. Metcalf (1980) Characterisation of a serum factor stimulating the differentiation of myelomonocytic leukemic cells. Int. J. Cancer, *26:*647–654.

Cohen, L., and L. Sachs (1981) Constitutive gene expression in myeloid leukemia and cell competence for induction of differentiation by the steroid dexamethasone. Proc. Natl. Acad. Sci. U.S.A., *78:*353–357.

Collins, J.S., R.C. Gallo, and R.E. Gallagher (1977) Continuous growth and differentiation of human myeloid leukemic cells in culture. Nature, *270:*347–349.

De Larco, J.E., and G.J. Todaro (1978) Growth factors from murine sarcome virus-transformed cells. Proc. Natl. Acad. Sci. U.S.A., *8:*4001–4005.

Dewey, M.J., D.W. Martin Jr., G. R. Martin, and B. Mintz (1977) Mosaic mice with teratocarcinoma-derived mutant cells deficient in hypoxanthine phosphoribosyltransferase. Proc. Natl. Acad. Sci. U.S.A., *74:*5564–5568.

Douer, D., and L. Sachs (1979) Production of human T cell colony inducing activity (TCI) by malignant cells from patients with lymphocytic and myeloid leukemia. J. Immunol., *122:*2473–2477.

Falk, A., and L. Sachs (1980) Clonal regulation of the induction of macrophage- and granulocyte-inducing proteins for normal and leukemic myeloid cells. Int. J. Cancer, *26:*595–601.

Fibach, E., M. Hayashi, and L. Sachs (1973) Control of normal differentiation of myeloid leukemic cells to macrophages and granulocytes. Proc. Natl. Acad. Sci. U.S.A., *70:*343–346.

Fibach, E., T. Landau, and L. Sachs (1972) Normal differentiation of myeloid leukemic cells induced by a differentiation-inducing protein. Nature (New Biol.), *237:*276–278.

Fibach, E., and L. Sachs (1974) Control of normal differentiation of myeloid leukemic cells. IV. Induction of differentiation by serum from endotoxin treated mice. J. Cell Physiol., *83:*177–185.

Fibach, E., and L. Sachs (1975) Control of normal differentiation of myeloid leukemic cells. VIII. Induction of differentiation to mature granulocytes in mass culture. J. Cell Physiol., *86:*221–230.

Fibach, E., and L. Sachs (1976) Control of normal differentiation of myeloid leukemic cells. XI. Induction of a specific requirement for cell viability and growth during the differentiation of myeloid leukemic cells. J. Cell Physiol., *89:*259–266.

Foulds, L. (1954) The experimental study of tumor progression: A review. Cancer Res. *14:*327–339.

Friend, C. (1978) The phenomenon of differentiation in murine erythroleukemic cells. Harvey Lectures 72. New York: Academic Press, pp. 253–281.

Ginsburg, H., and L. Sachs (1963) Formation of pure suspension of mast cells in tissue culture by differentiation of lymphoid cells from the mouse thymus. J. Natl. Cancer Inst., *31:*1–40.

Harris, H., O.J. Miller, G. Klein, P. Worst, and T. Tachibana (1969) Suppression of malignancy by cell fusion. Nature, *223:*363–368.

Hayashi, M., E. Fibach, and L. Sachs (1974) Control of normal differentiation of myeloid leukemic cells. V. Normal differentiation in aneuploid leukemic cells and the chromosome banding pattern of $D^+$ and $D^-$ clones. Int. J. Cancer, *14:*40–48.

Hitosumachi, S., Z. Rabinowitz, and L. Sachs (1971) Chromosomal control of reversion in transformed cells. Nature, *231:*511–514.

Hoffman-Liebermann, B., D. Liebermann, L. Sachs (1981a) Control mechanism regulating gene expression during normal differentiation of myeloid leukemic cells: Differentiation defective mutants blocked in mRNA production and mRNA translation. Dev. Biol., *81:*255–265.

Hoffman-Liebermann, B., D. Liebermann, and L. Sachs (1981b) Regulation of gene expression by tumor promoters. III. Complementation of the developmental program in myeloid leukemic cells by regulating mRNA production and mRNA translation. Int. J. Cancer, *28:*615–620.

Hoffman-Liebermann, B., and L. Sachs (1978) Regulation of actin and other proteins in the differentiation of myeloid leukemic cells. Cell, *14:*825–834.

Honma, Y., T. Kasukabe, J. Okabe, and M. Hozumi (1977) Glucocorticoid binding and mechanism of resistance in some clones of mouse myeloid leukemic cells resistant to induction of differentiation by dexamethasone. J. Cell Physiol., *93:*227–235.

Hozumi, M., T. Umezawa, K. Takenaga, T. Ohno, M. Shikita, and I. Yamane (1979) Characterization of factors stimulating differentiation of mouse myeloid leukemic cells from a Yoshida sarcoma cell line cultured in serum free medium. Cancer Res., *39:*5127–5131.

Ichikawa, Y. (1969) Differentiation of a cell line of myeloid leukemia. J. Cell Physiol., *74:*223–234.

Ichikawa, Y., M. Maeda, and M. Horiuchi (1976) In vitro differentiation of Rauscher virus induced myeloid leukemic cells. Int. J. Cancer, *17:*789–797.

Ichikawa, Y., M. Paran, and L. Sachs (1969) The stimulation of tumor cell growth by a substance produced by normal and tumor cells. J. Cell Physiol., *73:*43–48.

Ichikawa, Y., D.H. Pluznik, and L. Sachs (1966) In vitro control of the development of macrophage and granulocyte colonies. Proc. Natl. Acad. Sci. U.S.A., *56:*488–495.

Ichikawa, Y., D.H. Pluznik, and L. Sachs (1967) Feedback inhibition of the development of macrophage and granulocyte colonies. I. Inhibition by macrophages. Proc. Natl. Acad. Sci. U.S.A., *58:*1480–1486.

Klein, G. (1979) Lymphoma development in mice and humans: Diversity of initiation is followed by convergent cytogenetic evolution. Proc. Natl. Acad. Sci. U.S.A., *76:*2442–2446.

Klein, G., S. Frieberg, F. Wiener, and H. Harris (1973) Studies on hybrid cells derived from the fusion of the $TA_3$/Ha ascites carcinoma with normal fibroblasts. I. Malignancy, karyotype, and formation of isoantigenic variants. J. Natl. Cancer Inst., *50:*1259–1268.

Krystosek, A., and L. Sachs (1976) Control of lysozyme induction in the differentiation of myeloid leukemic cells. Cell, *9:*675–684.

Krystosek, A., and L. Sachs (1977) Steroid hormone receptors and the differentiation of myeloid leukemic cells. J. Cell Physiol., *92:*345–352.

Landau, T., and L. Sachs (1971) Characterization of the inducer required for the development of macrophage and granulocyte colonies. Proc. Natl. Acad. Sci. U.S.A., *68:*2540–2544.

Lehmann, F.G. (1979) Carcino-Embryonic Proteins, Amsterdam: Elsevier.

Liebermann, D., B. Hoffman-Liebermann, and L. Sachs (1980) Molecular dissection of differentiation in normal and leukemic myeloblasts: Separately programmed pathways of gene expression. Dev. Biol., *79:*46–63.

Liebermann, D., B. Hoffman-Liebermann, and L. Sachs (1981a) Regulation of gene expression by tumor promoters. II. Control of cell shape and developmental programs for macrophages and granulocytes in human myeloid leukemic cells. Int. J. Cancer, *28:*285–291.

Liebermann, D., B. Hoffman-Liebermann, and L. Sachs (1981b) Regulation and role of different macrophage and granulocyte inducing proteins in normal and leukemic myeloid cells. Int. J. Cancer (in press).

Liebermann, D., and L. Sachs (1979) Increase of normal myeloblast viability and multiplication without blocking differentiation by type C RNA virus from myeloid leukemic cells. Proc. Natl. Acad. Sci. U.S.A., *76:*3353–3357.

Lipton, J., and L. Sachs (1981) Characterization of macrophage- and granulocyte-inducing proteins for normal and leukemic myeloid cells produced by the Krebs ascites tumor. Biochim. Biophys. Acta, *673:*552–569.

Lotem, J., J. Lipton, and L. Sachs (1980) Separation of different molecular forms of macrophage- and granulocyte-inducing proteins for normal and leukemic myeloid cells. Int. J. Cancer, *25:*763–771.

Lotem, J., and L. Sachs (1974) Different blocks in the differentiation of myeloid leukemic cells. Proc. Natl. Acad. Sci. U.S.A., *71:*3507–3511.

Lotem, J., and L. Sachs (1975) Induction of specific changes in the surface membrane of myeloid leukemic cells by steroid hormones. Int. J. Cancer, *15:*731–740.

Lotem, J., and L. Sachs (1977a) Genetic dissection of the control of normal differentiation in myeloid leukemic cells. Proc. Natl. Acad. Sci. U.S.A., *74:*5554–5558.

Lotem, J., and L. Sachs (1977b) Control of normal differentiation of myeloid leukemic cells. XII. Isolation of normal myeloid colony-forming cells from bone marrow and the sequence of differentiation to mature granulocytes in normal and $D^+$ myeloid leukemic cells. J. Cell Physiol., *92:*97–108.

Lotem, J., and L. Sachs (1978a) In vivo induction of normal differentiation in myeloid leukemic cells. Proc. Natl. Acad. Sci. U.S.A., *75:*3781–3785.

Lotem, J., and L. Sachs (1978b) Genetic dissociation of different cellular effects of interferon on myeloid leukemic cells. Int. J. Cancer, *22:*214–220.

Lotem, J., and L. Sachs (1979) Regulation of normal differentiation in mouse and human myeloid leukemic cells by phorbol esters and the mechanism of tumor promotion. Proc. Natl. Acad. Sci. U.S.A., *76:*5158–5162.

Lotem, J., and L. Sachs (1980) Potential prescreening for therapeutic agents that induce differentiation in human myeloid leukemic cells. Int. J. Cancer, *25:*561–564.

Lotem, J., and L. Sachs (1981) In vivo inhibition of the development of myeloid leukemia by injection of macrophage- and granulocyte-inducing protein. Int. J. Cancer, *28:*375–386.

Maeda, M., M. Horiuchi, S. Numa, and Y. Ichikawa (1977) Characterization of a differentiation stimulating factor for mouse myeloid leukemic cells. Gann, *68:*435–447.

Maeda, S., and L. Sachs (1978) Control of normal differentiation of myeloid leukemic cells. XIII. Inducibility for some stages of differentiation by dimethylsulfoxide and its dissociation from inducibility by MGI. J. Cell Physiol., *94:*181–186.

Marks, P., and R.A. Rifkind (1978) Erythroleukemic differentiation. Annu. Rev. Biochem., *47:*419–448.

McCulloch, E.A. (1979) Abnormal myelopoietic clones in man. J. Natl. Cancer Inst., *63:*883–891.

Metcalf, D. (1969) Studies on colony formation in vitro by mouse bone marrow cells. I. Continuous cluster formation and relation of clusters to colonies. J. Cell Physiol., *74:*323–332.

Mintz, B., and K. Illmensee (1975) Normal genetically mosaic mice produced from malignant teratocarcinoma cells.

Proc. Natl. Acad. Sci. U.S.A., *72:*3585–3589.

Mintz, U., and L. Sachs (1973) Differences in inducing activity for human bone marrow colonies in normal serum and serum from patients with leukemia. Blood, *42:*331–340.

Natvig, J.B., and H.G. Kunkel (1973) Human immunoglobulins: Classes, subclasses, genetic variants, and idiotypes. Adv. Immunol., *16:*1–54.

Norgaard-Pederson, B., and N.H. Axelsen (1978) Carcinoembryonic proteins: Recent progress. Scand. J. Immunol. (Suppl.) *8:*1–683.

Oshima, G., M. Yamada, and T. Sugimura (1979) Changes in protease activity during differentiation of mouse myeloid leukemic cells. Biochem. Biophys. Res. Commun. *90:*158–163.

Paran, M., Y. Ichikawa, and L. Sachs (1968) Production of the inducer for macrophage and granulocyte colonies by leukemic cells. J. Cell Physiol., *72:*251–254.

Paran, M., L. Sachs, Y. Barak, and P. Resnitzky (1970) In vitro induction of granulocyte differentiation in hematopoietic cells from leukemic and non-leukemic patients. Proc. Natl. Acad. Sci. U.S.A., *67:*1542–1549.

Pluznik, D.H., and L. Sachs (1965) The cloning of normal "mast" cells in tissue culture. J. Cell Comp. Physiol., *66:*319–324.

Pluznik, D.H., and L. Sachs (1966) The induction of clones of normal "mast" cells by a substance from conditioned medium. Exp. Cell Res., *43:*553–563.

Potter, M., (1972) Immunoglobulin producing tumors and myeloma proteins of mice. Physiol. Rev. *52:*632–719.

Potter, V.R., (1978) Phenotypic diversity in experimental hepatomas: The concept of partially blocked ontogeny. Br. J. Cancer, *68:*1–23.

Rabinowitz, Z., and L. Sachs (1968) Reversion of properties in cells transformed by polyoma virus. Nature, *220:*1203–1206.

Rabinowitz, Z., and L. Sachs (1969) The formation of variants with a reversion of properties of transformed cells. II. In vitro formation of variants from polyoma-transformed cells. Virology, *38:*343–346.

Rabinowitz, Z., and L. Sachs (1970a) Control of the reversion of properties in transformed cells. Nature, *225:*136–139.

Rabinowitz, Z., and L. Sachs (1970b) The formation of variants with a reversion of properties of transformed cells. V. Reversion to a limited life span Int. J. Cancer, *6:*388–398.

Ringertz, N.R., and R.E. Savage (1976) Cell Hybrids. New York: Academic Press.

Rowley, J.D. (1977) Mapping of human chromosomal regions related to neoplasia: Evidence from chromosomes 1 and 17. Proc. Natl. Acad. Sci. U.S.A., *74:*5729–5733.

Sachs, L. (1974) Regulation of membrane changes, differentiation, and malignancy in carcinogenesis. Harvey Lectures, Vol. 68, New York: Academic Press, pp. 1–35.

Sachs, L. (1977) Control of normal cell differentiation in leukemia. Isr. J. Med. Sci., *13:*654–665.

Sachs, L. (1978) Control of normal cell differentiation and the phenotypic reversion of malignancy in myeloid leukemia. Nature, *274:*535–539.

Sachs, L. (1980) Constitutive uncoupling of pathways of gene expression that control growth and differentiation in myeloid leukemia: A model for the origin and progression of malignancy. Proc. Natl. Acad. Sci. U.S.A., *77:*6152–6156.

Sachs, L. (1981) Induction of normal differentiation in malignant cells as an approach to cancer therapy. In Molecular Actions and Targets for Cancer Chemotherapeutic Agents. New York: Academic Press, pp. 579–589.

Siden, E.J., D. Baltimore, D. Clark, and N.E. Rosenberg (1979) Immunoglobulin synthesis by lymphoid cells transformed in vitro by Abelson murine leukemia virus. Cell, *16:*389–396.

Simantov, R., and L. Sachs (1978) Differential desensitization of functional adrenergic receptors in normal and malignant myeloid cells: Relationship to receptor mediated hormone cytotoxicity. Proc. Natl. Acad. Sci. U.S.A., *75:*1805–1809.

Simantov, R., T. Shkolnik, and L. Sachs (1980) Desensitization of enucleated cells to hormones and the role of cytoskeleton in control of a normal hormonal response. Proc. Natl. Acad. Sci. U.S.A., *77:*4798–4802.

Symonds, G., and L. Sachs (1979) Activation of normal genes in malignant cells: Activation of chemotaxis in relation to other stages of normal differentiation in myeloid leukemia. Somat. Cell Genet., *5:*931–944.

Symonds, G., and L. Sachs (1982) Modulation of cell competence for induction of differentiation in myeloid leukemic cells. J. Cell. Physiol. (in press).

Unkeless, J.C., G. Kaplan, H. Plutner, and Z.A. Cohn (1979) Fc-receptor variants of a mouse macrophage cell line. Proc. Natl. Acad. Sci. U.S.A., *76:*1400–1404.

Weinhouse, S. (1972) Glycolysis, respiration and anomalous gene expression in experimental hepatomas: G.H.A. Clowes memorial lecture. Cancer Res., *32:*2007–2016.

Weiss, B., and L. Sachs (1978) Indirect induction of differentiation in myeloid leukemic cells by lipid A. Proc. Natl. Acad. Sci. U.S.A., *75:*1374–1378.

Wiener, F., G. Klein, and H. Harris (1974) The analysis of malignancy by cell fusion. V. Further evidence of the ability of normal diploid cells to suppress malignancy. J. Cell Sci., *15:*177–183.

Yamamoto, T., Z. Rabinowitz, and L. Sachs (1973) Identification of the chromosomes that control malignancy. Nature [New Biol.], *243:*247–250.

JOURNAL OF CELLULAR PHYSIOLOGY SUPPLEMENT 1:165–173 (1982)

# Gene Controlled Negative Regulation of DNA Synthesis in Erythropoietic Progenitor Cells

ARTHUR AXELRAD,* HELENA CROIZAT, DENISE ESKINAZI, SOLOMON STEWART, DUSHANDHAN VAITHILINGAM, AND HENK VAN DER GAAG
*Division of Histology, Department of Anatomy, University of Toronto, Toronto, Ontario, Canada M5S 1A8*

Congenic strains differing by a small segment of Chromosome 9 that bears the *Fv-2* locus have provided valuable material for investigating genetic resistance to Friend polycythemia virus (FV). C57BL/6 (B6) (*Fv-2*$^{rr}$) mice have been found to differ from B6.S (*Fv-2*$^{ss}$) mice not only in their response to this virus but also in the proliferative state of their erythropoietic progenitor cells BFU-E: in B6 mice the majority of BFU-E are normally quiescent, while in B6.S mice approximately 50% are actively synthesizing DNA at any time. We have shown that B6 but not B6.S marrow contains a macromolecule that negatively regulates DNA synthesis specifically of BFU-E in vitro. Evidence is presented that this macromolecule is a physiological negative regulator active in vivo in B6 mice. A new liquid culture system is described in which FV and erythropoietin (epo) act synergistically on Ficoll-Isopaque separated bone marrow cells to give rise after 7 days to large numbers of CFU-E detectable in 2-day plasma cultures with and without epo. Inclusion of concentrated B6 but not B6.S bone marrow supernatant in the liquid cultures drastically curtailed the amplification of CFU-E by FV and epo. These studies indicate that both DNA synthesis and the BFU-E stage of differentiation are necessary conditions for initiating the effects of polycythemia-inducing FV. Genetic resistance to FV appears not to reside within the target cells for the virus.

Two decades have passed since it was first realized that the development of polycythemia following exposure of mice to Friend virus (FV) was genetically controlled (Odaka and Yamamoto, 1962). The almost simultaneous production in 3 different laboratories (Odaka, 1969, 1970; Axelrad and Van der Gaag, 1968; Lilly, 1970) of strains that were congenic for resistance or susceptibility to this virus confirmed the validity of the notion that a single gene locus could be critically important in the host response to a tumor virus. Production of these congenic strains also provided animals of known genotype for further investigation. These strains theoretically differed only at the locus of interest and at some closely linked loci on the small chromosome segment transferred in the course of production of the congenic partners. Thus any differences uncovered between the strains might be leads to an understanding of the mechanism underlying the genetically controlled host response. Mapping of the locus responsible for dominant susceptibility to FV followed; it was on linkage group II (later assigned to Chromosome 9) and was named *Fv-2* (Lilly, 1970). [The name *Fv-1* (Lilly, 1970; Pincus et al., 1971) went to a gene dominant for relative resistance to a large variety of leukemia viruses, including FV].

In addition to the 3 pairs of congenic strains developed deliberately for the purpose of investigating the *Fv-2* locus, two other congenic strains were produced that differed at *H-7* (Snell and Bunker, 1965; Bailey, 1975). *H-7*

Dr. Croizat is a visiting scientist from the Institut de Radiobiologie Clinique La 207, Laboratoire de Cinetique Cellulaire, IGR, France and is supported by funds from the C.N.R.S.

*To whom correspondence/reprint requests should be addressed.

and *Fv-2* are closely linked to one another on Chromosome 9 but are clearly not identical.

Our FV-susceptible strain B6.S ($Fv\text{-}2^{ss}$), congenic with the FV-resistant C57BL/6 (B6) strain ($Fv\text{-}2^{rr}$), was produced from a cross with the FV-susceptible SIM strain ($Fv\text{-}2^{ss}$) by intercrosses and serial backcrosses to the B6 strain with selection for susceptibility to FV (Axelrad and Van der Gaag, 1972). The β-galactosidase locus *Bgs* (Felton et al., 1973; Paigen et al., 1976) is situated near *Fv-2* on Chromosome 9. Recently we examined the levels of enzymatic activity of β-galactosidase in these strains. We found that the activity, which is known to be high in B6, was also high in its congenic partner B6.S, but it was low in SIM, the parent strain from which the Chromosome 9 segment bearing the $Fv\text{-}2^{s}$ allele had been derived. This indicated that in the course of production of the B6.S strain congenic with B6 at *Fv-2*, recombination must have occurred at a point between the *Bgs* locus and the *Fv-2* locus (arrow in Fig. 1) (Eisen and Axelrad, unpublished observations). On the other side of *Fv-2*, the $Mod\text{-}1^{b}$ allele of B6.S was known to be derived from the SIM strain (Axelrad et al., 1978). Thus the segment of Chromosome 9 that differs in the congenic partners B6 and B6.S must be at least 10 cM long.

We are presently investigating this interesting genetic region which in addition to 1) determining host response to Friend polycythemia virus, also 2) controls the proportion of early erythropoietic progenitor cells BFU-E that are normally engaged in DNA synthesis in vivo (Suzuki and Axelrad, 1980), 3) regulates the expression of RNA sequences in hemopoietic tissues of normal FV-susceptible mice which show homology with sequences in the defective Spleen Focus-Forming virus component of Friend polycythemia virus (Mak et al., 1979), 4) encodes a pair of normal erythroid differentiation antigens on the late erythropoietic progenitor cells CFU-E (Vaithilingam and Axelrad, submitted), one of which appears to be *H-7*, and 5) controls the production of a macromolecule in normal marrow that negatively regulates DNA synthesis in the early erythropoietic progenitor cells BFU-E (Axelrad et al., 1981). The present paper deals with negative regulation of DNA synthesis in BFU-E and its relation to resistance to Friend polycythemia virus.

## Properties of the negative regulatory activity in normal B6 ($Fv\text{-}2^{rr}$) marrow that affect the proportion of BFU-E engaged in DNA synthesis

We have shown that mice which differ in their response to FV also differ with respect to the proliferative state of their erythropoietic progenitor cells BFU-E under normal, noninfective conditions (Suzuki and Axelrad, 1980). In FV-resistant mice of genotype $Fv\text{-}2^{rr}$, a majority of BFU-E are in a quiescent state, as judged by $^3$H-thymidine or hydroxyurea suicide experiments in vivo, while in FV-susceptible mice of genotype $Fv\text{-}2^{ss}$ approximately 50% of the BFU-E are engaged in DNA synthesis at any time. The difference is specific to the BFU-E; the proportion of CFU-S, CFU-E, and CFU-C in DNA synthesis did not differ between the congenic partner strains.

When cells from B6 mice were subjected to a single wash, their BFU-E became vulnerable to $^3$H-thymidine or hydroxyurea (Table 1; Axelrad et al., 1981a). These cycle-active agents are believed to operate by different mecha-

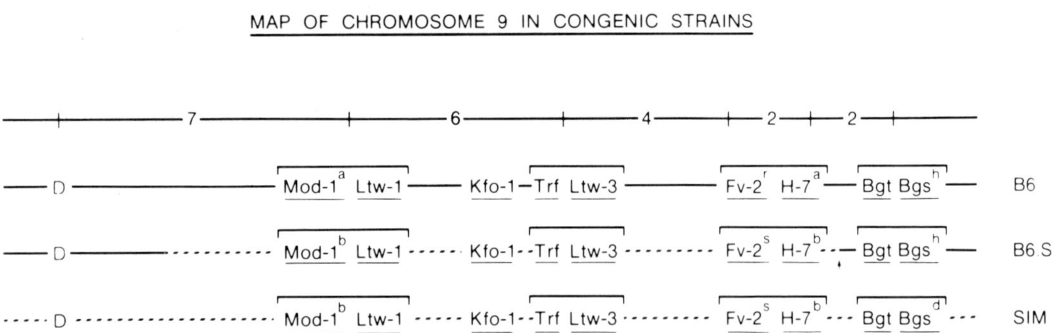


Fig. 1. Modified after Davisson and Roderick (1981). *Bgs*, β galactosidase structural; *Bgt*, β galactosidase temporal; *H-7*, Histocompatibility-7; *Fv-2*, Friend virus susceptibility; *Ltw-3,1*, Liver protein, molecular weight 20–30 K (Elliott, 1979); *Trf*, Transferrin (Schreffler, 1963); *Kfo-1*, Kidney protein, molecular weight 40–50 K (Elliott, 1979); *Mod-1*, Malic dehydrogenase; *D*, Nondilute.

TABLE 1. Washing renders BFU-E of B6 *(Fv-2^rr^)* mice vulnerable to $^3$H-thymidine or hydroxyurea

| | Percent of BFU-E killed by | |
|---|---|---|
| | $^3$H-thymidine | hydroxyurea |
| Unwashed | 17.7 ± 7.9 | 7.5 ± 9.7 |
| Washed | 49.7 ± 5.2 | 50.0 ± 6.0 |

nisms but have in common the ability to kill cells in DNA synthesis. We therefore concluded that washing of B6 bone marrow cells must remove something that prevents BFU-E from engaging in DNA synthesis.

We have recovered from bone marrow supernatants of B6 mice an activity with the unique capacity of reducing or maintaining at a low level the proportion of BFU-E normally engaged in DNA synthesis. A striking reduction in the proportion of BFU-E undergoing $^3$H-thymidine suicide was observed when washed B6.S or B6 marrow cells were suspended in supernatants from B6 marrow (Axelrad et al., 1981a). The activity appears to be loosely associated with the BFU-E, as it is readily washable from these cells.

When we found that supernatants from bone marrow cells of B6 mice contained something that could restrict the proportion of BFU-E killed by $^3$H-thymidine or hydroxyurea, we naturally wondered whether we had a nonspecific toxic agent or a truly physiological negative regulator of DNA synthesis. Five pieces of evidence lead us to believe that the negative regulatory activity in B6 marrow is indeed physiological.

1. The activity is under strict genetic control. In B6 (*Fv-2*$^{rr}$) mice, the proportion of BFU-E in DNA synthesis is naturally low, while in B6.S (*Fv-2*$^{ss}$) mice the proportion of BFU-E in DNA synthesis is high. Washing of B6 but not B6.S bone marrow cells substantially increased the proportion of BFU-E in DNA synthesis. Resuspension of washed B6 or B6.S bone marrow cells in supernatant from B6 marrow resulted in a low proportion of BFU-E undergoing DNA synthesis. Resuspension of washed B6 or B6.S bone marrow cells in superantant from B6.S marrow had no effect. Thus the *Fv-2* locus (or a closely linked gene on Chromosome 9) controls the negative regulatory activity detectable in B6 bone marrow supernatants. This locus does not control the response of the BFU-E to this activity, since the BFU-E in washed marrow cells from both B6 and B6.S mice were responsive. (B6.S × B6)$F_1$ hybrid mice were investigated to determine how the activity was inherited. No negative regulatory activity was detected in the marrows of the $F_1$ hybrid mice (Axelrad et al., 1981b). These observations are consistent with the hypothesis that the negative regulatory activity in B6 marrow is recessively inherited. Comparison of marrows from young adult and immature B6 mice showed that the negative regulatory activity was expressed only in the adult.

2. The activity is detectable at approximately the same low concentration at which it is found in B6 marrow, as shown by the rapidity with which it dilutes out in titration curves. Large doses are not necessary.

3. The activity appears to be specific for one particular target cell population in the hemopoietic system—the BFU-E. Washing of a marrow cell suspension followed by $^3$H-thymidine suicide assays showed that there was no significant effect of washing on DNA synthesis of CFU-S, CFU-E, or CFU-C. Only the DNA synthesis in BFU-E was affected by the washing. Recently we showed that exposure to B6 marrow supernatant had no effect on DNA synthesis of the bipotential BFU-ME (Croizat, Eskinazi, and Axelrad, unpublished observations). It thus seems to be not only lineage- but also stage-specific although the possibility of a quantitative rather than a qualitative restriction cannot at present be ruled out.

4. The activity is readily reversible without apparent toxicity. If after being maintained in the B6 marrow supernatant where DNA synthesis is kept at a low level for a period of time, the BFU-E are released by washing, within the time it takes to do the experiment, a high proportion of the BFU-E can be shown to be engaged in DNA synthesis. There is no reduction in burst-forming efficiency, and the proportion in DNA synthesis is not reduced by the cells having been in contact with the B6 marrow supernatant.

5. But the most convincing evidence for the physiological nature of the B6 negative regulator comes from an experiment with congenic monoclonal antibody (Vaithilingam and Axelrad, unpublished observations) specifically directed against the activity in B6 marrow supernatant.

Reciprocal immunization was carried out between the congenic strains B6 and B6.S. The antigen consisted of unwashed bone marrow and spleen cells from mice of the opposite strain. The cells were injected first in complete, then in incomplete Friend's adjuvant at weekly intervals. The sera obtained were reacted with bone marrow and spleen cells of the same and opposite strains and the cells were assayed for their residual ability to produce erythrocytic colonies in plasma culture. In this way congenic antisera were obtained. After 1 1/2 years of immunization, spleen cells from the immunized mice were fused with PEG to SP/1 myeloma cells that had lost the capacity to produce antibody. The media from hybrid colonies that grew out of HAT medium were screened for their ability to inhibit erythroid colony formation. Two colonies were found active and their cells were grown up and cloned. The monoclonal antibody produced by B6 cells against B6.S unwashed bone marrow and spleen cells was called anti B6.$S_8$, the monoclonal antibody produced by B6.S cells against B6 bone marrow and spleen cells was called anti $B6_{16}$.

Anti $B6_{16}$ was reacted with B6 marrow supernatant and the latter was tested at different dilutions against bone marrow cells, which were then assayed for $^3$H-thymidine suicide as a measure of the proportion of BFU-E engaged in DNA synthesis. Controls consisted of anti B6.$S_8$ antibody in the same system. It was found that the anti $B6_{16}$ antibody was capable of specifically neutralizing the B6 marrow supernatant's ability to lower the proportion of BFU-E in DNA synthesis. We reasoned that this antibody should afford us the opportunity of obtaining a definitive answer to the question of whether the material in B6 marrow supernatants is a physiological regulator of BFU-E DNA synthesis.

If the majority of BFU-E in B6 marrow are maintained in vivo in a quiescent state because they are coated by a physiological negative regulator, then neutralization of the negative regulator as it exists on unwashed bone marrow cells taken directly from the mouse should release the BFU-E from the restriction in DNA synthesis imposed by this material. The proportion of BFU-E engaged in DNA synthesis and therefore killed by $^3$H-thymidine should be seen to rise.

The results showed that this was indeed the case. Among unwashed B6 bone marrow cells as they came from the animal, only a small proportion of BFU-E were killed by $^3$H-thymidine, i.e., only a minority of BFU-E were engaged in DNA synthesis. Exposure of these cells to increasing concentrations of monoclonal antibody directed against the B6 negative regulator resulted in an increasing proportion of BFU-E undergoing $^3$H-thymidine suicide, signifying that an increasing proportion of BFU-E were engaged in DNA synthesis. No such effect was seen with the control antibody. To our knowledge this is the first time that exposure of cells to an antibody has been followed by an increase in the proportion of these cells undergoing DNA synthesis.

These observations convince us that the marrow supernatant from B6 mice must contain a physiological negative regulatory activity responsible for keeping low the proportion of BFU-E engaged in DNA synthesis. We believe that it is this activity that is responsible for the low proportion of BFU-E ordinarily engaged in DNA synthesis in B6 (*Fv-2*$^{rr}$) mice (Suzuki and Axelrad, 1980).

The negative regulatory activity in B6 marrow was shown to be stable to dialysis and lyophilization and to a temperature of 56°C for 30 min; it was excluded on G-25 Sephadex gel filtration, and it was retained by Amicon filters that permitted passage of particles of less than 50,000 but not less than 100,000 molecular weight. Thus the negative regulatory activity in B6 marrow must be macromolecular in nature (Axelrad et al., 1981).

The following evidence indicates that the negative regulator probably operates at the cell surface: 1) it is readily removed by a single wash; 2) it is immediately effective when added to the cells although its molecular weight appears to be high (>50,000); 3) it is accessible to antibody while associated with the cells; and 4) it appears to function at 0°C. When washed marrow cells are resuspended in B6 marrow supernatant at 0°C and then warmed to 37°C, at which temperature $^3$H-thymidine suicide assays are carried out, the proportion of BFU-E in DNA synthesis can be shown to have been lowered in relation to the concentration of supernatant during the period in the cold.

## The B6 negative regulator and Friend virus resistance

The negative regulator of BFU-E DNA synthesis was discovered in normal B6 mice during the course of investigations on the mechanism by which the *Fv-2*$^{rr}$ genotype confers solid resistance against the development of Friend virus-induced polycythemia in adult mice. It is therefore appropriate at this time to ask whether or not the B6 negative regulator in fact has anything to do with the genetic resistance to Friend polycythemia virus. Operationally this question translates into the related question: Does the presence of B6 negative regulator (which restricts the proportion of BFU-E engaged in DNA synthesis) inhibit the polycythemia-inducing action of Friend virus?

An experimental approach to this question had to await the development of a reliable system for examining early events in the induction of polycythemia by Friend virus in vitro.

### Friend Virus and erythropoietin act synergistically to amplify the CFU-E population of marrow cells in liquid culture

Several authors have reported the production of erythroid colonies and bursts without erythropoietin (epo) in cultures of hemopoietic cells infected with Friend virus (FV) in vitro (Clarke et al., 1975; Hankins et al., 1978, 1979). These culture systems reproduced the epo-independence characteristic of erythropoiesis in FV-induced polycythemia, but they failed to reproduce the massive increase in total number of CFU-E in the hemopoietic tissues of the affected animals (Liao and Axelrad, 1975), one of the cardinal features of this disease. We have recently found that FV and epo act synergistically in liquid culture to bring about a striking increase in the total number of CFU-E among marrow cells cultured over a 7-day period.

Femoral bone marrow cell suspensions were prepared from the femurs of B6.S female mice. The mononuclear cells were separated on-Ficoll-Isopaque (density 1.0773 gm/ml) and either kept as controls or infected with F-B virus, a polycythemic variant of FV ($6 \times 10^5$ FFU/$2.4 \times 10^7$ cells; 45 min, 0°C) and transferred to T-25 flasks (Falcon Plastics, Los Angeles, CA) where they were grown in media with and without epo (Connaught Step III; 3 units/ml). After 7 days the cells were harvested and assayed for CFU-E in plasma cultures with and without epo (0.25 units/ml) (Fig. 2).

The results (Table 2) showed that in liquid cultures containing cells treated with neither FV nor epo or with FV alone and assayed in the presence of epo, no CFU-E could be detected at 7 days, and in liquid cultures treated with epo alone moderate numbers of CFU-E were present. In contrast, cultures containing mixtures of FV with epo showed an almost ex-

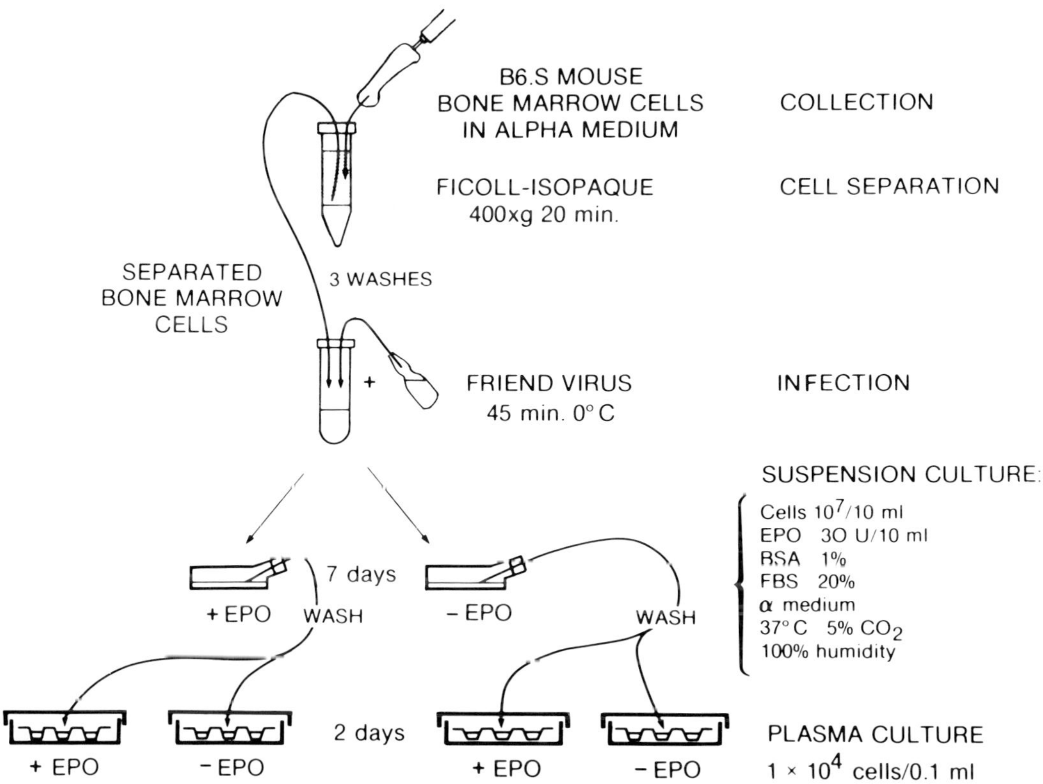


Fig. 2. Liquid culture for amplification of CFU-E by the synergistic action of Friend polycythemia virus and erythropoietin.

TABLE 2. Friend virus and erythropoietin act synergistically in vitro to amplify production of CFU-E

| Day | Virus | epo | Assayed in presence of added epo Number of CFU-E recovered per 5 × $10^4$ cells | per flask |
|---|---|---|---|---|
| 0 | | | 270.7 ± 32.4 | 65,000 |
| 7 | – | – | 0 | 0 |
| | + | – | 0.4 ± 0.4 | 13 |
| | – | + | 45.8 ± 6.0 | 1,830 |
| | + | + | 572.5 ± 40.5 | 22,900 |

TABLE 3. Friend virus and erythropoietin act synergistically in vitro to amplify production of "epo-independent" CFU-E

| In suspension culture Day | Virus | epo | Assayed in absence of added epo Number of CFU-E recovered per 5 × $10^4$ cells | per flask |
|---|---|---|---|---|
| 0 | | | 0.3 ± 0.3 | 76 |
| 7 | – | – | 0 | 0 |
| | + | – | 0 | 0 |
| | – | + | 0 | 0 |
| | + | + | 191.3 ± 18.0 | 7,650 |

plosive increase in the number of CFU-E. This increase in absolute number of CFU-E was accompanied by a fall in the total number of cells in the flasks during the 7 days of culture, resulting in a CFU-E enrichment of over 35-fold. Amplification of the CFU-E population under the influence of FV appears to require exposure to epo at an early stage.

Erythrocytic colonies were also produced in plasma cultures to which no epo was added (Table 3), but significant numbers of colonies developed only in those cultures seeded with cells from the marrow suspension culture flasks that had contained both FV and epo. These colonies appeared similar to those previously attributed to "epo-independent" CFU-E. Their number, though lower than those derived from "epo-dependent" CFU-E, also showed a substantial increase over the 7-day period.

The present observations should provide the basis of a useful culture system for investigation of the cellular and molecular mechanisms of interaction between FV and epo in the induction of polycythemia.

### Effect of B6 negative regulator of BFU-E DNA synthesis on amplifiction of CFU-E by Friend virus and erythropoietin in vitro: Test of a hypothesis

We have proposed the hypothesis that in order for FV to initiate a polycythemia, the erythropoietic progenitor cells BFU-E must be actively proliferating (Axelrad et al., 1978). This hypothesis should be testable in the present system: If we reduced the proportion of BFU-E engaged in DNA synthesis by the inclusion of B6 negative regulator in the liquid cultures, we should observe a decrease in the amplification of CFU-E and in the production of "epo-independent" CFU-E resulting from the synergistic action of FV and epo. These effects should be readily detectable in the semi-solid cultures.

Femoral bone marrow cells of B6.S female mice were separated on Ficoll-Isopaque, and exposed to FV in the cold or kept as controls. The cells were diluted to 1 × $10^7$ cells/ml, and 1 ml aliquots were added to tubes containing 3 ml of 10× concentrated bone marrow supernatant from B6 mice. Concentrated bone marrow supernatant from B6.S mice and PBS were used as controls. The contents of these tubes were added to flasks containing medium with or without epo (3 units/ml). All flasks also contained aprotinin (8.2 μg/ml). The cell suspensions in the flasks were harvested after 7 days and assayed for BFU-E and CFU-E in plasma cultures with and without epo (0.25 units/ml). For comparison, unseparated and separated bone marrow cells were also assayed at the start of the experiment. The results (Table 4) show that Ficoll-Isopaque separation was effective in reducing the number of CFU-E in the initial marrow cell preparations by over 90%, while maintaining or slightly enriching the suspensions for BFU-E. As before, the CFU-E population was greatly amplified in the cultures containing FV and epo and moderately amplified in those containing epo alone. No CFU-E were detectable in the cultures that did not contain epo.

The results in Table 4 show that inclusion of concentrated B6 marrow supernatant in the liquid cultures entirely prevented the amplification of the CFU-E population brought about by FV and epo; in contrast, B6.S marrow supernatant had little or no effect. The total number of cells was reduced by about 50% in the flasks that contained bone marrow supernatants, whether of B6.S or of B6 origin. B6 marrow supernatant had no consistent effect on the number of BFU-E recoverable from the suspension cultures after 7 days. These observations show that the action of the B6 negative regulator in restricting amplification of CFU-E was genetically specific and not caused by toxicity to the BFU-E or CFU-E.

Erythrocytic colonies were also produced from the liquid cultured bone marrow cells ex-

TABLE 4. B6 negative regulator of BFU-E DNA synthesis specifically inhibits amplification of CFU-E production by Friend virus and epo

| Bone marrow cells | Number of CFU-E per $5 \times 10^4$ | Number of CFU-E per flask | Number of BFU-E per $5 \times 10^4$ | Number of BFU-E per flask |
|---|---|---|---|---|
| Unseparated | 341.5 ± 22.8 | 68,300 | 10.3 ± 1.5 | 2,060 |
| Separated | 28.8 ± 3.1 | 5,750 | 19.4 ± 2.3 | 3,870 |

| In suspension culture | | 7 days Number of CFU-E recovered per $1 \times 10^4$ | per flask |
|---|---|---|---|
| − Virus − epo | — | 0 | 0 |
| + Virus − epo | — | 0 | 0 |
| − Virus + epo | — | 28.8 ± 2.9 | 12,960 |
| + Virus + epo | — | 105.8 ± 5.6 | 58,190 |
| + Virus + epo + B6.S | marrow supernatant | 90.0 ± 1.3 | 23,400 |
| + Virus + epo + B6 | marrow supernatant | 26.0 ± 3.6 | 7,280 |

posed to FV and epo in plasma cultures to which no epo had been added (Table 5). However, the number of such "epo-independent" erythroid colonies seen in plasma cultures seeded with cells harvested from the flask cultures containing B6 marrow supernatant was greatly reduced. Thus both "epo-dependent" and "epo-independent" CFU-E amplification were curtailed when B6 marrow supernatant was present in the liquid cultures where FV-sensitive B6.S bone marrow cells were exposed to FV and epo.

TABLE 5. Negative regulator of BFU-E DNA synthesis inhibits production of "epo-dependent" and "epo-independent" CFU-E by Friend virus and epo

| In suspension culture: Day | Virus | epo | B6 sup | Number recovered per flask assayed in plasma culture: CFU-E + epo | CFU-E − epo | BFU-E + epo | BFU-E − epo |
|---|---|---|---|---|---|---|---|
| 0 | | | | 1,560 | 0 | 2,800 | 0 |
| 7 | + | + | − | 21,054 | 2,596 | 343 | 0 |
| | + | + | + | 4,194 | 864 | 245 | 0 |

## DISCUSSION

The salient findings in the present work are 1) that DNA synthesis BFU-E is under negative control and this is mediated by a macromolecule present in B6 bone marrow, and 2) that B6 marrow supernatant, which possesses the negative regulatory macromolecule, strikingly inhibited amplification of the CFU-E populations induced by FV and epo; little increase in the number of either "epo-dependent" or "epo-independent" CFU-E occurred in liquid cultures containing supernatant from B6 marrow.

In contrast, B6.S marrow supernatant was incapable of inhibiting either DNA synthesis of normal BFU-E or FV and epo-induced amplification of CFU-E. We therefore conclude that both functions are determined at the region of Chromosome 9 by which the congenic strains B6 and B6.S differ. Resistance to the action of Friend polycythemia virus imposed by the *Fv-2*$^{rr}$ genotype thus appears to be mediated by the B6 negative regulator's restriction of DNA synthesis in BFU-E.

Our hypothesis that proliferation of BFU-E is a necessary condition for sensitivity to Friend polycythemia virus (Axelrad et al., 1978) has thus survived the test presented to it in the present work. The behavior of Friend virus in this system would be in line with that of other retroviruses, which also require a proliferative state of the host cells for expression of their viral genes (Humphries and Temin, 1974; Varmus et al, 1977). Since the action of the B6 negative regulator on DNA synthesis seems to be specific for the BFU-E, the results of the present work indicate that another crucial condition for expression of the polycythemic effects of Friend virus is the stage of differentiation of the cells.

What these studies have further shown us is that genetic resistance to FV, like genetic control of DNA synthesis, does not reside in the BFU-E itself but is exerted upon the BFU-E via a negative regulatory macromolecule present in marrow of B6 mice. This would also account for the fact that the genetic resistance in these mice is relative in that it is not fully developed in infancy and that it can be overcome in the adult by radiation and repopulation of the hemopoietic system with syngeneic bone marrow cells of B6 origin (Axelrad et al., 1978).

Dewey and Eldridge also report (this conference) that genetic resistance to Friend virus does not reside in the target cells themselves, since it could be conferred on otherwise sensitive cells from DBA/2 parents by cohabitation with cells from genetically resistant B6 parents in allophenic mice. The resistance was also shown to be relative, since animals with equal proportions of blood cells from the two strains showed sensitivity to FV, while those with 90% or more B6 blood cells were unaffected by high doses of the virus.

Our earlier transplantation studies (Axelrad and Van der Gaag, 1968; Axelrad et al., 1978) and those of Odaka and Matsukura (1969) showed that both genetic susceptibility and genetic resistance to FV were of donor origin in radiation chimeras; it was therefore concluded that those qualities reside in the target cells themselves. These studies must now be reinterpreted in the light of the present findings. The bone marrow grafts of pluripotential stem cells no doubt transferred not only potential target cells for FV but also potential producers of negative regulator. Therefore the system was incapable of distinguishing one from the other.

The same difficulty applies in the case of the radiation chimeras produced by Silver and Teich (1981) in which 1:1 mixtures of DBA/2 (*Fv-2*$^{ss}$, *Hbb*$^{dd}$) and B10.D2 (*Fv-2*$^{rr}$, *Hbb*$^{oo}$) bone marrow cells were infused into irradiated hosts of either strain. Challenge with Friend polycythemia virus showed that the majority of the newly synthesized hemoglobin one month following infection was of the FV-sensitive DBA/2 type. Their conclusion was that the resistance of B10.D2 mice to FV is inherent in the cells. But no attempt was made to vary the proportions of marrow from the 2 strains as the proportions of cells in the allophenic mice of Dewey and Eldrige were varied. In fact there appeared to be a slight selective advantage in favor of the DBA/2 cells in the chimeras even without infection. This experiment can therefore not be regarded as conclusively demonstrating inherent resistance in B10.D2 cells, which might have been producing negative regulator in inadequate amounts to be detected in this system.

For the amplification of CFU-E in the presence of FV in liquid culture, as in vivo (Liao and Axelrad, 1975), epo must have been essential no matter what was in the medium, because the absence of epo led to no CFU-E production or survival in the same medium. It is interesting also that it was necessary to expose the marrow cells to epo in the early liquid culture phase of the experiment in order for "epo-independent" CFU-E to be detected in the later semisolid cultures. While CFU-E amplification was a regular consequence of the combined action of FV and epo, the proportion of "epo-independent" CFU-E varied over a wide range and was always a small fraction of the total. B6 marrow supernatant inhibited the production of both. We therefore believe that "epo-independent" CFU-E represent a variable subpopulation of the total CFU-E population of infected marrow that have previously been exposed to epo. Whether or not this turns out to be the case, the system described here should provide a useful tool for molecular dissection of the synergism between Friend virus and epo that leads to unregulated amplification and "epo-independence" of the marrow CFU-E population.

## ACKNOWLEDGMENTS

We thank Mrs. Lottie Viezel and Mr. Bruce Smith for excellent technical assistance. This work was supported by grants from the National Cancer Institute of Canada and from the Medical Research Council of Canada.

## LITERATURE CITED

Axelrad, A., and H.C. Van der Gaag (1968) Genetic and cellular basis of susceptibility or resistance to Friend leukemia virus in mice. Can. Cancer Conf. *8*:313–343.

Axelrad, A., M. Ware, and H.C. Van der Gaag (1972) Host cell susceptibility and resistance to murine leukemia viruses and their genetic control. In Emmelot, P., and Bentveltzen, P. eds.: Proceedings of the Conference on RNA Viruses and Host Genome in Oncogenesis. North-Holland, Amsterdam: pp. 239–254.

Axelrad, A.A., S. Suzuki, H. Van der Gaag, B.J. Clarke, and D.L. McLeod (1978) Friend virus and malignant erythropoiesis: A genetic analysis. In Golde, D.W., Cline, M.J., Metcalf, D., Fox, C.F. eds.: ICN-UCLA Symposia on Molecular and Cellular Biology, New York: Academic Press, pp. 69–90.

Axelrad, A.A., H. Croizat, and D. Eskinazi (1981a) A washable macromolecule from *Fv-2*$^{rr}$ marrow negatively regulates the proportion of erythropoietic progenitor cells BFU-E in DNA synthesis. Cell *26*:233–244 (Part 2).

Axelrad, A.A., H. Croizat, and D. Eskinazi (1981b) Gene control of progenitor cell proliferation during erythropoietic differentiation. In Stamatoyannopoulos, G., Nienhuis, A. eds.: Hemoglobins in Development and Differentiation. New York: Alan R. Liss, Inc., pp. 45–55.

Bailey, D.W. (1975) Genetics of histocompatibility in mice. I. New loci and congenic lines. Immunogenetics, *2*:249–256.

Clarke, B.J., A.A. Axelrad, M.M. Shreeve, and D.L. McLeod (1975) Erythroid colony induction without erythropoietin by Friend leukemia virus in vitro. Proc. Natl. Acad. Sci.

U.S.A., *72:*3556–3560.

Davisson, M.T., and T.H. Roderick (1981) The linkage map: 52nd Annual Report. Bar Harbor, The Jackson Laboratory (in press).

Elliott, R.W. (1979) Use of two-dimensional electrophoresis to identify and map new mouse genes. Genetics, *91:*295–308.

Felton, J., M. Meisler, and K. Paigen (1973) A locus determining β-galactosidase activity in the mouse. J. Biol. Chem., *249:*3267–3272.

Hankins, W.D., T.A. Kost, M.J. Koury, and S.B. Krantz (1978) Erythroid bursts produced by Friend leukemia virus in vitro. Nature, *276:*506–508.

Hankins, W.D., T.A. Kost, M.J. Koury and S.B. Krantz (1979) An in vitro system for erythroid transformation by Friend leukemia virus. In Baum, S.J., Ledney G.D., eds.: Experimental Hematology Today. New York: Springer, pp. 89–97.

Humphries, E.H., and H.M. Temin (1974) Requirement for cell division for initiation of transcription of Rous sarcoma virus RNA. J. Virol., *14:*531–546.

Liao, S.-K., and A.A. Axelrad (1975) Erythropoietin-independent erythroid colony formation in vitro by hemopoietic cells of mice infected with Friend virus. Int. J. Cancer, *15:*467–482.

Lilly, F. (1970) *Fv-2:* Identification and location of a second gene governing the spleen focus response to Friend leukemia virus in mice. J. Natl. Cancer Inst., *45:*163–169.

Mak, T., A. Axelrad, and A. Bernstein (1979) The *Fv-2* locus controls expression of Friend erythroleukemia virus-specific sequences in normal and infected mice. Proc. Natl. Acad. Sci. U.S.A., *76:*5809–5812.

Odaka, T., and T. Yamamoto (1962) Inheritance of susceptibility to Friend mouse leukemia virus. Jpn. J. Exp. Med., *32:*405–413.

Odaka, T. (1969) Inheritance of susceptibility to Friend mouse leukemia virus. V. Introduction of a gene responsible for susceptibility in the genetic complement of resistant mice. J. Virol., *3:* 543–548.

Odaka, T., and M. Matsukura (1969) Inheritance of susceptibility to Friend mouse leukemia virus. VI. Reciprocal alteration of innate resistance or susceptibility by bone marrow transplantation between congenic strains. J. Virol., *4:*837–843.

Odaka, T. (1970) Inheritance of susceptibility to Friend mouse leukemia virus. VII. Establishment of a resistant strain. Int. J. Cancer, *6:*18–23.

Paigen, K., M. Meister, J. Felton, and V. Chapman (1976) Genetic determination of the β-galactosidase developmental program in mouse liver. Cell, *9:*533–539.

Pincus, T., W.P. Rowe, and F. Lilly (1971) A major genetic locus affecting resistance to infection with murine leukemia viruses. II. Apparent identity to a major locus described for resistance to Friend murine leukemia virus. J. Exp. Med., *133:*1234–1241.

Shreffler, D.C. (1963) Linkage of the mouse transferrin locus. J. Hered., *54:*127–129.

Silver, J., and N. Teich (1981) Expression of resistance to Friend virus-stimulated erythropoiesis in bone marrow chimeras containing $Fv\text{-}2^{rr}$ and $Fv\text{-}2^{ss}$ bone marrow. J. Exp. Med. (in press).

Snell, G.D., and H.P. Bunker (1965) Histocompatibility genes of mice. V. Five new histocompatibility loci identified by congenic resistant lines on a C57BL/10 background. Transplantation, *3:*235–252.

Suzuki, S., and A.A. Axelrad (1980) *Fv-2* locus controls the proportion of erythropoietic progenitor cells (BFU-E) synthesizing DNA in normal mice. Cell, *19:*225–236.

Vaithilingam, D.S., and A.A. Axelrad Novel differentiation antigens on late erythrocytic progenitor cells CFU-E detected by alloantisera made between mouse strains congenic at the *Fv-2* locus. (submitted for publication).

Varmus, H.E., T. Podgett, S. Heasley, G. Simon, and J.M. Bishop (1977) Cellular functions are required for the synthesis and integration of avian sarcoma virus-specific DNA. Cell, *11:*307–319.

JOURNAL OF CELLULAR PHYSIOLOGY SUPPLEMENT 1:175–183 (1982)

# Sources and Biology of Regulatory Factors Active on Mouse Myeloid Leukemic Cells

**D. METCALF***

*Cancer Research Unit, Walter and Eliza Hall Institute of Medical Research, Royal Melbourne Hospital, Melbourne, 3050 Victoria, Australia*

The action of serum or cells in enforcing differentiation in mouse myelomonocytic leukemic cells was monitored in agar cultures of WEHI-3B leukemic cells. The repeated intravenous injection of 5 μg endotoxin initially increased serum differentiating activity but after the third injection responses to further injections decreased markedly. Congenitally athymic (nude) mice exhibited normal rises in serum differentiating activity when injected with endotoxin but C3H HeJ mice failed to respond to challenge with purified lipid A. Whole body irradiation up to 1,200 rads did not increase serum differentiating activity but did not suppress responses to challenge injection of endotoxin. Coculture of WEHI-3B cells with peritoneal cells from normal or irradiated BALB/c mice caused marked granulocytic differentiation in WEHI-3B colonies. This effect was not seen if leukemic cells were cultured with thymus, spleen, or bone marrow cells. The serum halflife of the factor in postendotoxin serum enforcing differentiation of WEHI-3B cells was shown to be 1.5–2.3 hr.

The first demonstration that undifferentiated leukemic cells could be induced to differentiate in vitro was achieved by adding DMSO to cultures of Friend erythroleukemic cells (see review, Friend, 1980). Subsequent studies showed that similar effects could be achieved by the use of a wide variety of unrelated chemicals. More recent studies have shown that undifferentiated mouse and human myeloid leukemic cell lines could be induced to differentiate by many of the same agents to form polymorphs and macrophages (Lotem and Sachs, 1974, 1980; Okabe et al., 1977; Sugiyama et al., 1979; Collins et al., 1978; de Both et al., 1981).

While these studies have been useful in providing simple systems for the induction of differentiation in leukemic cells, it has yet to be established whether such agents can exert similar effects on leukemic cells in vivo. Furthermore, this approach provides little information on whether leukemic cells can be induced to differentiate by specific factors occurring within the body, and, if so, on the nature and biology of these factors.

Ichikawa (1969, 1970) was the first to demonstrate that natural biological substances could induce undifferentiated Ml myeloid leukemic cells to differentiate, a process that was accompanied by a significant reduction in leukemogenicity of such cells. In subsequent studies using this leukemia and two similar mouse myeloid leukemias (R453 and WEHI-3B), it has been demonstrated that a variety of biological materials can a) induce differentiation of leukemic cells to form polymorphs and macrophages, b) suppress stem cell self-replication within the leukemic population, and c) reduce the leukemogenicity of pretreated leukemic cells.

The biological materials used to induce these effects include embryo and fibroblast-conditioned medium (Ichikawa, 1969 1970; Fibach and Sachs, 1976), spleen cell or macrophage-conditioned medium (Ichikawa, 1969; Hozumi et al., 1977; Lotem et al., 1980), lung-conditioned medium (Maeda and Ichikawa, 1973; Fibach and Sachs, 1976; Metcalf, 1979), ascites fluid from tumor-bearing animals (Hozumi et al., 1979), and serum from mice injected with endotoxin (Fibach and Sachs, 1974; Metcalf, 1980; Lotem et al., 1980; Burgess and Metcalf, 1980) or injected with a variety of antigens (Metcalf, 1981).

*To whom correspondence/reprint requests should be addressed.

As a background to a discussion of the nature of the active factors in such materials, it is necessary to summarise what is known about the factors controlling normal granulocyte (G) and macrophage (M) cells. Three types of factor can stimulate the proliferation and differentiation of normal mouse GM cells. a) M-CSF, present in L-cell-conditioned medium, yolk sac-conditioned medium or pregnant mouse uterus extract, has a selective capacity to stimulate M formation. The prototype M-CSF has been purified from L-cell-conditioned medium and is a glycoprotein of M.W. 70,000 (Stanley and Heard, 1977). b) GM-CSF, produced in vitro by all mouse tissues, has the capacity to stimulate both G and M formation. The prototype GM-CSF has been purified from mouse-lung-conditioned medium and is a glycoprotein of M.W. 23,000 (Burgess et al., 1977). Similar molecules have been characterized in a variety of other mouse-organ-conditioned media (Nicola et al., 1979), in mitogen-stimulated lymphocyte-conditioned medium (Burgess et al., 1980) and in medium conditioned by WEHI-3B cells (Williams et al., 1978). c) G-CSF was originally described in postendotoxin serum (Lotem et al., 1980; Burgess and Metcalf, 1980), and embryo-conditioned medium (Maeda et al., 1977). This molecule has yet to be purified but has a M.W. of 22,000–28,000 and has a selective capacity to stimulate G formation.

Tests on myeloid leukemic cells have shown M-CSF to be inactive in enforcing differentiation (Metcalf, 1981), purified GM-CSF to be active (Lotem and Sachs, 1978, Metcalf, 1979), and G-CSF-containing materials to have an exceptionally high capacity to induce differentiation—hence the alternative term for this factor, "differentiation factor DF" (Maeda et al., 1977; Lotem et al., 1980; Burgess and Metcalf, 1980; Nicola and Metcalf, 1981).

All of the crude biological materials so far reported to induce differentiation in mouse myeloid leukemic cells contain GM-CSF and/or G-CSF and their activity appears to be ascribable to their content of these two factors.

In the mouse, the sources of GM-CSF and the factors modifying GM-CSF levels have been extensively analysed (see reviews, Burgess and Metcalf, 1980a; Metcalf, 1981a). With the demonstration of the exceptional efficiency of G-CSF in inducing differentiation in leukemic cells, it has become necessary to carry out similar studies on G-CSF.

In a recent study (Nicola and Metcalf, 1981), it was demonstrated that fragments of lung, muscle, bone shaft, kidney, heart, thymus, and spleen all produce detectable levels of G-CSF when cultured in vitro. Prior injection of endotoxin before removal of the tissues not only led to an immediate increase in serum G-CSF levels (Metcalf, 1980; Burgess and Metcalf, 1980) but also to increases in the capacity of the above tissues to produce G-CSF in vitro. The production of G-CSF required active protein synthesis and levels of G-CSF rose progressively during a 7-day incubation period. The active materials produced by these different tissues exhibited comparable biological activity in stimulating G colony formation by normal marrow cells and inducing granulocytic differentiation in WEHI-3B leukemic cells. All appeared to be macromolecules of M.W. 22,000–28,000 with similar behavior on $(NH_4)_2SO_4$ fractionation and affinity or hydrophobic column chromatography.

The provisional conclusions from these studies are that G-CSF exhibits a generally similar biology to GM-CSF, being produced by many tissues and with serum levels reflecting fluctuations in tissue content and production of the factor.

The present studies were undertaken to obtain further information on the cell types producing G-CSF, the half-life in the serum of G-CSF, and the effects of endotoxin in modifying serum G-CSF levels.

## MATERIALS AND METHODS

### Mice

Mice used were 3-month-old male and female mice of the strains C57BL and BALB/c.

### Production of post-endotoxin serum

Mice were injected intravenously with 5 μg endotoxin (*E. coli* W, lipopolysaccharide WO III, B4 Difco) in 0.2 ml saline. For routine purposes, mice were bled from the axilla 3 hours after injection and, after separation, the serum stored at −15°C prior to use.

### Whole body irradiation

Mice were irradiated using a Philips RT250 unit operating at 250 Kv and 15 mA. HVL was 0.8 mm copper and a focal skin distance of 50 cm was used with full backscatter conditions. The dose rate was 127 rad/min.

### Collection of peritoneal and other cells

BALB/c mice were killed by cervical dislocation then injected intraperitoneally with 4 ml of sterile balanced salt solution. The abdomen was opened and the peritoneal cell suspension harvested using a sterile Pasteur pipette. Bone marrow cell suspensions were prepared in balanced salt solution by oral pipetting of femur marrow plugs. Spleen and thymus cell suspensions were prepared using a fine metal sieve. All cell suspensions were washed by centrifugation and resuspended in Eisen's balanced salt solution. Viable cell counts were performed using eosin dye exclusion.

### WEHI-3B leukemia

The WEHI-3B subline was derived by serial transplantation from a myelomonocytic leukemia developing in a mineral oil-injected BALB/c mouse (Warner et al., 1969). The cell line was adapted to continuous culture in vitro and cloned. The WEHI-3B subline maintained in this laboratory grows in suspension culture as a population of undifferentiated blast cells and promyelocytes. In agar medium the cells form large colonies composed of similar cells, the plating efficiency varying from 30–70% (Metcalf, 1979, 1980).

The present studies were performed using WEHI-3B cells grown in continuous liquid culture in Dulbecco's modified Eagle's medium with 10% FCS and containing from 0.5 to 2.0 $\times 10^6$ cells/ml.

### Agar cultures

All semisolid cultures were prepared as 1 ml cultures in 35 mm plastic Petri dishes using an equal volume mixture of double-strength Dulbecco's modified Eagle's medium and 0.6% agar in distilled water.

The formula of the modified Eagle's medium was Dulbecco's modified Eagle's medium HG-16 instant tissue culture powder (Grand Island Biological Co., New York) 10g; 390 ml double glass-distilled water; 3 ml L-asparagine, 6.7 mg/ml (final concentration in medium 20 μg/ml; 1.5 ml DEAE-dextran, 50 mg/ml (final concentration in medium 75 μg/ml) (Pharmacia, Uppsala, Sweden, M.W. 500,000); 0.575 ml penicillin ($2 \times 10^5$ IU/ml); 0.375 ml streptomycin (200 mg/ml); 4.9 g $NaHCO_3$; and 250 ml fetal calf serum.

WEHI-3B cells were added to the liquid agar medium (300 cells/ml) and 1 ml volumes of the agar medium pipetted into each culture dish containing 0.1 ml of test serum or other additives. After mixing, the cultures were allowed to gel, then incubated for 7 days at 37°C in a fully humidified atmosphere of 10% $CO_2$ in air.

Colonies were typed using an Olympus dissection microscope at × 35 magnification. At 7 days, cultures of WEHI-3B cells were free of any surviving cells in intercolony areas and all colonies were discrete aggregates containing from 50 to 3,000 cells. Based on previous studies (Metcalf, 1979, 1980) colonies were classified as "undifferentiated" if they consisted of tight spherical aggregates with no outlying cells, or "differentiated" if they exhibited a halo of loosely dispersed cells or were composed wholly of loosely dispersed cells.

Serum differentiating activity was expressed either as the percentage of differentiated colonies obtained with different serum dilutions or expressed in arbitrary units obtained by using data from the linear portion of the dose response curve and multiplying by the serum dilution. Thus if 0.1 ml of 1:10 dilution serum stimulated 40% differentiated colonies then the level of differentiating activity was expressed as 10 × 40 = 400 units.

## RESULTS

### Effects of repeated injections of endotoxin

Following a single intravenous injection of 5 μg endotoxin, the capacity of the serum to induce differentiation in colonies of WEHI-3B cells rises rapidly to a peak at 3–6 hours. However, within 24–48 hours, serum activity falls to the very low levels of activity seen in normal serum. To determine whether sustained elevations of serum differentiating activity could be achieved by repeated injections of endotoxin, groups of C57BL mice were injected intravenously with 5 μg endotoxin four times per week.

On each occasion, two sets of four mice were withdrawn from the group that had received previous injections of endotoxin. One set of four mice were pool bled without endotoxin injection and the other set injected with endotoxin, then pool bled three hours later. This procedure was repeated on each of 13 occasions. Following completion of the injection protocol, all sera were assayed for differentiating activity at dilutions of 1:1, 1:6, 1:60, and 1:600 in a single assay experiment using cultures of 300 WEHI-3B cells.

As shown in Figure 1, good responses were obtained following the first three injections of endotoxin but thereafter successive responses were weaker and after the seventh injection

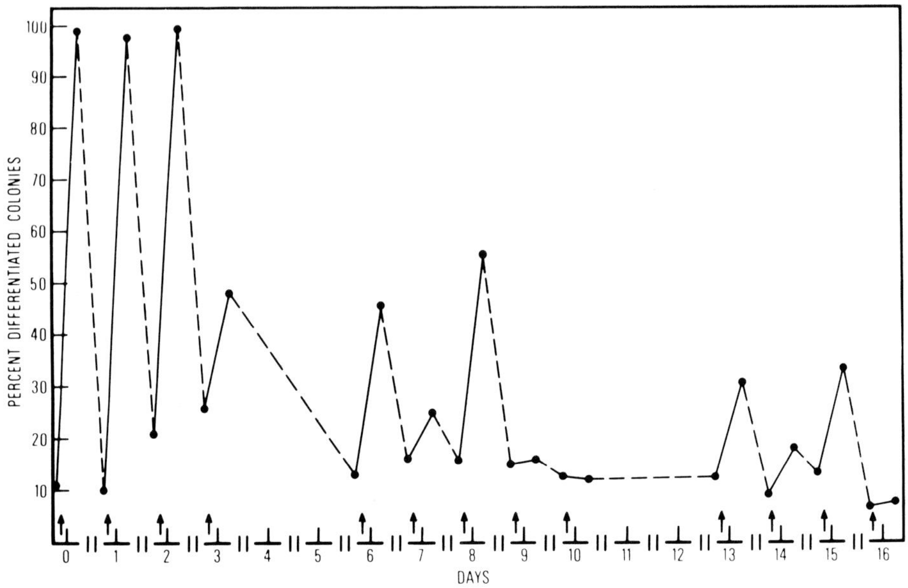


Fig. 1. Progressive fall in serum differentiating activity in mice given repeated intravenous injections of 5 μg endotoxin at times indicated by arrows. On each occasion, serum was assayed from four mice from the group prior to and 3 hours after the injection of endotoxin.

no significant rises in activity were observed when the three hour postinjection sera were assayed. It will also be noted from the preinjection data that at no time was there a sustained elevation of serum activity 24 hours after the preceding injection.

A similar sequence of depressed responsiveness to repeated endotoxin injections has been described previously for serum GM-CSF levels (Metcalf, 1974; Quesenberry et al., 1975), and this was confirmed in the present experiment when each serum was assayed in parallel for colony-stimulating activity in cultures of 75,000 C57BL bone marrow cells.

### Effect of whole body irradiation on serum differentiating activity

To determine the effects of whole body irradiation on serum differentiating activity, groups of 4-6 C57BL mice were irradiated with doses ranging from 150–1200 rads. Three hours after irradiation, these mice were pool bled and the sera assayed for differentiating activity in cultures of 300 WEHI-3B cells. As shown in Table 1, prior irradiation had no significant influence on the detectable level of serum differentiating activity.

However, irradiation did not prevent mice from responding to a subsequent intravenous injection of 5 μg endotoxin made immediately after irradiation. When sera were collected 3 hours after the challenge injection of endotoxin, levels of differentiating activity were as high as in control unirradiated mice injected with endotoxin.

### Response to endotoxin of genetically abnormal mice

In view of the fact that T-lymphocytes have been shown to be able to synthesise GM-CSF and similar regulatory molecules (Parker and Metcalf, 1974, 1974a), it was possible that T-lymphocytes might play a significant role in the rise in serum levels of differentiating activity following the injection of endotoxin.

To test this possibility, groups of BALB/c *nu/nu* (congenitally athymic) mice were injected intravenously with 5 μg endotoxin or 0.2 ml saline. The sera were collected 3 hours after injection and assayed in cultures of 300 WEHI-3B cells. As shown in Figure 2, *nu/nu* mice exhibited a normal response to the injection of endotoxin and exhibited obvious elevations in serum differentiating activity. These elevations were essentially identical in magnitude to those seen in C57BL or BALB/c mice following the injection of endotoxin (Metcalf, 1980).

In a previous study, the injection of 0.5 mg lipid A, purified from endotoxin, was shown to be as effective as endotoxin in elevating serum differentiating activity (Metcalf, 1980). C3H/

TABLE 1. Effect of whole-body irradiation on capacity of serum to induce differentiation in WEHI-3B colonies[1]

| Irradiation dose (rads) | Injected with | GM colony stimulating activity[2] | Percentage differentiated colonies with serum diluted[3] 1:6 | 1:60 | 1:600 |
|---|---|---|---|---|---|
| 0 | Saline (0.2 ml) | 0 | 13 | 4 | 2 |
| 150 | | 0 | 28 | 2 | 4 |
| 250 | | 3 | 35 | 2 | 5 |
| 450 | | 0 | 11 | 2 | 4 |
| 650 | | 2 | 27 | 4 | 1 |
| 850 | | 2 | 9 | 0 | 4 |
| 1,200 | | 4 | 24 | 0 | 0 |
| 0 | Endotoxin (5 μg) | 90 | 100 | 29 | 1 |
| 150 | | 54 | 100 | 48 | 11 |
| 250 | | 72 | 100 | 62 | 5 |
| 450 | | 90 | 100 | 65 | 5 |
| 650 | | 96 | 100 | 63 | 18 |
| 850 | | 57 | 100 | 78 | 13 |
| 1,200 | | 72 | 100 | 92 | 19 |

[1]Groups of 4 C57BL mice injected I.V. with 0.2 ml saline or 5 μg endotoxin immediately after irradiation. Serum collected and pooled 3 hours after injection.
[2]Mean number of granulocyte-macrophage colonies stimulated by 0.1 ml of 1:6 dilution serum in duplicate cultures of 75,000 C57BL bone marrow cells.
[3]Mean percentage of differentiated colonies in duplicate cultures of WEHI-3B cells containing 0.1 ml serum at dilutions indicated.

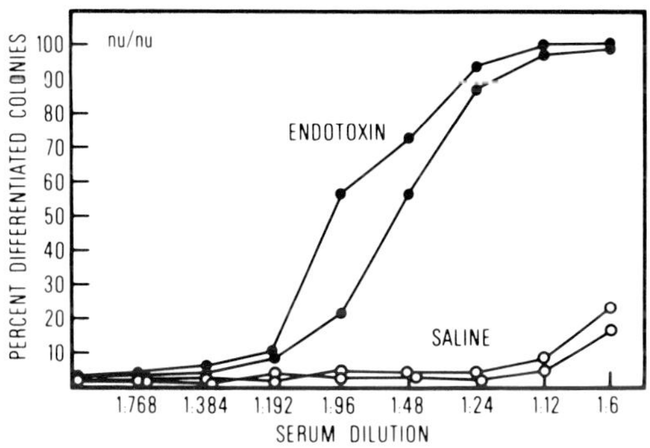


Fig. 2. Elevated differentiating activity for WEHI-3B cells of serum from athymic BALB/C nu/nu mice injected intravenously 3 hours previously with 5 μg entotoxin. Mean data from six mice per group.

TABLE 2. Effect of lipid A on serum differentiating activity[1]

| Strain | Percent differentiated WEHI-3B colonies with serum diluted 1:6 | 1:60 | 1:600 |
|---|---|---|---|
| C3H/WEHI | 99 | 98 | 18 |
| C3H/HeJ | 16 | 0 | 0 |

[1]Groups of 4–6 mice injected intravenously with 0.5 mg purified lipid A and serum collected and pooled 3 hours after injection. Sera assayed in serial dilutions in cultures of 300 WEHI-3B cells.

HeJ mice are genetically unresponsive to challenge with endotoxin or lipid A as assessed by rises in GM-CSF levels (Apte and Pluznik, 1976). As shown in Table 2, when C3H HeJ mice were injected with 0.5 mg lipid A, they also failed to exhibit elevated levels of serum differentiating activity for WEHI-3B cells compared with the level of activity in sera from normally responsive C3H/WEHI mice injected with 0.5 mg lipid A.

## Effects of cocultivation of WEHI-3B cells with normal cells

Agar cultures were prepared in which $10^6$ normal BALB/c thymus or peritoneal cells were cocultured with 300 WEHI-3B cells. Control cultures showed that neither normal cell population formed colonies in the type of agar medium used. As shown in Table 3, addition of thymus cells did not induce differentiation in WEHI-3B colonies but did not interfere with the differentiation occurring if 0.1 ml of 1:6 dilution postendotoxin serum was also added to the cultures. Indeed, differentiation to purely granulocytic colonies appeared to be slightly increased in the presence of the thymus cells.

Addition of $10^6$ BALB/c peritoneal cells caused a marked differentiation of WEHI-3B colonies. Addition of postendotoxin serum had little further effect on this differentiation. A notable feature of the differentiation occurring in the presence of peritoneal cells was the very high proportion of colonies wholly composed of differentiating granulocytes. When decreasing numbers of peritoneal cells were cocultured with the WEHI-3B cells, it was found that ad-

TABLE 3. Influence of added thymus or peritoneal cells on differentiation of WEHI-3B colonies[1]

| Added to cultures | Percent differentiated colonies | Percent colonies | | | | | |
|---|---|---|---|---|---|---|---|
| | | U | U + M | U + M + G | M | G + M | G |
| Saline 0.1 ml | 0 | 100 | 0 | 0 | 0 | 0 | 0 |
| Endotoxin serum 0.1 ml | 89 | 11 | 11 | 17 | 7 | 21 | 33 |
| $10^6$ thymus cells | 0 | 100 | 0 | 0 | 0 | 0 | 0 |
| $10^6$ thymus cells + endotoxin serum 0.1 ml | 99 | 0 | 0 | 8 | 0 | 17 | 75 |
| $10^6$ peritoneal cells | 95 | 3 | 0 | 3 | 0 | 10 | 84 |
| $10^6$ peritoneal cells + endotoxin serum 0.1 ml | 100 | 0 | 0 | 0 | 4 | 7 | 89 |

[1]Cultures contained 300 WEHI-3B cells with or without added BALB/c thymus or peritoneal cells. Differentiated colonies assessed by gross morphology and composition determined from stained preparations of 45 sequential colonies. U, undifferentiated blast and promyelocytic cells; G, granulocytes at varying stages of differentiation; M, macrophages.

dition of fewer than $10^5$ cells usually resulted in no significant differentiation in WEHI-3B colonies. Comparison of the effectiveness of the peritoneal cells from normal mice versus cells from mice injected 3 hours previously with endotoxin produced variable results. In the example shown in Figure 3, normal cells appeared more efficient than cells from endotoxin-injected mice, but in other experiments the results were reversed.

Preirradiation of donor mice with 850 rads whole-body irradiation did not suppress the ability of peritoneal cells collected 1 hour later to induce differentiation in WEHI-3B colony cells. When $10^6$ bone marrow, spleen, or thymus cells from preirradiated mice were cocultured with WEHI-3B cells, no colony differentiation was observed, but again the addition of these cells did not suppress the ability of postendotoxin serum to induce colony differentiation in the same cultures.

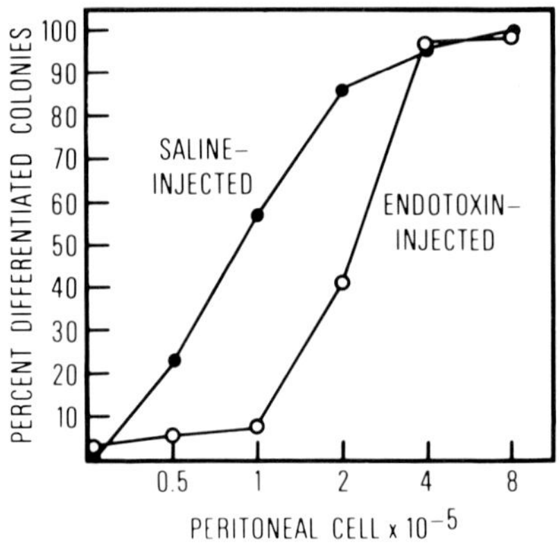


Fig. 3. Effects of coculture of various numbers of BALB/c peritoneal cells on colony differentiation in cultures of 300 WEHI-3B cells. Donor mice were injected 3 hours previously with 5 μg endotoxin or 0.2 ml saline. Cells pooled from 12 donor mice. Mean data from four replicate cultures.

## Half-life of serum differentiating activity

The rapid fall in serum differentiating activity in the interval from 6 to 24 hours after endotoxin injection suggested that the half-life of the active material in the serum must be relatively short. To measure this activity, C57BL mice were injected intravenously with 0.5 ml of pooled C57BL serum collected 3 hours after the injection of endotoxin. At intervals from 1–6 hours after injection, groups of four mice were bled and each serum assayed separately for differentiating activity in cultures of 300 WEHI-3B cells.

The results of one such experiment are shown in Figure 4 in which it can be seen that levels of detectable differentiating activity fell with an apparent half-life of 1.8 hours. In replicate experiments using different pools of

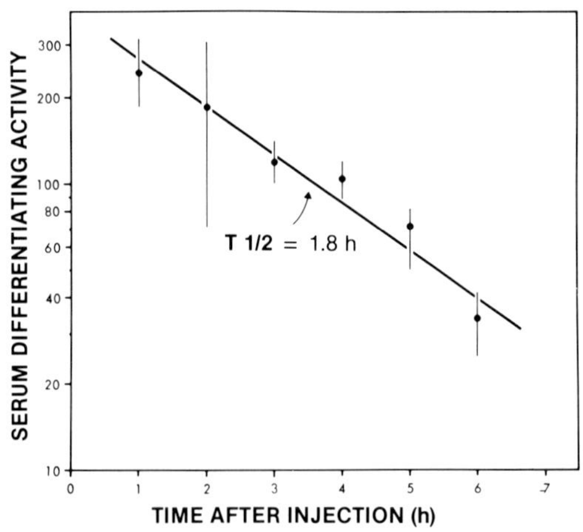


Fig. 4. Fall with time in level of differentiating activity in serum of C57BL mice injected intravenously with 0.5 ml of 3 hour postendotoxin serum. Mean data from 3–4 mice per timepoint. Vertical bars are standard deviations for four replicate cultures.

postendotoxin serum the observed halflife varied from 1.5–2.3 hours.

### Differentiating activity in media conditioned by tumor cells or in the serum of tumor-bearing mice

WEHI-3B cells constitutively produce GM-CSF and factors stimulating the proliferation in vitro of eosinophil, megakaryocyte, and erythroid cells. Assays of concentrated medium conditioned by WEHI-3B cells indicated that the media were able to induce differentiation in WEHI-3B cells.

In contrast, media harvested from cultures of 14 other neoplastic mouse cell lines (including T-lymphomas, fibrosarcoma, erythroleukemic, and macrophage cell lines) failed to exhibit differentiating activity.

In an effort to determine whether the functional activity of WEHI-3B cells would result in vivo in elevated serum levels of differentiating activity, groups of BALB/c mice were injected subcutaneously with $10^6$ WEHI-3B cells. Control BALB/c mice were injected with similar numbers of three other mouse tumor cell lines (2 T-lymphomas and 1 macrophage tumor). Recipient mice developed subcutaneous tumors 1–2 cm in diameter within 3–4 weeks of transplantation and at this time they were killed and the serum collected. Sera were assayed at dilutions of 1:6, 1:60, and 1:600 for differentiating activity in cultures of 300 WEHI-3B cells.

In no case were elevated levels of differentiating activity observed in sera from these tumor-bearing mice compared with levels in normal BALB/c serum.

## DISCUSSION

Previous evidence has indicated that G-CSF, the factor stimulating G colony formation by normal marrow cells, is the most active factor in inducing differentiation in mouse myeloid leukemic cells. This factor is synthesized by many different tissues (Nicola and Metcalf, 1981), but it is not known which specific cell types are involved in G-CSF synthesis. The present evidence confirms previous studies (Ichikawa, 1969; Hozumi et al., 1977) indicating that macrophages could be one important source of the G-CSF appearing in the serum following the injection of endotoxin. This conclusion is supported by the observed radioresistance of the response to endotoxin, its independence of T-lymphocytes and the evidence of Hultner et al. (1981) that serum GM-CSF responses to endotoxin can occur if unresponsive C3H/HeJ mice are injected with responsive marrow cells (containing macrophage precursors).

The present evidence does not of course exclude other radioresistant cells as sources of G-CSF. It indicates only that, in part, the serum rises in G-CSF after endotoxin injection may either be due to the synthetic activity of macrophages (Sullivan et al., 1981) or to a necessary involvement of macrophages in the process.

While endotoxin is highly efficient in elevating serum levels of differentiating activity, the present results show that the repeated use of this agent rapidly leads to a state of nonresponsiveness. The tissues involved in G-CSF synthesis are clearly capable of synthesizing elevated amounts of this factor over extended periods since sustained elevations of serum differentiating activity have been observed during inections in both mouse and man (Metcalf, 1981, 1981a). The failure of repeated injections of endotoxin to cause repeated elevations in G-CSF levels seems more likely to be due to an alteration in the distribution or half-life of repeatedly injected endotoxin such that the molecules no longer make effective contact with the cells producing G-CSF.

Freund's adjuvant does not appear to be effective in elevating serum G-CSF levels (Metcalf, 1981) and as shown in the present experiments injected G-CSF has a very short serum half-life. There is an urgent need therefore to identify or develop some agent capable of inducing sustained elevations in G-CSF levels so that the effects of this situation can be determined on the behavior in vivo of myeloid leukemic cells.

Fractionation studies on postendotoxin serum and organ conditioned media (Burgess and Metcalf, 1980; Nicola and Metcalf, 1981) have shown that, with all fractionation procedures, the factor inducing differentiation in myeloid leukemic cells cofractionates with the factor (G-CSF) stimulating granulocyte colony formation by normal mouse marrow cells. The conclusion that these factors are identical must remain provisional until chemically pure preparations become available. With this reservation, the data extend earlier evidence from studies on human myeloid leukemic cells (Metcalf et al., 1974; Francis et al., 1979) that the proliferation and/or differentiation of these leukemic populations retain a surprising degree of responsiveness to normal regulatory factors.

From what is so far known of the biology of GM-CSF and G-CSF production, microbial infections appear to be by far the most important modulators of CSF levels. It should follow therefore that infections, whether clinical or subclinical, must have a significant influence on the initiation and emergence of myeloid leukemic clones. Because local production of CSFs is not necessarily reflected in varying circulating CSF levels, such an influence might vary from tissue to tissue, e.g., bone marrow or spleen, according to which tissues respond most markedly to a particular infection.

CSFs have two potentially opposing effects on myeloid leukemic cells. On the one hand CSFs are likely to be necessary for the proliferation of leukemic cells and if CSF levels were very low, a leukemic clone might fail to proliferate. Conversely, high CSF levels would promote the proliferation of a leukemic clone. However, high CSF levels would tend to suppress a leukemic clone by restricting stem cell self-replication and enforcing differentiation. The interaction between these two opposing effects could be very subtle if it was also influenced by the relative ratios of GM-CSF to G-CSF since both are proliferative stimuli but GM-CSF appears relatively inefficient in inducing the differentiation of leukemic cells (Metcalf, 1979, 1980).

There is little available epidemiological information on the relation between infections and myeloid leukemia development. One early study in Rf mice indicated that germ-free mice failed to develop myeloid leukemia after whole body irradiation, but did so when the mice were subsequently "conventionalised" by bacterial contamination (Upton et al., 1966). Conversely there is anecdotal evidence from the prechemotherapy era (Forkner, 1938) that remissions in acute (? myeloid) leukemia sometimes occurred following severe infections.

In view of the recent in vitro studies on the effects of CSF on myeloid leukemic cells, it seems worthwhile to attempt careful epidemiological studies on the possible connection between infections and myeloid leukemia development. One group worthy of careful study are those patients with myeloproliferative disorders, some but not all of whom will subsequently develop myeloid leukemia. With the wider availability of germ-free mice and improved methods for inducing myeloid leukemia it also appears feasible to determine more precisely whether infections play a significant role in myeloid leukemia development.

In both types of study it will be important to assess CSF levels during the preleukemic period and the types of CSF involved since present evidence indicates that these are the molecules mediating the effects of infections on target granolopoietic populations.

## ACKNOWLEDGMENTS

This work was supported by the Carden Fellowship Fund of the Anti-Cancer Council of Victoria, the National Health and Medical Research Council, Canberra, the National Cancer Institute, Bethesda, grants CA-22556 and CA-25972, and a Royal Society Exchange Fellowship.

## LITERATURE CITED

Apte, R.N., and D.H. Pluznik (1976) Genetic control of lipopolysaccharide-induced generation of serum colony stimulating factor and proliferation of splenic granulocyte/macrophage precursor cells. J. Cell Physiol., *89:*313–324.

Burgess, A.W., and D. Metcalf (1980) Characterisation of a serum factor stimulating the differentiation of myelomonocytic leukemic cells. Int. J. Cancer, *26:*647–654.

Burgess, A.W., and D. Metcalf (1980a) The nature and action of granulocyte-macrophage colony-stimulating factor. Blood, *56:*947–958.

Burgess, A.W., J. Camakaris, and D. Metcalf (1977) Purification and properties of colony-stimulating factor from mouse lung-conditioned medium. J. Biol. Chem., *252:*1998–2003.

Burgess, A.W., D. Metcalf, S.H.M. Russell, and N.A. Nicola (1980) Granolocyte/macrophage-, megakaryocyte-eosinophil-, and erythroid- colony-stimulating factors produced by mouse spleen cells. Biochem. J., *185:*301–314.

Collins, S.J., F.W. Ruscetti, R.E. Gallagher, and R.C. Gallo (1978) Terminal differentiation of human promyelocytic leukemias induced by dimethylsulphoxide and other polar compounds. Proc. Natl. Acad. Sci. U.S.A., *75:*2458–2462.

De Both, N.J., A. Hagemeijer, E.H. Rhijnsburger, M. Vermey, E. Van T Hull, and E.M.E. Smit (1981) DMSO-induced terminal differentiation and trisomy 15 in myeloid cell line transformed by the Rauscher murine leukemia virus. Cell Differ. *10:*13–21.

Fibach, E., and L. Sachs (1974) Control of normal differentiation of myeloid leukemic cells. IV. Induction of differentiation by serum from endotoxin treated mice. J. Cell Physiol., *83:*177–186.

Fibach, E., and L. Sachs (1976) Control of normal differentiation of myeloid leukemic cells. XI. Induction of a specific requirement for cell viability and growth during the differentiation of myeloid leukemic cells. J. Cell Physiol., *89:*259–266.

Forkner, L.E. (1938) Leukemia and Allied Disorders. New York: Macmillan.

Francis, G.W., J.J. Berney, P.M. Chipping, and A.V. Hoffbrand (1979) Stimulation of human haemopoietic cells by colony-stimulating factors. Sensitivity of leukaemic cells. Br. J. Haematol., *41:*545–562.

Friend, C. (1980) Differentiation and Neoplasia. New York: Springer.

Hozumi, M., K. Takenaga, M. Tornida, and J. Okabe (1977) Role of leucocytes in ascites in the production of factor(s) stimulating differentiation of mouse myeloid leukemia cells. Gann, *68:*643–652.

Hozumi, M., T. Umezawa, K. Takenaga, T. Ohno, M. Shikita, and I. Yamane (1979) Characterization of factors stimulating differentiation of mouse myeloid leukemia cells from a Yoshida sarcoma cell line cultured in serum-free medium. Cancer Res., *39:*5127–5131.

Hultner, L., F.G. Staber, H.G. Mergenthaler, and P. Dormer (1981) Production of murine granulocyte-macrophage colony-stimulating factors (GM-CSF) by bone marrow-derived cells in vivo. Exp. Haematol. Suppl., *9:*10.

Ichikawa, Y. (1969) Differentiation of a cell line of myeloid leukemia. J. Cell Physiol., *74:*223–234.

Ichikawa, Y. (1970) Further studies on the differentiation of a cell line of myeloid leukemia. J. Cell Physiol., *76:*175–184.

Lotem, J., and L. Sachs (1974) Different blocks in the differentiation of myeloid leukemic cells. Proc. Natl. Acad. Sci. U.S.A., *71:*3507–3511.

Lotem, J., and L. Sachs (1978) In vivo induction of normal differentiation in myeloid leukemic cell. Proc. Natl. Acad. Sci. U.S.A. *75:*3781–3785.

Lotem, J., and L. Sachs (1980) Potential prescreening for therapeutic agents that induce differentiation in human myeloid leukemia cells. Int. J. Cancer, *25:*561–564.

Lotem, J., J.H. Lipton, and L. Sachs (1980) Separation of different molecular forms of macrophage- and granulocyte-inducing proteins for normal and leukemic myeloid cells. Int. J. Cancer, *25:*763–771.

Maeda, M., and Y. Ichikawa (1973) Production of growth- and differentiation-stimulating factors for mouse leukemia cells by different cell species. Gann, *64:*257–263.

Maeda, M., M. Horiuchi, S. Numa, and Y. Ichikawa (1977) Characterisation of a differentiation-stimulating factor for mouse leukemia cells. Gann, *68:*435–447.

Metcalf, D. (1974) Depressed responses of the granulocyte-macrophage system to bacterial antigens following preimmunisation. Immunology, *26:*115–125.

Metcalf, D. (1979) Clonal analysis of the action of GM-CSF on the proliferation and differentiation of myelomonocytic leukemic cells. Int. J. Cancer, *24:*616–623.

Metcalf, D. (1980) Clonal extinction of myelomonocytic leukemic cells by serum from mice injected with endotoxin. Int. J. Cancer, *25:*225–233.

Metcalf, D. (1981) Regulatory control of the proliferation and differentiation of normal and leukemia cells. Natl. Cancer Inst. Monogr. (in press).

Metcalf, D. (1981a) Hemopoietic colony stimulating factors. In Baserga, R. ed.: Tissue Growth Factors. New York: Springer pp. 343–384.

Metcalf, D., M.A.S. Moore, J.W. Sheridan, and G. Spitzer (1974) Responsiveness of human granulocytic leukemic cells to colony-stimulating factors. Blood, *43:*847–859.

Nicola, N.A., and D. Metcalf (1981) Biochemical properties of differentiation factors for murine myelomonocytic leukemic cells in organs conditioned media-separation from colony-stimulating factors. J. Cell Physiol, *109:*253–264.

Nicola, N.A., A.W. Burgess, and D. Metcalf (1979) Similar molecular properties of granolocyte-macrophage colony stimulating factors produced by different mouse organs. J. Biol. Chem., *254:*5290–5299.

Okabe, J., Y. Honma, and M. Hozumi (1977) Inhibition of RNA and protein synthesis makes nondifferentiating mouse myeloid leukemia cells sensitive to a factor(s) stimulating differentiation. Int. J. Cancer, *20:*933–940.

Parker, J.W., and D. Metcalf (1974) Production of colony-stimulating factor in mixed leucocyte cultures. Immunology, *26:*1039–1049.

Parker, J.W., and D. Metcalf (1974a) Production of colony-stimulating factor in mitogen-stimulated lymphocyte cultures. J. Immunol. *112:*502–510.

Quesenberry, P., J. Halperin, M. Ryan, and E. Stohlman (1975) Tolerance to the granulocyte-releasing and colony-stimulating factor elevating effects of endotoxin. Blood, *45:*789–800.

Stanley, E.R., and P.M. Heard (1977) Factors regulating macrophage production and growth. Purification and some properties of the colony stimulating factor from medium conditioned by mouse L cells. J. Biol. Chem., *252:*4305–4312.

Sugiyama, K., M. Hozumi, and J. Okabe (1979) Induction of differentiation of Rauscher virus-induced mouse myeloid leukemia cells with a factor(s) in ascitic fluid and inhibitors of nucleic acid and protein syntheses. Cancer Res., *39:*1056–1062.

Sullivan, F.R. P.J. Gans, and L.A. McCarroll (1981) Kinetics of colony-stimulating activity production by human monocytes in vitro. Exp. Hematol. Suppl., *9:*7.

Upton, A.C., V.K. Jenkins, H.E. Walburg, R.L. Tyndall, J.W. Conklin, and N. Wald (1966) Observations on viral chemical and radiation-induced myeloid and lymphoid leukemias in Rf mice. Natl. Cancer Inst. Monogr. *22:*329–347.

Warner, N.L., M.A.S. Moore, and D. Metcalf (1969) A transplantable myelomonocytic leukemia in BALB/c mice: Cytology, karyotype and muramidase content. J. Natl. Cancer Inst., *43:*963–968.

Williams, N., R.R. Eger, M.A.S. Moore, and N. Mendelsohn (1978) Differentiation of mouse bone marrow precursor cells into neutrophil granulocytes by an activity separation from WEHI-3 cell-conditioned medium. Differentiation, *11:*59–63.

JOURNAL OF CELLULAR PHYSIOLOGY SUPPLEMENT 1:185–194 (1982)

# Induction of Erythroid Tumorigenic Colonies by Friend Helper Virus F-MuLV Alone and Isolation of a New Class of Friend Erythroleukemic Cells

TSUNEFUMI SHIBUYA AND TAK W. MAK*
*The Ontario Cancer Institute, and Department of Medical Biophysics, University of Toronto, Toronto, Ontario M4X 1K9, Canada*

Friend erythroleukemia virus (FLV) is a complex of two viruses, a defective spleen focus virus component (SFFV) and a helper independent virus, F-MuLV. The virus complex induces in susceptible mice a multistage disease, an early stage (1–3 weeks) of rapid erythropoietic changes and a late stage, from which erythroleukemic cell lines (FLC) can be established. While the F-MuLV alone does not induce any detectable changes in adult mice, it has been found recently to induce, in newborn BALB/c and NIH/Swiss mice, splenomegaly and certain early erythropoietic changes. The present study indicates that if the infected mice were kept alive by red cell transfusion and the cells from these F-MuLV-induced enlarged spleens were transplanted repeatedly in adult mice, erythroid tumorigenic colonies (CFU-FV) can be detected. These helper virus of induced CFU-FV (CFU-FV-H) have properties that differed from those CFU-FVs induced by the FLV complexes. The colonies are more diffuse in morphology and do not release any detectable SFFV activities. In addition, FLC lines can also be established from these F-MuLV infected spleens and CFU-FV-H and have been in culture for eight months. They are found to be distinct from those established from CFU-FV induced by the FLV complexes. First, these FLC lines are more selective in the type of medium they can grow in, for while FLC lines induced by FLV complexes will grow in both α-medium and Iscove medium, the FLC induced by F-MuLV will only grow in Iscove medium and not in α-medium. Secondly, the growth of the F-MuLV induced FLC lines are cell density dependent and will not proliferate below $10^5$ cells/ml. Thirdly, the growth of the F-MuLV-induced FLC lines can be greatly enhanced by the addition of erythropoietin (epo). Finally, the F-MuLV induced FLC lines will not respond to either dimethylsulphoxide (DMSO) or epo to induce along the erythroid pathway. These studies indicate that similar to the disease induced by FLV complexes, the erythroleukemia induced by F-MuLV is also a multistage disease and a new class of erythroid tumorigenic cells and FLC lines can also be established from the spleens of these mice infected by the F-MuLV alone.

The mechanism of transformation by avian and murine type-C leukemia virus has been the subject of intense research for many years. Recent research in molecular biology has indicated that there are two classes of leukemia viruses, the weakly transforming leukemia virus and the rapidly transforming acute leukemia virus. The acute leukemia viruses are complexes of two viruses: a helper virus component that provides the replication functions, and a replication defective component that carries unique cellular sequences (Stephenson, 1980). It has been postulated that the weakly transforming viruses act by recombining with endogenous viral sequences to become more oncogenic agents (Hartley et al., 1977; Lerner, 1978), or that they can function by activation of cellular transforming genes as a consequence of integrating adjacent to these genes (Hayward et al., 1981; Noori-Daloir et al., 1981). The acute leukemia viruses, however, have been postulated to initiate transformation by activation of unique cellular sequences

*To whom correspondence/reprint requests should be addressed.

within the viral genomes (Stehelin and Graf, 1978; Mak et al., 1978; Shields et al., 1979). It is not known, however, what function, if any, the helper viruses play in the mechanism of transformation by these acute leukemia viruses.

Friend erythroleukemia virus (FLV) (Friend, 1957), like other acute leukemia viruses, is a complex of two viruses consisting of a replication defective spleen focus forming virus (SFFV) (Alexrad and Steeves, 1964) and a replication competent F-MuLV (Tambourin and Wendling, 1974; Steeves, 1975; Bernstein et al., 1977). The erythroleukemia induced by these FLV complexes involves rapid changes in erythropoiesis and splenomegaly (Friend, 1957; Axelrad and Steeves, 1964; Tambourin and Wendling, 1974; Steeves, 1975; Bernstein et al., 1977; MacDonald et al., 1980a). However, in spite of these rapid pathological changes, the early phase is not tumorigenic and the cells from the infected spleens cannot be established as permanent Friend erythroleukemia cell lines (FLC). It is not until late after infection (> 4 weeks) that spleen fragments can give rise to tumor in vivo and FLC in vitro (Friend and Haddad, 1960; Ostertag et al., 1972; Mager et al., 1980, 1981a; Wendling et al., 1981). Based on these observations, it appears that the erythroleukemia induced by FLV is a multistage disease, and it has been suggested that the late stage is the malignant phase (Mager et al., 1980, 1981a; Wendling et al., 1981).

Recent studies have also indicated that, like the FLV complexes, the helper virus F-MuLV can also induce splenomegaly in newborn BALB/c and NIH/Swiss mice but not in adult mice (Troxler and Scolnick, 1978; MacDonald et al., 1980b; Niho et al., 1982). The capability of the F-MuLV, however, to induce the tumorigenic phase and form established FLC cells is not known. In a recent study, Oliff et al. (1981) reported that if the lives of the F-MuLV infected newborn mice were prolonged by red cell transfusion, erythroleukemic cell lines can be established from the spleens of these infected animals. In the present studies, we have utilized a newly developed quantitative colony assay (Mager et al., 1981a), which can detect tumorigenic cells transformed by FLV (CFU-FV) to determine if the F-MuLV alone can induce a malignant transformation in newborn BALB/c mice. The results showed that erythroleukemic cell lines and tumorigenic colonies (CFU-FV) (Mager et al., 1981a) can be induced in the spleens of these newborn mice. In addition, continuous erythroleukemic cell cultures could be established from these CFU-FVs. Moreover, they were found to be distinct from those established from spleens infected with FLV complexes of both the anemia (FV-A) or polycythemia strain (FV-P) (Mager et al., 1981a,b).

## MATERIALS AND METHODS

### Viruses

The replication competent NB-tropic Friend F-MuLV was cloned by limiting dilution as described previously (Bernstein et al., 1977; Mak et al., 1978). Polycythemia and anemia stocks of Friend virus complexes used were described by MacDonald et al. (1980a).

### Mice

BALB/c mice were obtained from Jackson Laboratory (Bar Harbor ME). Newborn mice were obtained from the animal colony facilities of the Ontario Cancer Institute, Toronto.

### Colony assay for Friend virus induced tumorigenic cells (CFU-FV)

The assay for CFU-FV was performed as described previously (Mager et al., 1981a,b). Briefly, cells ($1 \times 10^6$/ml) from individual spleens were cultured in 1.3% methylcellulose in Iscove modified Dulbecco medium with 30% heat-inactivated fetal calf serum (Flow Laboratory, Inc., Rockville, MD) and $1 \times 10^{-5}$ M 2-mercaptoethanol. Dishes were incubated at 37°C for 12–14 days at which time macroscopic colonies (containing $10^4$–$10^5$ cells) were scored by eye.

### Assay of hematocrit

Blood samples were taken from tail vein every week to heparinized hematocrit tubes (DADE Division, American Hospital Supply Co., Miami, FL). This amount of bleeding was confirmed not to affect the hematocrit of normal mice.

## RESULTS

### Induction of pathological and cellular changes by clone F-MuLV

Previous studies have indicated that F-MuLV can induce in newborn BALB/c mice splenomegaly and certain erythropoietic changes (Troxler and Scolnick, 1978; MacDonald et al., 1980b; Niho et al., 1982). In order to examine whether these F-MuLV infected animals only exhibit certain erythropoietic changes or whether they can progress into the malignant stage, we injected 3 × 10$^5$ XC units of cloned F-MuLV (Bernstein et al., 1977) into one day old newborn BALB/mice. The spleen weight and hematocrit values of these mice and the ability of their spleen cells to form tumorigenic colonies were monitored. These stocks of cloned F-MuLV had been shown repeatedly for several years by biological or biochemical assays to contain only the helper virus (Bernstein et al., 1977; Mak et al., 1978; MacDonald et al., 1980a,b). The data obtained from these studies, summarized in Figure 1, indicate that, as previously reported, increased spleen weight can be detected 5 to 6 weeks after injection of the cloned F-MuLV (Fig. 1a).

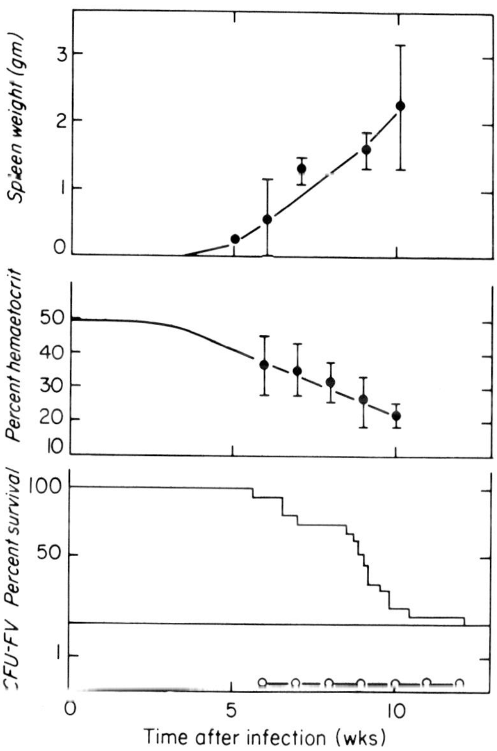


Fig. 1. Pathological and biological changes following infection of newborn BALB/c mice by cloned F-MuLV. The 3 × 10$^5$ XC units of cloned F-MuLV were injected into one day old BALB/c mice and their spleen weights, hematocrits, survival rates, and CFU-FV colonies were monitored. Top panel: mice were sacrificed and spleen weights were determined. Second panel: hematocrits were measured through tail veins. Third panel: survival rates were determined. Bottom panel: the spleens were removed and 5 × 10$^6$ cells each were plated and the CFU-FVs were counted 14 days later. For experimental details, see Mager et al. (1981a,b).

This increase in spleen weight is also accompanied by an onset of anemia with hematocrits dropping to 20% or less (Fig. 1b). This decrease in hematocrits was accompanied by an increase in the mortality rate in these infected mice (Fig. 1c). However, with this splenomegaly and severe reduction of hematocrits, no tumorigenic colonies (CFU-FV) can be detected in spleens of these mice before their deaths, usually before 12 weeks (Fig. 1d).

### Induction of tumorigenic erythroleukemic cells and CFU-FV by cloned F-MuLV

It is possible that no tumorigenic cells can be established because of early deaths of these infected mice. In order to increase the time that may be necessary for the evolution of these tumorigenic cell clones, we have repeatedly passaged the cells in the enlarged spleens to uninfected adult BALB/c mice. Figure 2 illustrates one such experimental scheme of transplantations of the F-MuLV infected spleen cells. As can be seen in Figure 2, with the injection of F-MuLV into the newborn BALB/c mouse, splenomegaly was detected. In an attempt to prolong the life of this mouse, we initiated transfusion of red blood cells from uninfected mice, similar to those described by Oliff et al. (1981). At biweekly intervals from 10–14 weeks after infection, 0.5 ml of packed red blood cells from uninfected BALB/c mice were injected intravenously into the infected mouse. The cells from this enlarged spleen was then transplanted into uninfected BALB/c adult mice. Results (Fig. 2) showed that splenomegaly developed in these recipient mice within 3 weeks of transplantation. After two such serial transplants, the spleens were cultured for FLC lines and were found to contain these tumorigenic erythroleukemic cells (Fig. 2). In an attempt to examine if the tumorigenic colonies CFU-FV can be detected in these F-MuLV infected mice, 5 × 10$^6$ spleen cells each were plated for CFU-FV as described before (Mager et al., 1980a). Results in Figures 2 and 3 indicate that macroscopic tumorigenic CFU-FV can be detected in these spleens. The frequency of these colonies from one such mouse was about 8 (6,7,10,11, and 7) CFU-FVs per 5 × 10$^6$ spleen cells. A picture showing the methylcellulose Petri dish with CFU-FV colonies and a photograph of the edge of an individual colony are shown in Figure 3 upper left and upper right panels, respectively. These results indicate that CFU-FV cells developed in F-MuLV infected newborn susceptible mice

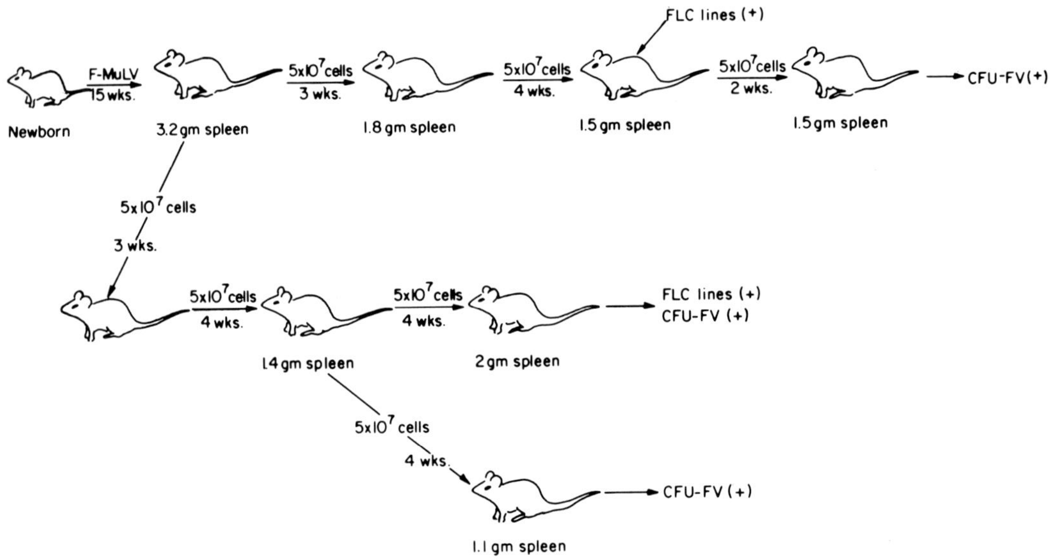


Fig. 2. Detection of Friend erythroleukemic cell lines and CFU-FVs in mice transplanted repeatedly with spleen cells from F-MuLV infected newborn mice. A newborn BALB/c mouse was infected with $3 \times 10^5$ XC plaque units of cloned F-MuLV. At the times noted, $5 \times 10^7$ cells from the enlarged spleens were transplanted to uninfected adult (6–8-week-old) BALB/c mice. The spleens from these mice were assayed for CFU-FV and cultured for FLC lines as described in Materials and Methods. The plus signs denote when these tumorigenic cells were detected.

TABLE 1. Properties of erythroid tumorigenic colonies (CFU-FV) induced in spleens of newborn BALB/c mice infected with cloned F-MuLV[1]

| Type of tumorigenic colonies (CFU-FV) | Susceptible host | Colony morphology | Viruses released | | |
|---|---|---|---|---|---|
| | | | FV-A | FV-P | F-MuLV |
| CFU-FV-H[1] (F-MuLV) | newborn only | loose or tight | – | – | + |
| CFU-FV-A[2] | newborn or adult | tight | + | – | + |
| CFU-FV-P[2] | newborn or adult | tight | – | + | + |

[1]Newborn 1 day old BALB/c mice were injected with $3 \times 10^5$ XC-units (Rowe et al., 1970) of cloned F-MuLV (Bernstein et al., 1977). After 8–10 weeks the infected mice were transfused with 0.5 ml of erythrocytes from uninfected BALB/c mice twice weekly. Five weeks later the mice were sacrificed and the spleens were plated for CFU-FV as described (Mager et al., 1981a,b). Colonies with loose morphology were picked and cultured in Iscove modified medium (Iscove et al., 1980) with 20% fetal calf serum for 5 days. Then 0.5 ml of the supernatants were injected into adult DBA/2 mice and the presence of spleen foci and hematocrit levels in the mice were examined. XC plaque assays were determined as described (Rowe et al., 1970).

[2]Data taken from present study and Mager et al., 1981a,b.

and emerged after a prolonged period in the spleens. To determine if these cells were tumorigenic, $3.5 \times 10^7$ spleen cells from the infected female BLAB/c mice were injected intraperitonally into an eight-week-old adult male BALB/c mouse. Six weeks later, the mouse was sacrificed and found to have splenomegaly. Moreover, the spleen cells from this recipient adult mouse also formed CFU-FV in methylcellulose at a markedly higher frequency of about 50 CFU-FV/$5 \times 10^6$ spleen cells. These results indicate that the CFU-FV from the F-MuLV infected BALB/c mice were tumorigenic.

### Comparison of the properties of CFU-FVs obtained from F-MuLV, FV-A, or FV-P infected mice

We next studied the nature of these CFU-FVs and compared them with those obtained from spleens of adult mice infected with the FLV complexes, FV-P or FV-A. Results of these studies are illustrated in Figure 3 and Table 1. The morphology of the colonies and

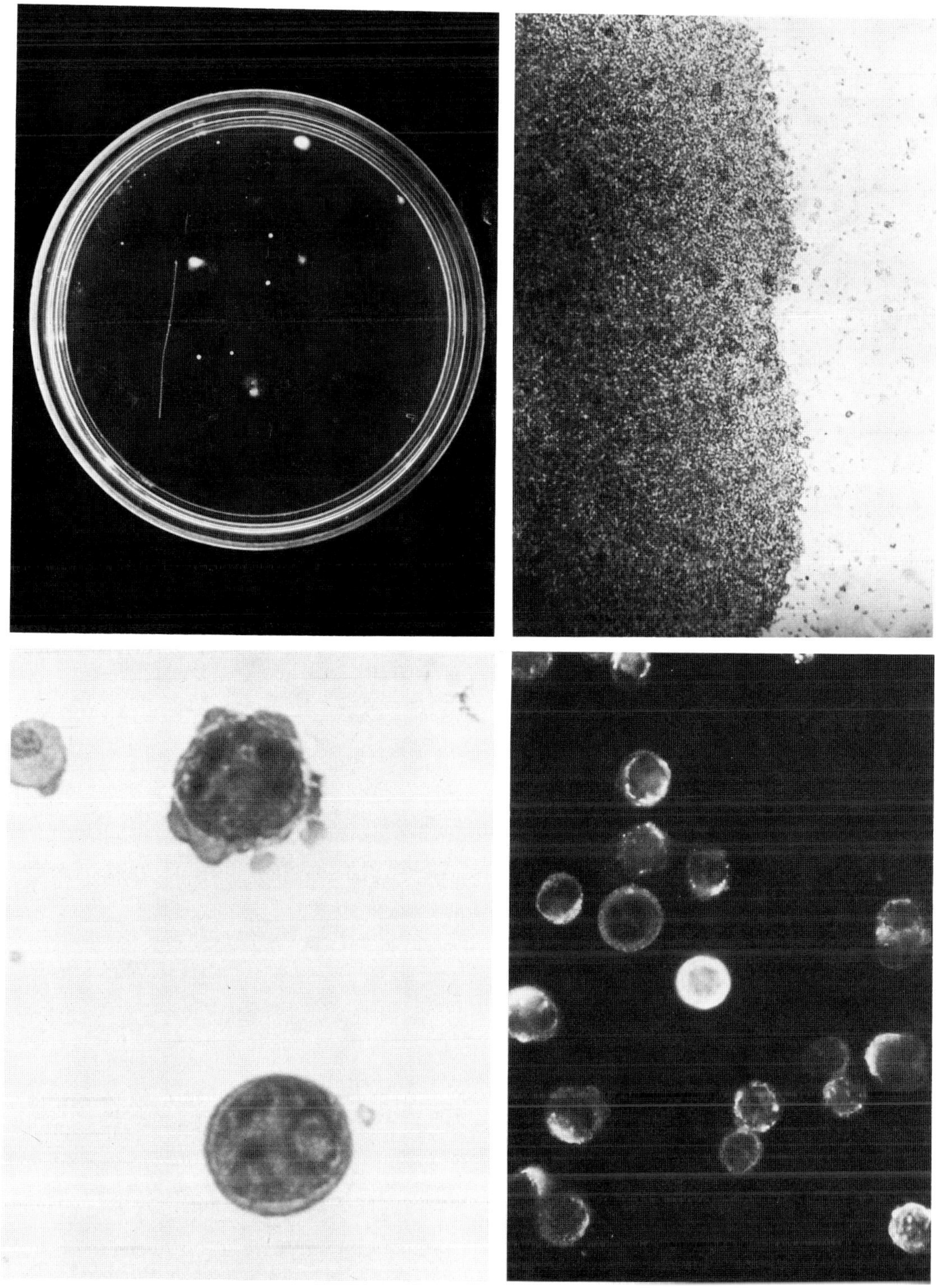

Fig. 3. Detection of F-MuLV induced erythroleukemic CFU-FVs and morphological and immunological characterization of these cells. Upper left panel: a 60 mm Petri dish 14 days after plating $5 \times 10^6$ spleen cells from a mouse infected with cloned F-MuLV at birth. Upper right panel: high magnification of a colony derived from the same experiment as in upper left panel. Lower left panel: morphology of cells established from these colonies stained with Wright-Giemsa stain. Lower right panel: detection of spectrin by immunofluorescence on cells established from CFU-FV as described by Mager et al. (1981b).

TABLE 2. Properties of friend erythroleukemic cells established

| Erythroleukemic cells established from CFU-FV | Growth in medium | | Spectrin positive cells | Benzidine positive cells | Inducibility by | |
|---|---|---|---|---|---|---|
| | IMDM[3] | α-medium[4] | | | DMSO | Epo |
| CFU-FV-H[1] (F-MuLV) | + | − | + + | − | − | − |
| CFU-FV-A[2] (FV-A) | + | + | + + | − | − | + |
| CFU-FV-P[2] (FV-P) | + | + | + + + | + | + | − |

[1]CFU-FV induced by F-MuLV in spleens of BALB/c mice were cultured in Iscove modified Dulbecco medium (IMDM) supplemented with 20% fetal calf serum for 3 months. The percent of spectrin positive cells was determined by immunofluorescence staining with antimouse spectrin obtained from Dr. H. Eisen (Eisen et al., 1977; Mager et al., 1981b). Benzidine positive cells were determined by staining with 1% benzidine. Studies of inducibility and growth in erythropoietin were carried out as described in Figure 3 with 1% dimethylsulphoxide (DMSO) or 1 unit/ml of erythropoietin (epo) (Mager et al., 1981b).
[2]Data obtained from this study and taken from Mager et al., (1981a,b).
[3]IMDM, Iscove modified Dulbecco medium (Iscove et al., 1980).
[4]Stanners et al., 1971.

of the cells in these colonies was examined. Unlike the CFU-FV obtained from FV-P or FV-A infected spleens, which had a tight and compact appearance (Mager et al., 1981a,b), the morphology of these F-MuLV induced CFU-FVs was more diffuse in appearance (Fig. 3, upper right panel). In morphology, these cells resembled erythroblasts when examined with Wright-Giemsa stain (Fig. 3, lower left panel). Studies of the CFU-FV colonies induced by FLV complexes, FV-A and FV-P, indicate that the tumorigenic colonies released FV-A and FV-P, respectively (Mager et al., 1981a,b). In order to determine whether the F-MuLV induced colonies also release $SFFV_A$ or $SFFV_P$, we picked five colonies and cultured them in microwells for a period of 5 to 7 days and the supernatants of these cultures were assayed for FV-A and FV-P in adult DBA/2 mice (Axelrad and Steeves, 1964; MacDonald et al., 1980a) as well as for XC plaque-forming units (Rowe et al., 1970). Results indicated that none of the colonies examined released any detectable SFFVs although XC plaque-forming viruses were found in all the cultures. We next examined if these colonies were erythroid in nature. To this end, we stained the cells in the colonies for the presence of benzidine positive cells and the erythroid specific protein spectrin (Mager et al., 1981b). Results indicate that although no benzidine positive cells were detected, a high percentage of these cells contained spectrin. Thus, like the CFU-FVs induced by the Friend complexes, FV-A and FV-P, these newly described CFU-FVs are also erythroid in nature. Previous studies of the CFU-FV indicated that erythroid permanent cell lines could be established from a high frequency of these tumorigenic colonies (Mager et al., 1981a,b). In an attempt to establish cell lines from the F-MuLV induced CFU-FV, individual colonies were picked and cultured in two different media: α-medium (Stanners et al., 1971) supplemented with 20% fetal calf serum (FCS) and Iscove modified Dubecco medium (Iscove et al., 1980) supplemented with 20% FCS. We found that although some of the cells from these CFU-FV grew in the Iscove modified medium, they uniformly failed to grow in the α-medium supplemented with FCS after 11 trials (Table 2). This is in contrast to the tumorigenic cells from FV-A or FV-P induced CFU-FV, which grew well in this medium (Mager et al., 1981a,b). The cells from these F-MuLV induced CFU-FV were continuously cultured in Iscove medium for more than 8 months and had remained predominantly erythroid (0% benzidine positive and 95% spectrin positive, Fig. 3, lower right panel). However, although these cells have some properties similar to FLC cells established from FV-P and FV-A infected spleens, they differ with respect to some other properties.

One of the properties of FLC cells that has been the subject of intense investigation for many years is their ability to respond in culture to certain inducers to differentiate along the erythroid pathway. Previous studies have shown that FLC cell lines established from FV-P infected spleens will respond to dimethylsulfoxide (DMSO) (Friend et al., 1971) and other inducers (Marks and Rifkind, 1978), while FLC cell lines established from FV-A infected spleens will respond to the hormone erythropoietin (epo) (Mager et al., 1981b). To investigate if these F-MuLV induced cells will respond to induction by DMSO or epo, 1% DMSO or 1 unit/ml of epo were added to one such cell culture at $1 \times 10^5$ or $2 \times 10^5$ cells per ml and the total number of viable cells and percent of hemoglobin containing cells, as measured by benzidine staining, were monitored. The results of these experiments are illustrated in Figure 4. The data showed that after 3 or 5 days of incubation in the presence of DMSO or epo, no benzidine positive cells were detected, indicating that these cells did not respond to induction by these inducers (Table 2, Fig. 4). However, when the total viable cells were examined, it was found that their growth was cell density dependent (growth was detected only at cell densities of $2 \times 10^5$ cell/ml or more (Fig. 4). In addition, at $2 \times 10^5$ cells/ml they grew at a significantly faster rate in the presence of 1 unit/ml of epo than in the absence of the hormone. At $1 \times 10^5$ cells/ml, their growth is dependent on added epo (Fig. 4). This phenomenon of cell density dependence of this cell clone was further examined by in vitro culture of these cells at initial seeding densities of $5 \times 10^4$, $10^5$, $2 \times 10^5$, and $3 \times 10^5$ cells/ml. As can be seen in Figure 5, similar to the results in Figure 4, only cultures with initial cell densities of $2 \times 10^5$ or higher grew.

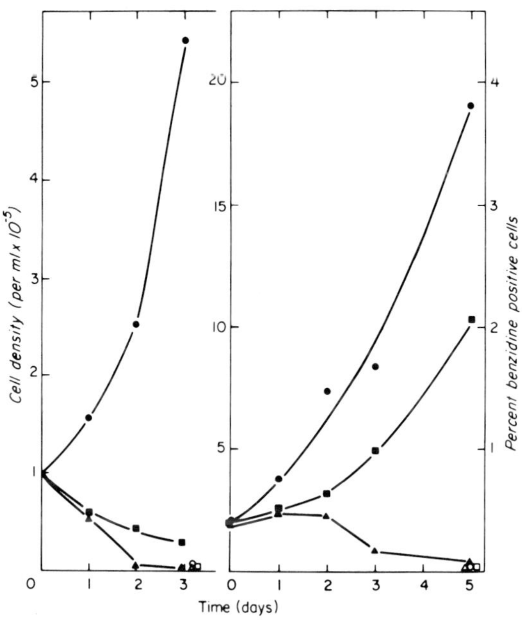


Fig. 4. Growth and inducibility of F-MuLV induced Friend erythroleukemic cells in the presence or absence of dimethylsulphoxide (DMSO) or erythropoietin (epo). Erythroleukemic cells established from CFU-FV induced by F-MuLV as described in Figure 2 and Tables 1 and 2 were cultured in Iscove conditioned medium (Iscove et al., 1980) supplemented with 20% fetal calf serum at cell concentrations of $10^5$/ml or $2 \times 10^5$/ml in the presence or absence of 1% DMSO or 1 unit/ml of epo (Mager et al., 1981b). The viable cells in the cultures were monitored by trypan blue stains and the benzidine positive cells were determined by staining with 1% benzidine as described in Mager et al., (1981b). Left panel: cultures seeded at initial cell concentrations of $10^5$/ml. Right panel: cultures seeded at initial cell concentrations of $2 \times 10^5$/ml. Symbols: cell densities in the absence (■) or presence of 1 unit/ml of epo (●); or 1% DMSO (▲). Percent benzidine positive cells in the absence (□), or presence of 1 unit/ml of epo (○) or 1% DMSO (△).

## DISCUSSION

The data in the present study and those recently reported by Oliff et al. (1981) indicate that the competent Friend helper virus F-MuLV alone, without the SFFV component, can induce a malignant phase of the Friend disease in newborn BALB/c or NIH Swiss mice. These results throw additional light on the possible mechanisms of transformation by Friend leukemia virus as well as the heterogenous nature of this neoplastic disease. We have shown that similar to those studies reported by Oliff et al. (1981), if the newborn BALB/c mice infected with F-MuLV alone can be maintained alive by red blood cell transfusion, tumorigenic cells can be detected by a newly developed colony assay (CFU-FV) (Mager et al., 1981a), which is specific for the Friend virus infected mice. These colonies were similar to CFU-FVs induced in the spleens of FV-P or FV-A infected adult mice in that they are specific for the viruses, independent of added growth factors (other than FCS), and

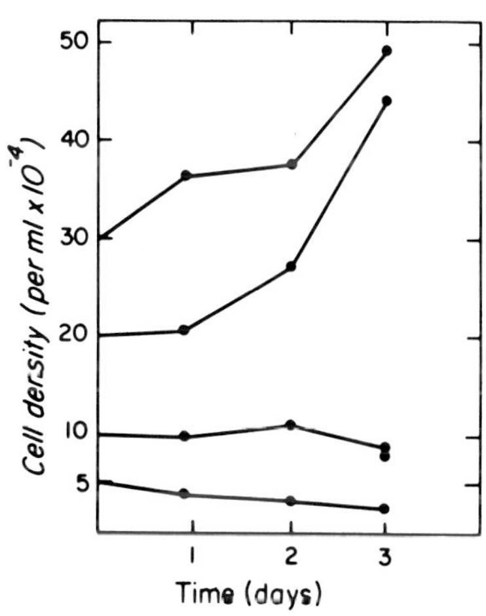


Fig. 5. Growth of F-MuLV induced FLC line at different initial seeding cell densities. A Friend erythroleukemic line obtained by F-MuLV infection of BALB/c newborn mice described in Figure 4 was seeded at different cell densities in IMDM medium as described in Figure 4. At various times, the viable cell numbers were measured.

they are erythroid in origin. They are, however, distinct from those induced by the Friend virus complexes in several aspects. First, these newly described CFUs did not release either $SFFV_A$ or $SFFV_P$ component. Secondly, in colony morphology these CFU-FVs were more diffuse in appearance when compared to the compact colonies induced by FLV complexes; and finally the cells from these colonies were more limited (selective) in their growth medium. For example, they do not grow in α-medium.

Like the CFU-FV induced by the FLV complexes (Mager et al., 1981a,b), these CFU-FVs could be established into long-term cultures at high frequency. These F-MuLV induced erythroleukemic cells, however, had slightly different properties. One property is that their growth is more cell density dependent than the cell lines established from FV-A or FV-P infected spleens. Another property of interest was that these cells did not respond to either DMSO or epo. Instead, although they grew at high cell density in the absence of epo, their growth was enhanced by the presence of the hormone when seeded at high cell concentration and dependent on the hormone when seeded at low cell concentrations. These data suggest that the F-MuLV transformed spleen cells belong to a distinct class of erythroleukemic cells. Further, these results and those reported previously (Mager et al., 1981a,b) illustrate that different types of erythroleukemic cells can emerge as a consequence of infection with different strains of FLV complexes or the F-MuLV helper alone. At the present time, it is not possible to determine the stage of differentiation these cells belong to, nor the relative stage of differentiation they are at with respect to the cell lines derived from FV-A or FV-P infected spleens. Based on the levels of spectrin and benzidine positive cells one might speculate that the cells derived from the FV-P infected spleens are probably more differentiated than those from FV-A or F-MuLV infected spleens. This hypothesis is based on the following two findings: First, while the cells established from the F-MuLV and FV-A infected spleen cells had high levels of spectrin and no benzidine positive cells (Mager et al., 1981a,b), the FV-P FLC cells had both spectrin and benzidine positive cells (Mager et al., 1981a,b). Second, spectrin is a marker for more immature erythroid cells than is hemoglobin (Eisen et al., 1977). One additional piece of information suggests that the F-MuLV induced Friend disease may involve more immature cells than those affected by FV-A or FV-P infection. Niho et al. (1982) have recently reported that in mice infected as newborns by F-MuLV, a dramatic rise (about 3,-000-fold) in the number erythroid bursts were observed instead of the increase in the CFU-E-like colonies induced by FV-A or FV-P infection.

The mechanisms by which these viruses transform cells or the reasons for the heterogenous nature of the FLC cells in response in to infection by the different viruses are also not clear. It is possible that these viruses infect or transform different target cells, probably erythroid progenitor cells at different stages of differentiation. This view is supported by the findings that early after infection with the FLV complexes, FV-A and FV-P, induce certain erythropoietic changes that correlated with the FLC cell lines derived from the corresponding CFU-FVs. For example, the epo dependence of erythroid progenitor cells present in the these early stages after infection with FV-A or FV-P correlate with the responsiveness of the cell lines to induction by epo (Mager et al., 1981b).

Distinct genetic and biological (Mager et al., 1980; Wendling et al., 1981) differences can be detected between the early and late leukemic cells induced by the FLV complexes, although some similarities also exist. Little is known about their evolution or progress. Because the early erythropoietic changes in these susceptible adult mice are dependent on the presence of SFFV genomes, they have been postulated to be responsible for the early changes in the infected mice. The role(s) of the F-MuLV, if any, in the early or late stages is not known. It is likely however that, subsequent to early viral infection, alterations are necessary for the evolution of these truly malignant cell clones. These alterations could occur in cells other than those affected in the early stages of the disease (a "two population" hypothesis) or additional events could occur in the early virus-infected cells leading to a process of clonal evolution or progression as would be predicted by a "one population model" (Mager et al., 1981b).

Regardless which model is correct, it is of interest to consider these possible additional events. One possibility is that the clonal progression occurs by in vivo selection of clones with high self-renewal capabilities (Mager et al., 1981b). Another possibility is that further viral infections (SFFV or F-MuLV or both) or rearrangements of these viral genomes are necessary to bring about this progression.

The data in the present study and those reported by Oliff et al. (1981) showing that truly malignant erythroleukemic cells can evolve from enlarged F-MuLV infected spleens after a prolonged period indicates that the erythroleukemia induced by F-MuLV, like those of the FLV complexes, is also a multistage disease. Furthermore, two additional implications should be considered. First, unlike other acute leukemia viruses, the helper virus in the FLV complex, F-MuLV, probably plays a more important role in the development of Friend erythroleukemia than had been previously believed. If the progression of this disease requires further viral involvements, the present study indicates that the F-MuLV alone can fulfill such functions. Secondly, we are forced to consider the possibility that the SFFV components may only serve as a promoter in the development of the neoplastic disease, possibly by the expansion of the erythroid precursor cell populations in infected adult mice (Suzuki and Axelrad, 1980). Such erythropoietic changes, which were also detected in newborn BALB/c mice infected with F-MuLV (Niho et al., 1982), would favour a subsequent erythroid transformation event, either by F-MuLV or SFFV. An attractive working hypothesis to explain such a transformation event by weakly transforming retroviruses has recently been advanced for two avian leukemia viruses (Hayward et al., 1981; Noori-Daloir et al., 1981).

## ACKNOWLEDGMENTS

We would like to thank Dr. H. Eisen for the generous gift of anti-spectrin antiserum used in these studies and for Dr. A. Howatson for critical review of the manuscript. This work was supported by a grant from the Medical Research Council of Canada.

## LITERATURE CITED

Axelrad, A.A., and R.A. Steeves (1964) Assay for Friend leukemia virus. Rapid quantitative method based on enumeration of macroscopic spleen foci. Virology, *24*:513–516.

Bernstein, A., T.W. Mak, and J.R. Stephenson (1977) The Friend virus genome: Evidence for the stable association of MuLV sequences and sequences involved in erythroleukemic transformation. Cell, *12*:287–294.

Eisen, H., R. Back, and R. Emery (1977) Induction of spectrin in erythroleukemic cells transformed by Friend virus. Proc. Natl. Acad. Sci. U.S.A., *74*:3898–3902.

Friend, C. (1957) Cell-free transmission in adult Swiss mice of a disease having the character of a leukemia. J. Exp. Med., *105*:307–318.

Friend, C., and J.R. Haddad (1960) Tumor formation with transplants of spleen and liver from mice with virus-induced leukemia. J. Natl. Cancer Inst., *25*:1279–1289.

Friend, C., W. Scher, J.G. Holland, and T. Sato (1971) Hemoglobin synthesis in murine virus induced leukemic cells in vitro: Stimulation of erythroid differentiation by dimethylsulphoxide. Proc. Natl. Acad. Sci. U.S.A., *68*:378–382.

Hartley, J.W., N.K. Wolford, L.J. Old, and W.P. Rowe (1977) A new class of murine leukemia virus associated with development of spontaneous lymphomas. Proc. Natl. Acad. Sci. U.S.A., *74*:789–792.

Hayward, W.S., B.G. Neel, and S.M. Astrin (1981) Activation of a cellular *onc* gene by promoter insertion in ALV-induced lymphoid leukosis. Nature, *90*:475–480.

Iscove, N.N., L.J. Guilbert, and C. Wayman (1980) Complete replacement of serum in primary cultures of erythropoietin dependent red cell precursors (CFU-E) by albumin, transferrin ion, unsaturated fatty acid, lecithin, and cholesterol. Exp. Cell Res., *126*:121–126.

Lerner, R.A. (1978) Recombinant origins of leukemogenic murine viruses. Am. J. Pathol., *93*:10–26.

MacDonald, M.E., F.J. Reynolds Jr., W.J.M. Van de Ven, J.R. Stephenson, T.W. Mak, and A. Bernstein (1980a) Anemia- and polycythemia-inducing isolates of Friend spleen focus-forming virus: Biological and molecular evidence for two distinct viral genomes. J. Exp. Med., *151*:1477–1492.

MacDonald, M.E., T.W. Mak, and A. Bernstein (1980b) Erythroleukemia induction by replication-competent type C viruses cloned from the anemia- and polycythemia-inducing isolates of Friend leukemia virus. J. Exp. Med., *151*:1493–1503.

Mager, D., T.W. Mak, and A. Bernstein (1980) Friend leukemia virus-transformed cells, unlike normal stem cells form spleen colonies is $Sl/Sl^d$ mice. Nature, *288*:592–594.

Mager, D., T.W. Mak, and A. Bernstein (1981a) Quantitative colony method for tumorigenic cells transformed by two distinct strains of Friend erythroleukemia virus. Proc. Natl. Acad. Sci. U.S.A., *78*:1703–1707.

Mager, D., M.E. MacDonald, I.B. Robson, T.W. Mak, and A. Bernstein (1981b) Clonal analysis of the late stages of erythroleukemia induced by two distinct strains of Friend leukemia virus. Mol. Cell Biol., *1*:721–730.

Mak, T.W., D. Penrose, C. Gamble, and A. Bernstein (1978) The Friend spleen focus-forming virus (SFFV) genome: Fractionation and analysis of SFFV and helper virus-related sequences. Virology, *87*:73–80.

Marks, P.A. and R.A. Rifkind (1978) Erythroleukemic differentiation. Annu. Rev. Biochem., *47*:419–448.

Niho, Y., T. Shibuya and T.W. Mak (1982) Modulation of erythropoiesis by the helper-independent Friend leukemia virus F-MuLV. (submitted for publication).

Noori-Daloir, M.R., H.J. Kung, L.B. Crittenden, and R.L. Witter (1981) Specific integration of REV provirus in avian Bursa lymphoma. Nature (in press).

Oliff, A.A., S. Ruscetti, E.C. Douglass, and E. Scolnick (1981) Isolation of transplantable erythroleukemia cells from mice infected with helper-independent Friend murine leukemia virus. Blood, *58*:244–254.

Ostertag, W., H. Melderis, G. Steinheider, N. Kluge, and S. Dube (1972) Synthetis of mouse hemoglobin and lgobin mRNA in leukemic cell cultures. Nature (New Biol.), *236*:231–234.

Rowe, W.P., W.E. Pugh, and J.W. Hartley (1970) Plaque assay techniques for murine leukemia viruses. Virology, *42*:1130–1139.

Shields, A., S. Goff, M. Paskind, G. Otto, and D. Baltimore (1979) Structure of the Abelson murine leukemia virus genome. Cell, *18*:955–962.

Stanners, C.P., G.L. Eliceiri, and H. Green (1971) Two types of ribosomes in mouse hamster hybrid cells. Nature (New Biol.), *230:*52–54.

Steeves, R.A. (1975) Spleen focus-forming virus in Friend and Rauscher leukemia virus preparations. J. Natl. Cancer Inst., *54:*289–297.

Stehelin, D., and T. Graf (1978) Avian myelocytomatosis and erythroblastosis viruses lack the transforming gene src of avian sarcoma viruses. Cell, *13:*745–750.

Stephenson, R.J. (1980) Molecular Biology of RNA Tumor Viruses. Academic Press, New York.

Suzuki, S., and A.A. Axelrad (1980) Fv-2 locus controls the proportion of erythropoietic progenitor cells (BFU-E) synthesizing DNA in normal mice. Cell, *19:*225–236.

Tambourin, P., and F. Wendling (1971) Malignant transformation and erythroid differentiation by polycythemia-inducing Friend virus. Nature (New Biol.), *234:*230–233.

Troxler, D., and E.M. Scolnick (1978) Rapid leukemia induced by cloned Friend strain of replicating murine type-C virus. Virology, *85:*17–27.

Wendling, F., F. Moreau-Gachelin, and P. Tambourin (1981) Emergence of tumorigenic cells during the course of Friend virus leukemias. Proc. Natl. Acad. Sci. U.S.A., *78:*3614–3618.

JOURNAL OF CELLULAR PHYSIOLOGY SUPPLEMENT 1:195–207 (1982)

# Erythroblast Cell Lines Transformed by a Temperature-Sensitive Mutant of Avian Erythroblastosis Virus: A Model System To Study Erythroid Differentiation In Vitro

**HARTMUT BEUG,* GABI DOEDERLEIN, CHRISTA FREUDENSTEIN, AND THOMAS GRAF**
*Institute of Virology, German Cancer Research Center, Im Neuenheimer Feld 280, 6900 Heidelberg, Federal Republic of Germany*

A continuous chicken erythroblast cell line transformed by the temperature-sensitive mutant *ts*34 of avian erythroblastosis virus was developed. This cell line, designated HD3, could be induced to terminally differentiate by shift to the nonpermissive temperature. The differentiated cells resembled erythrocytes as judged by morphology, expression of hemoglobin as determined by benzidine staining and radioimmunoassay, and by the expression of differentiation-specific cell surface antigens. Terminal differentiation was dependent on an erythropoietin-like activity present in anemic chicken serum. In contrast, induction of differentiation in the same cells by butyric acid was erythropoietin independent and did not lead to the formation of erythrocytes. In addition, we found that the responsiveness to temperature inducibility and to butyric acid could be dissociated in variant sublines of HD3 and that both types of differentiation inducers appear to act via different pathways.

Hematopoietic cell systems have for a long time been exploited to study the mechanisms underlying different aspects of cell differentiation. On the one hand, using in vitro colony assays, techniques have been developed that promote terminal differentiation in vitro of committed precursor cells from different lineages (for review, see Metcalf and Moore, 1971) or even of pluripotent hematopoietic stem cells (Dexter et al., 1977). On the other hand, cell lines of virus-transformed or "spontaneously" transformed hematopoietic cells have been studied, in which differentiation can be induced in tissue culture. The first type of approach has the advantage that processes which occurr in vivo can be faithfully duplicated in culture but has the disadvantage that the progenitor cells are difficult to characterize and cannot be grown to any large numbers. The second approach circumvents this problem since virus-transformed continuous cell lines can be grown to virtually unlimited numbers without losing their capacity to differentiate (for review, see Till and McCulloch, 1980), thus allowing more detailed biochemical studies requiring large amounts of homogeneous cells. The best studied system of this type are the Friend leukemia virus-transformed erythroleukemia cell lines that can be induced to partially differentiate after the addition of either DMSO, butyric acid or a number of other chemical substances (Marks and Rifkind, 1978). A possible drawback of this system is that it is artificial, since the cells appear to be aberrant in that they do not require erythropoietin for differentiation (Hankins et al., 1978; Housman and Volloch, 1980). Also, no mutants of the virus are available in which the putative transforming gene product can be temperature inactivated to probe its effect on differentiation.

In studying the mechanism of leukemogenesis by avian erythroblastosis virus (AEV), we have developed temperature sensitive *(ts)* mutants of the virus which at permissive (36°) but not at nonpermissive (42°) are capable of transforming bone marrow cells with the phenotype of erythroblasts. These mutants also exhibit a greatly reduced leukemogenic potential in chicks (Graf et al., 1978a). If *ts*-AEV-transformed cells are shifted from 36° to 42°,

*To whom correspondence/reprint requests should be addressed.

0021-9541/82/SUPPL. 1-0195$04.00 

thus probably inactivating the transforming activity of the virus, they begin to accumulate hemoglobin and to partially differentiate (Graf et al., 1978a). These results suggested that AEV induces a block of differentiation in its target cells, known to be immature cells of the erythroid lineage (Gazzolo et al., 1980; Graf et al., 1981). At the same time our findings provided a new system which permitted the study of erythroid differentiation under conditions very close to the in vivo situation. However, using clonal strains of erythroblasts grown from *ts* AEV-infected bone marrow, we encountered the problem that the cells could not be induced to terminally differentiate to erythrocytes under the culture conditions applied. Furthermore, because of the limited in vitro lifespan of the AEV-transformed erythroblast strains (20–40 population doublings), studies requiring large scale cultures could not be performed.

In this communication, we will describe the establishment and characterization of a continuous erythroblast cell line (HD3) that retained its capacity to differentiate in response to a temperature shift in an erythropoietin-like factor dependent fashion. We will also describe the characterization of variant sublines of HD3 that have lost either their responsiveness to temperature shift, to butyric acid treatment, or to both. These and additional studies suggest that both types of inducers act by different mechanisms.

## MATERIALS AND METHODS

### Viruses and cells

The ES4-strain of AEV (Graf et al., 1976) and its temperature sensitive mutant *ts*34 AEV (Graf et al., 1978a) were used as RAV-2 pseudotypes. Normal bone marrow cells were prepared from 1–4 week-old Spafas chicks as described by Graf (1973). The growth medium used consisted of Dulbecco's modified Eagle's medium supplemented with 8% fetal calf serum, 2% chicken serum, 10 mM Hepes pH 7.3, and $10^{-5}$ M thioglycerol.

### Isolation of erythroblast cell lines

All cell lines were isolated from clonal strains of AEV-transformed erythroblasts generated by infecting bone marrow cells with *wt* or *ts*34 AEV. These strains were obtained from colonies grown in Methocel for 10 days (Graf et al., 1981). Trials to isolate a cell line from an erythroblast strain involved the following strategy. After propagation of the clonal cell strain to cell numbers between $2 \times 10^8$ and $2 \times 10^9$ cells (Beug et al., 1979), cells were passaged until cellular senescence was evident from a decrease in growth rate, the appearance of irregularly sized, large cells and cell disintegration. During the crisis the cells were kept at cell densities of $1–3 \times 10^6$/ml. Complete medium changes and centrifugation were avoided. The nomenclature adopted for the lines follows the proposal made by Witter et al. (1979). LSCC HD4 was isolated by passaging the virus-producing 6C2 clone of *wt*-AEV transformed erythroblasts at 41°. After a crisis of about 2 1/2 months a few cells started to grow out which could be developed into a rapidly dividing cell line. The HD4 cell line has been kept in continuous culture for more than 3 years and produces high titers (> $10^4$ colony forming units per ml) of *wt* AEV (RAV-2).

The LSCC HD2 line was isolated at 42° from a *ts*34-AEV-transformed erythroblast nonproducer cell clone (A6). This clone had an unusually long in vitro lifespan of about 45 generations. The cell line obtained after a crisis period of about 2 months has been kept in culture for 2 1/2 years. It produces high titers (> $10^4$ colony forming units per ml) of *ts*34-AEV (RAV-2), as a consequence of an accidental superinfection with RAV-2.

The LSCC HD3 cell line was also isolated from clone A6, but in this case by passage at the permissive temperature (36°) (see Results). Starting from approx. $2 \times 10^9$ senescent cells grown in 600 mm² Roux flasks, cultures were maintained at 36°. They were fed by partial medium changes at twice weekly to weekly intervals and kept at a density of approximately $1–3 \times 10^6$ ml.

After about 4 months the cultures had to be reduced to 3 100 mm dishes. In one of the dishes cells started to grow again, the cell number doubling only every 4–6 days. A gradual increase in growth rate was then observed, finally reaching a doubling time of approximately 24 hours. At this time, the cell line was recloned in Methocel, using a bone marrow feeder layer (Graf et al., 1981). All further experiments were done with one subclone (Cl), which has been kept in continuous culture for about 2 years and was designated LSCC HD3.

### Isolation of HD3 sublines

Early passage HD3 cells were seeded in Methocel using a bone marrow feeder layer (Graf et al., 1981). After incubation for 8–10 days at 36°, single colonies were isolated with a drawn-out Pasteur pipet and seeded into 0.5 ml of growth medium supplemented with 30% of erythroblast conditioned medium (a 2 day harvest from dense cultures of HD3 cells). Subclones were grown up to 1–5 × $10^7$ cells and frozen in 0.5 ml aliquots (containing approximately 1–2 × $10^6$ cells) in liquid nitrogen. Subline 28 C1 was isolated by passaging subline 28 cells at 42° and splitting the culture 1:5 every third day for 2 months. Therafter, the cells were cloned in Methocel at 42°, colonies isolated and grown up at 36° and frozen in liquid nitrogen. After rethawing, each clone was assayed for its inducibility by temperature and by butyric acid using the acid benzidine staining technique (see below).

### Differentiation induction in growth medium

For experiments in growth medium cells were incubated at 36° or 42° in the presence of 5 × $10^{-6}$ M hemin prepared according to Ross and Sautner (1976) as a $10^{-2}$ M stock. During the experimental period cells were maintained at densities between 1.5 and 2.5 × $10^6$ cells/ml, avoiding acidification of the medium by partial medium changes. Induction of HD3 cells with butyric acid was performed by incubating 3–4 × $10^6$ cells/ml in growth medium supplemented with hemin and 1.5 × $10^{-3}$ M butyric acid (from a $10^{-1}$ M stock solution neutralized with 1 N NaOH). Half of the medium was replaced 1 1/2 days later with fresh growth medium plus butyric acid.

### Differentiation induction in differentiation medium

The "differentiation medium" used (Beug et al., 1982) consisted of Dulbecco's modified Eagle's medium (4 parts), sterile bidistilled water (1 part), fetal calf serum (Gibco, 1 part), normal chicken serum (Gibco, 1 part), anemic chicken serum prepared as described in Beug et al., 1982, 1 part), $10^{-3}$ M mercaptoethanol (Sigma Chemical Co., St. Louis, MO) in PBS (1 part) and detoxified BSA (100 mg/ml in PBS prepared as described in McLeod et al., 1974, 1 part). Cells were seeded in this medium at concentrations of 0.5–1 × $10^6$ cells/ml and incubated in an atmosphere of 0.5–1% $CO_2$ in air, resulting in a pH of the medium between 8.0 and 8.2. (A lower pH inhibited terminal differentiation; a higher one caused extensive cell death). In control experiments to assay for erythropoietin dependence, the anemic serum was replaced by an aliquot of normal serum taken from the same animal before it was made anemic (Beug et al., 1982). Butyric acid was added to differentiation medium at a final concentration of 4 × $10^{-3}$ M.

### Detection of hemoglobin by benzidine staining

Two techniques were used. The acid benzidine staining technique was performed according to Orkin et al. (1975). In this method, the benzidine solution was directly added to cells suspended in medium. Benzidine positive cells are stained blue. The neutral benzidine staining technique was essentially performed as described by McLeod et al. (1974). Samples of 1–4 × $10^4$ cells in 200 μl of growth medium were cytocentrifuged on precleaned glass slides, air dried and fixed for 5 min in methanol. They were then incubated in a 1% solution of 3,3′ methoxybenzidine in methanol for 1.5 min, 1% $H_2O_2$ in 50% ethanol for 1.5 min, bidistilled $H_2O$ for 30 sec, and counterstained with a Giemsa-like stain (Diff-Quik, Harleco, The Netherlands). The cytoplasm of the benzidine positive cells stains yellow and that of the negative cells blue.

### Evaluation of erythroid differentiation in cytospin preparations stained with benzidine

Morphology and nucleus-cytoplasm ratio was evaluated microscopically under white or green light (546 nm), whereas the yellow-brown color of the benzidine stain was selectively evaluated under illumination with dark-blue light (Leitz, BG 12, 480 nm). Strongly benzidine positive, oval cells with a nuclear diameter less than half of the diameter of the cell were scored as erythrocytes (Ery); similarly stained, round cells with a larger nucleus as late reticulocytes (LR); weakly benzidine positive cells with a large nucleus as early reticulocytes (ER); and large benzidine negative cells as erythroblasts (Ebl). For details, see Beug et al., 1982 and Fig. 2A.

### Radioimmunoassay for the determination of hemoglobin

Cell extracts were prepared from 20–100 × $10^6$ cells according to the procedure of Conscience et al. (1977). Hemoglobin was assayed in these extracts by radioimmunoassay as described earlier (Graf et al., 1978a). In addition, two modifications of the original assay were used. In the first (HbA-RIA) mainly adult hemoglobin(s) (HbA) were measured. HbA was prepared from peripheral erythrocytes of adult chickens according to the method of Bruns and Ingram (1973) and further purified by preparative isoelectric focusing (pH 7–9; Freudenstein et al., in preparation). It was used both as $^{125}$I-labeled antigen and as "cold" standard for derivation of the calibration curve. With the polyspecific antihemoglobin antibody used (Cappel Laboratories) this assay detected embryonic hemoglobin (HbP) with an at least three-fold less sensitivity than HbA. In the second modification of the assay (HbP-RIA), HbP purified from the blood of five-day-old embryos according to Bruns and Ingram (1973) and by isoelectric focusing (Freudenstein et al., in preparation) was used as $^{125}$I-labeled antigen. The same material was used as a cold standard to obtain a calibration curve. In this assay, the amount of antihemoglobin antibody had to be raised to 12.5 μg as compared to 2.5 μg in the original procedure (Graf et al., 1978a). This assay detected HbA and HbP at about the same sensitivity (50% inhibition was obtained with 40 ng of HbA and 35 ng of HbP.

### Detection of differentiation specific cell surface antigens

This method was performed as described earlier (Beug et al., 1979), except that the cells were double labeled with antierythrocyte serum (Beug et al., 1979) and a monoclonal antibody to immature erythroid cells (4.5.A.5; Hayman et al., 1981). They were then stained with a 1:1 mixture of TRITC-labeled goat antirabbit-IgG antibody (0.25 mg/ml) and FITC-labeled goat antimouse-IgG antibody (0.25 mg/ml). These latter reagents were prepared as described earlier (Beug et al., 1979, 1981b). The antierythrocyte serum was adsorbed to living peripheral chicken erythrocytes and eluted at acid pH (Kirchbach, 1981) to remove a residual activity against normal and AEV-transformed erythroblasts.

## RESULTS

### Establishment of erythroblast cell lines transformed with *ts*34 AEV

Attempts to isolate continuous lines of *ts*34-AEV-transformed erythroblasts were first performed at 41° because continuous cell lines of wildtype AEV-transformed erythroblasts (e.g., LSCC HD4; Beug et al., 1979) were obtained with relative ease under these conditions (Table 1; T. Graf, unpublished). With *ts*34-AEV-erythroblasts a cell line could also be isolated at 41° from a particular erythroblast strain with an unusually long in vitro life span (Table 1). This line, designated LSCC HD2, could, however, no longer be induced to differentiate by temperature shift. HD2 cells grown at either 36° or 42° contained less than 0.5% of benzidine positive cells and did not produce hemoglobin as detected by radioimmunoassay (Table 2). They were, however, clearly identified as erythroblasts by staining with specific antisera and produced high titers of a virus indistinguishable from the input *ts*34-AEV virus (unpublished data).

Because of these findings, further trials to isolate cell lines were performed at 36°. This proved to be more difficult because the "crisis" period, during which the number of living cells continued to decrease, was prolonged as compared to cells grown at 42° (see Material and Methods). However, after two unsuccessful trials we were able to obtain a continuous cell line from a culture of *ts*34-AEV-transformed clone A6 nonproducer cells that had been in crisis for more than 4 months (Table 1). This cell line, designated LSCC HD3, has been kept in continuous culture for more than 2 years. The cloned line grows in suspension with a population doubling time at 36° of about 18 hours. It can be cultivated to densities of up to 5–6 × $10^6$ cells per ml of culture fluid and had retained the nonproducer phenotype of its parent erythroblast clone. It does not synthesize helper virus-related proteins but synthesizes large amounts of p75 AEV, the putative transforming protein of AEV (Beug et al., 1980, 1981a).

### Induction of hemoglobin in sublines of HD3 erythroblasts

Experiments in which HD3 cells were shifted to the nonpermissive temperature under the same conditions as described for *ts*-34-AEV-erythroblast strains, that is, normal growth medium (Graf et al., 1978a), showed that only 20 to 40% of the cells became benzidine positive. In addition, the ability of HD3 cells to differentiate decreased during prolonged in vitro passage (data not shown). To circumvent these problems, two approaches were employed. First, HD3 cells were shifted to 42° in the presence of hemin, since this compound is known to increase hemoglobin accumulation in DMSO-induced Friend erythroleukemia cells (Ross and Sautner, 1976). At concentrations of 5–10 × $10^{-6}$ M, hemin stimulated hemoglobin synthesis also in our system as measured both by benzidine staining and by radioimmunoassay. We also tried to derive sublines of the HD3 cell line that would exhibit

TABLE 1. Derivation of *ts*AEV-transformed erythroblast cell lines

| Cell type | Parental cell type | Transforming virus | Conditions of establishment |
|---|---|---|---|
| LSCC HD2 | *ts*34 AEV clone A6 | *ts*34 AEV | grown at 42°. Duration of crisis approximately 2 months |
| LSCC HD3 | *ts*34 AEV clone A6 | *ts*34 AEV | grown at 36°. Duration of crisis: 4–5 months |
| HD3 sublines 2, 28, 32 37, 40, 41 | HD3 | *ts*34 AEV | derived by subcloning in Methocel at 36° |
| HD3 subline 28 C1 | HD3 subline 28 | *ts*34 AEV | derived by cultivation at 42° for 2 months and subcloning in Methocel at 42° |
| LSCC HD4 | *wt* AEV clone 6C2 | *wt* AEV | grown at 42°. Duration of crisis: more than 2 months |

TABLE 2. Induction of hemoglobin synthesis in *ts*34 AEV-transformed erythroblast clones and cell lines

| Cell type | Percent benzidine positive cells[1] | | μg/Hb/mg protein as determined by Hb-RIA | |
|---|---|---|---|---|
| | 36° | 42°[2] | 36° | 42°[1] |
| Cell lines | | | | |
| LSCC HD 3 (*ts*34 AEV) | | | | |
| Subline 2 | 5 | 76 | 1.0 | 23 |
| 32 | 14 | 77 | 0.4 | 20 |
| 37 | 0.6 | 6 | 0.4 | 1.2 |
| 40 | 9 | 82 | 0.6 | 14 |
| 41 | 15 | 84 | 3.5 | 31 |
| 28 C1 | 12 | 6 | 0.1 | 0.3 |
| LSCC HD 2 (*ts*34 AEV) | 0.3 | 0.5 | <0.02 | <0.02 |
| LSCC HD 4 (*wt* AEV) | 3 | 2 | 0.6 | 0.9 |
| Fresh clones | | | | |
| *ts*34 AEV clone C4 | 0.1 | 39 | 0.5 | 12.5 |
| *ts*34 AEV clone A4 | 24 | 99 | 3.0 | 48 |
| *wt* AEV clone A4 | 0.8 | 2 | 0.7 | 2.4 |

[1]Cells stained at acid pH by the method of Orkin et al. (1975).
[2]Cells were shifted in the presence of $10^{-5}$ M hemin for 5 days.

a more complete and homogeneous differentiation response to temperature shift. For this, 25 subclones were isolated from early-passage HD3 cells by cloning in Methocel (see Materials and Methods) and frozen in aliquots shortly after isolation. Each clone was then thawed and tested for its proportion of benzidine positive cells after temperature shift in the presence of hemin. After passaging for 10 days (equivalent to eight population doublings), the assay was repeated.

Three types of sublines were obtained. A small group (four sublines) were of the type which we had searched for. In these sublines, 75–85% of the cells became strongly benzidine positive after shift to 42° (Table 2). The amounts of hemoglobin synthesized by these sublines corresponded to those obtained from *ts*34-AEV-erythroblast strains analyzed in parallel (Table 2). Furthermore, the ability of these sublines to differentiate did not (sublines 32 and 41) or slightly (sublines 2 and 40) decrease during in vitro passaging (data not shown). A second group, consisting of two sublines (35, 37), had completely lost their ability to synthesize hemoglobin in response to temperature shift (Table 2). The rest of the sublines were intermediate in response and resembled the parental HD3 cells. An example of this type is subline 28.

These results show that it is indeed possible to isolate HD3 sublines that resemble freshly transformed *ts*34-AEV-erythroblast strains in their potential to differentiate. The time course of hemoglobin induction in the presence of added hemin as measured by benzidine staining and radioimmunoassay was determined for one of the inducible sublines (subline 2, Fig.

1A), as well as for one of the noninducible variants (subline 37,Fig. 1B) and a *wt*-AEV-transformed cell line (Fig. 1C). It is interesting that in the absence of added hemin the inducible subline shifted to 42° still showed a significant proportion of benzidine positive cells but hardly expressed any hemoglobin as measured by radioimmunoassay (Fig. 1D).

### Induction of terminal differentiation in sublines of HD3

In the experiments described so far, none of the HD3 sublines could be induced to complete terminal differentiation even when hernin was added to the cultures. As described elsewhere in greater detail (Beug et al., 1982), we were successful, however, in inducing terminal differentiation to essentially mature erythrocytes when *ts*-AEV-erythroblast strains were grown under conditions known to be favorable for the development of normal erythroid colonies (Samarut et al., 1976, 1980). These conditions comprised incubation of the cells at a pH of 8.2 in Dulbecco's medium containing serum from anemic chickens (as a source of erythropoietin) as well as bovine serum albumin, fetal calf serum and mercaptoethanol (to stabilize the cells).

Cells from the various HD3 sublines were shifted in the presence of this "differentiation medium" for 3 to 5 days. They were then analyzed after cytocentrifugation by staining with benzidine at a neutral pH with a subsequent counterstain. The method of benzidine staining employed here is much less sensitivie than the acid pH staining technique of Orkin et al. (1975), which has been used for the assays described so far. The stage of differentiation of different cells was evaluated according to their shape, nucleus to cytoplasm ratio, and staining properties. Four cell types could be distinguished according to these criteria (Fig. 2A): Erythrocytes (Ery), late reticulocytes (LR), early reticulocytes (ER) and erythroblasts (Ebl) (see Materials and Methods). As can be seen in Table 3, a significant proportion of the cells from sublines 2, 32, 40, and 41 differentiated into the more mature forms, including erythrocyte-like cells (see Fig. 2A and 4H). In contrast, cells from sublines 37, 28 C1, and from the *wt*-AEV-transformed cell line HD4 did not differentiate to a significant degree. The same cells were also tested for the expression of erythroid cell surface antigens, using rabbit sera described previously (Beug et al., 1979). To remove a residual reactivity towards normal and AEV-transformed erythroblasts, the antierythrocyte serum was purified by adsorption to and elution from viable erythrocytes before use. The results shown in Table 4 are in agreement with the morphological and cytochemical results (Table 3). Cells of sublines 2, 32, and 41 expressed exclusively erythroblast antigen if grown at 36°, whereas a high proportion of the same cells strongly expressed erythrocyte antigen when shifted to 42°. The noninducible variant sublines 37 and 28 C1, as well as HD4 cells expressed only erythroblast antigen both at 36° and after shift to 42°. Further characterization of the terminally differentiated HD3 cells revealed that they exhibited the ultrastructure of erythrocytes (Beug et al., 1982) and contained almost the levels of adult hemoglobin as found in normal erythrocytes (Beug et al., 1982; Freudenstein et al., in preparation). In addition, terminal differentiation of the HD3 cells could also be shown to depend on erythropoietin-like factors present in the serum from anemic but not from normal chickens (see Fig. 2A, B; Table 6, and Beug et al., 1982).

### Inducibility of hemoglobin production by temperature shift and by butyric acid differs in various sublines of *ts*AEV-transformed erythroblasts

The above results (Tables 2–4) showed that the same HD3 variant sublines that were unable to produce hemoglobin upon temperature shift in growth medium plus $10^{-5}$ M hemin were also refractory to temperature-induced terminal differentiation. In the following experiments we studied whether or not there is a correlation between the responsiveness to temperature shifts and the ability to be induced for hemoglobin production by the addition of a chemical inducer (Graf et al., 1978b). For this purpose, sublines of HD3 were either treated at 36° with $10^{-3}$ M butyric acid or shifted to 42° in the absence of the chemical inducer. Extracts prepared from these cells were then tested in a radioimmunoassay, detecting either predominantly hemoglobins of adult erythrocytes (HbA) or determinants of the primitive erythrocyte hemoglobin (HbP) in addition to adult hemoglobins. Surprisingly, we observed no correlation between the temperature inducibility and the inducibility by treatment with butyric acid in the variant sublines tested (Table 5): Three sublines (2, 32, 41) were inducible by both butyric acid and temperature shift and were therefore assigned the phenotype $ts^+BA^+$. A second type of variant ($ts^-BA^-$, exemplified by subline 37) could not be induced to synthesize hemoglobin by either temperature shift or butyric acid treat-

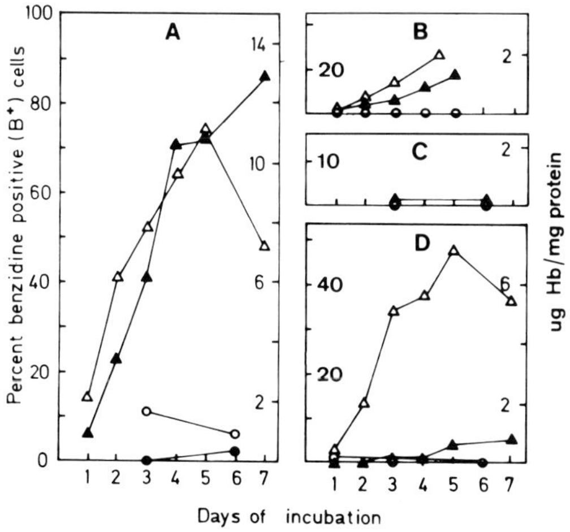


Fig. 1. Time course of hemoglobin induction in HD3 erythroblast cell lines. A–C: Cells of HD3 subline 2 (A), subline 37 (B), or of the *wt* AEV-transformed cell line HD4 (C). HD3 cells were kept at 36° (○, ●) or shifted to 42° (△, ▲). At daily intervals, aliquots were removed from the cultures and assayed for benzidine positive cells by the acid pH staining technique (○, △) or for hemoglobin content by radioimmunoassay (●, ▲) A–C, HD3 sublines 2, 37 and HD4 cells, respectively, grown in the presence of 5 × $10^{-6}$ M hemin. D, Subline 2 cells grown in the absence of added hemin.

ment (Table 5). The third type of variant ($ts^+$ $BA^-$, represented by subline 40), retained its ability to produce hemoglobin in response to temperature shift but was unresponsive to butyric acid (Table 5).

Variants of the fourth possible phenotype ($ts^-BA^+$) could not be found among the approximately 25 clones initially screened. Such a variant could, however, be selected by cultivation of HD3 subline 28 cells for 2 months at nonpermissive temperature and subcloning in Methocel at 42°. From the 30 subclones tested, one subline (28 Cl) was obtained that was still strongly inducible by butyric acid, but failed to express increased levels of hemoglobin after shift to 42° (see Table 5, Fig. 4A–C). The remaining sublines were either noninducible by butyric acid (24 sublines) or partially inducible by both butyic acid and elevated temperature (five sublines).

The data in Table 5 also show that in all sublines inducible with butyric acid the amounts of hemoglobin detected with the "HbP" assay (which measures HbP plus HbA)

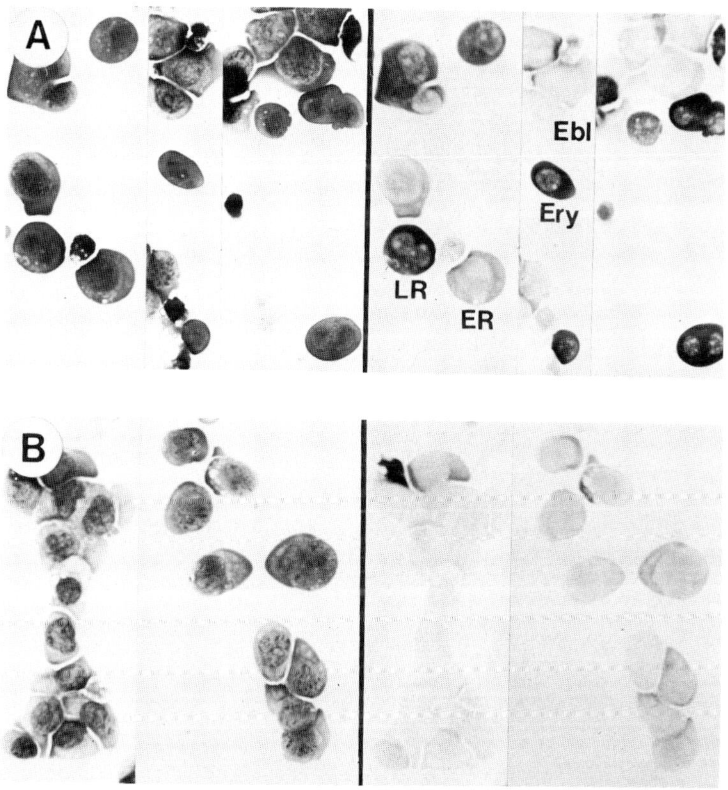


Fig. 2. Erythropoietin-dependence of terminal differentiation of HD3 erythroblasts. HD3 subline 41 cells were incubated at 42° in differentiation medium containing anemic chicken serum (A) or normal serum from the same chicken (B). Each pair of panels represents the same fields of cells photographed under illumination of green (left) or dark-blue light (right). Typical examples of the four phenotypes distinguished (see Material and Methods) are marked on the right panel of A.

TABLE 3. Temperature-induced terminal differentiation of *ts*34 AEV-transformed cell lines

| | Incidence of different cell types[1] | | | | | | | |
|---|---|---|---|---|---|---|---|---|
| | 36° | | | | 42°[2] | | | |
| Cell line | Ery | LR | ER | Ebl | Ery | LR | ER | Ebl |
| LSCC HD 3 | | | | | | | | |
| Subline 2 | 0 | 0 | 1 | 99 | 9 | 13 | 37 | 23 |
| Subline 32 | 0 | 0.3 | 1.4 | 98 | 8 | 22 | 38 | 32 |
| Subline 37 | 0 | 0 | 0 | 100 | 0.5 | 0.7 | 1.3 | 97 |
| Subline 40 | 0 | 0 | 1 | 99 | 5 | 17 | 29 | 50 |
| Subline 41 | 0 | 0.2 | 1 | 99 | 18 | 29 | 32 | 21 |
| Subline 28 C1 | 0 | 0 | 0 | 100 | 0 | 0.5 | 1 | 98 |
| LSCC HD 4 | 0 | 0 | 0 | 100 | 0 | 0 | 1 | 99 |

[1]Percent of total cell population. Ery, eythrocyte; LR, late reticulocyte; ER, early reticulocyte; Ebl, erythroblast.
[2]Cells were shifted for 5 days in the presence of differentiation medium.

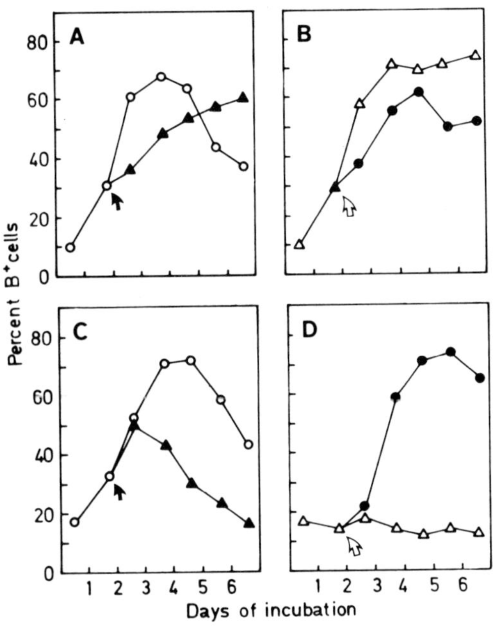


Fig. 3. Hemoglobin induction following a switch from a temperature shift to butyric acid treatment and viceversa. Cultures of HD3 subline 41 (A,B) and subline 28 C1 (C,D) in growth medium containing hemin were either preincubated at 36° with 1.5 × $10^{-3}$ M butyric acid or subjected to a preshift to 42°. After washing, half of the cells preincubated with butyric acid (A,C) were transferred to growth medium without butyric acid (solid arrows) and incubated at 42° (▲). The other half was resuspended in growth medium with butyric acid and incubated further at 36° (○). Similarly, half of the preshifted cells (B,D) were transferred to medium with 1.5 M butyric acid (open arrows) and incubated at 36° (●) whereas the other half was further incubated at 42° without butyric acid (△). At daily intervals, cells were assayed for benzidine positive (B⁺) cells by the acid-pH staining technique.

were much larger than those detected with the "HbA"-assay (which predominantly detects adult hemoglobins). This suggests that butyric acid induction leads to accumulation of hemoglobins possibly different from adult HbA, presumably resembling an embryonic, HbP-like hemoglobin species.

### Differentiation induction by temperature shift and by butyric acid treatment is based on different mechanisms

The results described above indicate that the differentiation induced by temperature shift on the one hand, and by treatment with chemical inducers on the other, can be affected independently of each other in different cellular variants, suggesting different mechanisms for these two types of differentiation induction. If this idea is correct, the effects of the two treatments should be nonadditive, i.e., treatment with butyric acid should not affect differentiation induction by temperature shift and vice versa. To test this, HD3 subline 41 cells were either incubated first with butyric acid for 36 hours, washed and shifted to 42°, or first shifted to 42° for the same time and then treated with butyric acid at 36°. In each case, control dishes were continuously treated with butyric acid or shifted to 42°. After different times of incubation, cell cultures were assayed for their proportion of benzidine positive cells as determined by the acid pH method. The results in Fig. 3A and B clearly show that the pretreatment of HD3 cells with butyric acid could not replace the effect of temperature shift to 42°, resulting in a delayed response to temperature shift in the butyric acid pretreated culture. Similarly, preincubation of HD3 cells at 42° could not replace the effect of butyric acid treatment, leading to a similar delay of differ-

Fig. 4. Induction of differentiation in HD3 sublines shifted to 42° or treated with butyric acid. Panels A–D show micrographs of cells incubated in growth medium plus hemin and stained with the acid benzidine technique. Panels E–H show cells incubated in differentiation medium and stained with the neutral benzidine technique. Pictures at left were photographed under green light, at right under blue light. Panels I–M are indirect immunofluorescence micrographs of the same cells stained with antierythrocyte serum and TRITC-labeled second antibody (left side) and antierythroblast (4.5.A.5) antibodies stained with FITC-labeled second antibody (right side). The upper three rows represent HD3 subline 28 C1 cells; the last row (D,H,M) are HD3 subline 41 cells. The first row (A,E,I) shows subline 28 C1 cells grown at 36°; the second row (B,F,K) shows 28 C1 cells treated with butyric acid for 3 days at 36° and the third row (C,G.L) shows 28 C1 cells shifted for 5 days to 42°. The fourth row shows subline 41 cells shifted for 5 days to 42° in differentiation medium.

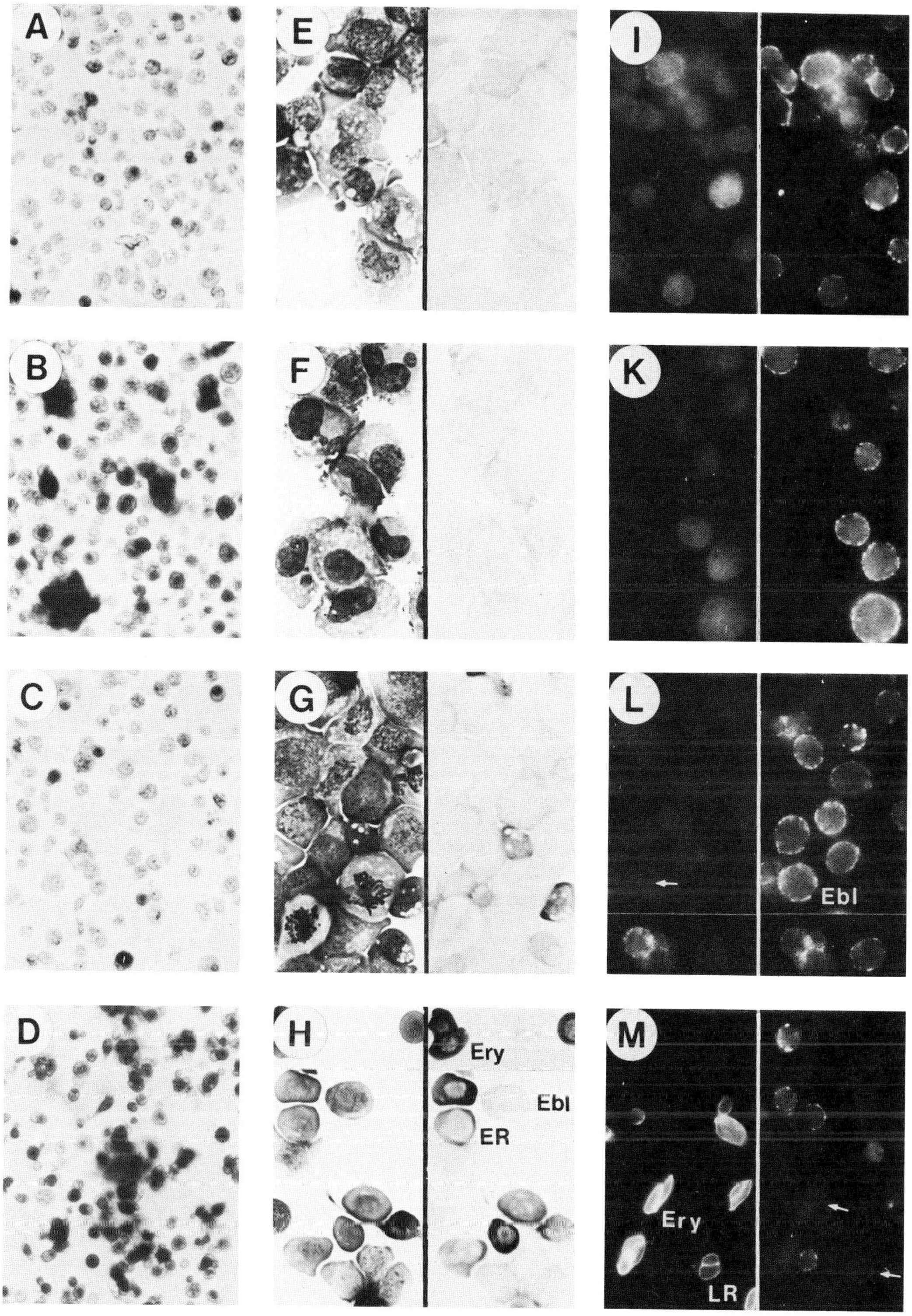


Figure 4

entiation induction by this inducer. In a parallel experiment, employing HD3 subline 28 C1 cells (of the $ts^-BA^+$ phenotype), pretreatment with butyric acid could not "complement" the inability of the cells to respond to a temperature shift. Instead, the percentage of benzidine positive cells dropped to the uninduced level (Fig. 3C). Conversely, the same cells, when first preshifted to 42° and then treated with butyric acid at 36°, exhibited a differentiation response delayed for a similar period of time as was employed to preincubate the cells at 42° in the absence of inducer (Fig. 3D).

Next, we asked the question whether or not induction of susceptible cells with butyric acid results in a terminal differentiation and if the differentiation induction seen with this inducer is dependent on erythropoietin. For this purpose, two $ts^+BA^+$ sublines (32 and 41) and the $ts^-BA^+$ subline 28 C1 were transferred to differentiation medium containing either anemic or normal chicken serum and treated either with butyric acid at 36° or shifted to 42°. Three to five days later they were assayed for benzidine staining at acid pH and for the proportion of terminally differentiated cells determined by the evaluation of stained cytospins and immunofluorescent staining with differentiation-specific antisera. The results shown in Figure 4 and Table 6 demonstrate that the butyric acid treated cells failed to terminally differentiate (Fig. 4F, K; Table 6). Fur-

TABLE 4. Expression of erythroid cell surface antigens in *ts*34-AEV-transformed cell lines

| | Erythrocyte antigen(s) | | Erythroblast antigen(s) | |
|---|---|---|---|---|
| Cell line | 36° | 42°[1] | 36° | 42°[1] |
| LSCC HD 3 | | | | |
| subline 2 | < 1[2] | 87 | > 99 | 13 |
| subline 32 | < 1 | 60 | > 99 | 40 |
| subline 37 | < 1 | 9 | > 99 | 91 |
| subline 41 | < 1 | 65 | > 99 | 35 |
| subline 28 C1 | < 1 | 1.5 | > 99 | 98.5 |
| LSCC HD4 | < 1 | < 1 | > 99 | > 99 |

[1]Cells were shifted for 5 days in the presence of differentiation medium.
[2]Percent fluorescence positive cells.

TABLE 5. Hemoglobin production in *ts*34-AEV-transformed erythroblast cell lines: Effect of temperature shift and of butyric acid

| | µg hemoglobin/mg protein as determined by [1] | | | | | | |
|---|---|---|---|---|---|---|---|
| | Hb-A-RIA | | | Hb-P-RIA | | | |
| Cell line | 36° | 36° + BA | 42° | 36° | 36° + BA | 42° | Phenotype |
| LSCC HD 2 | 0.02 | 0.1 | 0.02 | ND | ND | ND | $ts^-$ $BA^-$ |
| LSCC HD 3 | | | | | | | |
| subline 2 | 0.9 | 3.4 | 28 | ND | ND | ND | $ts^+$ $BA^+$ |
| subline 32 | 0.4 | 6.9 | 24 | 0.5 | 22 | 68 | $ts^+$ $BA^+$ |
| subline 37 | 0.4 | 0.6 | 1.2 | ND | ND | ND | $ts^-$ $BA^-$ |
| subline 40 | 0.6;2.0[2] | 0.5;1.6 | 38;41 | 0.9;3.4 | 1.4;3.1 | 68;78 | $ts^+$ $BA^-$ |
| subline 41 | 3.6 | 13.3 | 78 | 13.6 | 28 | 89 | $ts^+$ $BA^+$ |
| subline 28 C1 | 0.1 | 3.2 | 0.3 | 0.1 | 10 | 0.5 | $ts^-$ $BA^+$ |
| LSCC HD 4 | 0.2 | 0.2 | 0.15 | ND | ND | ND | $ts^-$ $BA^-$ |

[1]BA, butyric acid; ND, not determined.
[2]Determinations from two independent experiments.

TABLE 6. Effect of anemic chicken serum of differentiation induced by butyric acid treatment or by temperature shift in differentiation medium

| | | 36° | | 36° + BA[4] | | 42° | |
|---|---|---|---|---|---|---|---|
| Parameter assayed | LSCC HD3 subline | NCS[3] | ACS[3] | NCS | ACS | NCS | ACS |
| Percent benzidine positive cells[1] | subline 32 | ND | 10 | 39 | 40 | 67 | 78 |
| | subline 41 | ND | 21 | 55 | 56 | 67 | 87 |
| | subline 28 C1 | ND | 6 | ND | 45 | 48 | 45 |
| Percent terminally differentiated cells[2] | subline 32 | 0 | 0 | 0 | 0 | 5 | 21 |
| | subline 41 | 0 | 0 | 0 | 0 | 4 | 22 |
| | subline 28 C1 | 0 | 0 | 0 | 0 | 1 | 1 |

[1]Cells stained at acid pH by the method of Orkin et al. (1975).
[2]Cells with morphological and staining properties of erythrocytes and late reticuloctyes (Fig. 2) were scored as terminally differentiated.
[3]Cells incubated in differentiation medium containing either anemic chicken serum (ACS) or homologous normal chicken serum (NCS) instead.
[4]BA; butyric used at $4 \times 10^{-3}$ M.

thermore, they were induced to produce hemoglobin in an erythropoietin-independent manner (Table 6). However, shift to 42° resulted in the expected terminal differentiation in both $ts^+BA^+$ sublines, which was also clearly dependent on erythropoietin (Fig. 4H, M; Table 6).

## DISCUSSION

In this paper we have described a system to study terminal differentiation in vitro using a newly developed cell line of chicken erythroblasts transformed with the *ts*34 mutant of AEV. It was shown that after shift to the nonpermissive temperature (42°), a substantial fraction of the cells could be induced to differentiate into essentially mature erythrocytes, that terminal differentiation was dependent on erythropoietin, and that cellular variants could be isolated that are altered in their ability to differentiate in response to temperature shift and/or chemical inducers.

The main advantage of the *ts*-AEV-erythroblast system over the Friend cell system is that a terminal differentiation can be induced under physiological conditions and that this process is dependent on the normal differentiation hormone. However, at its present stage our system has the disadvantage that a maximum of only about 25% of the HD3 cells can be induced to differentiate into erythrocytes after shift to 42°. This might at least in part be due to the properties of the particular *ts*-AEV mutant used for infection since we have recently found that more than 95% of the cells can be induced to terminally differentiate when another *ts* mutant is used to transform erythroblasts (Beug et al., 1982). We are currently trying to develop a cell line transformed by this *ts* mutant. Another disadvantage (which, however, is also known for Friend cell lines) is that our *ts*-AEV cell line and sublines are somewhat unstable with respect to their ability to differentiate. This problem can be essentially overcome by freezing aliquots of freshly recloned cells and using them only for a limited time after thawing. A similar instability of inducibility by chemical inducers has also been observed in the Friend cell system (Housman et al., 1980).

The development of continuous chicken cell lines appears to be dependent largely on the cell type. For example, it is very difficult to obtain fibroblast cell lines (see Witter et al., 1979 for references) while almost every lymphoblastoid cell clone transformed by the T-strain of reticuloendotheliosis virus develops into a cell line (Levis et al., 1981; Beug et al., 1981). Erythroblasts transformed by AEV are intermediate in this respect. It is relatively easy to obtain cell lines from cell clones if they are grown at 42° but much more difficult when grown at 36°. The chances of being successful can be increased if clones are chosen that have a high population doubling potential before entering the phase of cellular senescence.

That the addition of hemin to *ts*34-AEV-transformed HD3 erythroblasts leads to an improved inducibility of these cells by temperature shift was also observed in fresh strains of *ts*34-AEV erythroblasts (N. Ade, unpublished). Hemin itself, however, did not function as an inducer of differentiation, because addition of hemin to *ts*34-AEV erythroblasts grown at 36° or to *wt* AEV erythroblasts as both 36° and 42° had no effect. A similar stimulating effect of hemin on globin expression has also been observed in dimethyl sulfoxide-treated Friend erythroleukemia cells (Ross and Sautner, 1976). As no stimulatory effect of hemin was observed when added to differentiation medium, it seems possible that hemin can partially compensate a deficiency in growth medium of factors (such as transferrin) necessary for iron uptake.

One of the questions that could be successfully studied with the experimental system described was whether induction of differentiation by temperature shift proceeds via a pathway similar to that of induction by chemicals such as butyric acid. Several lines of evidence indicate that these two modes of differentiation induction utilize different mechanisms.

1. Cellular variants could be isolated, in which the responsiveness to temperature shift on the one hand and to butyric acid on the other was selectively lost. 2. The effect of temperature shift and of butyric acid is not additive, that is, *ts*-AEV-transformed erythroblasts do not memorize the effects of these two modes of induction in the same "compartments." In contrast, in experiments performed with Friend erythroleukemia cells, Housman et al. (1978) have found that different types of chemical inducers seem to act in an additive fashion. These authors have, however, not included butyric acid in their studies and it remains to be seen whether or not this drug differs in this respect from other chemicals. 3. The induction of differentiation with butyric acid is, in contrast to the induction by temperature shift, incomplete and not erythropoietin dependent. This could be a consequence

of the toxic effects of the compound since its addition to HD3 cells shifted to 42° inhibited their terminal differentiation (H.B., unpublished data). Alternatively, the butyric acid-induced cells may be triggered along an aberrant pathway of differentiation as suggested by the finding that they may express predominantly embryonic-type hemoglobin. Experiments are in progress to test this possibility in greater detail.

Unfortunately, we have not been able to find other potent chemical inducers for AEV-transformed chick erythroblasts besides butyric acid. Treatment with polar agents such as dimethyl-sulfoxide, hexamethyl-benzanthracence and related compounds were without effect (Graf et al., 1978b; S. Palmieri, unpublished). Therefore, it was not possible to test whether or not insensitivity to butyric acid of some of our variant cell lines is due to a general insensitivity to chemical inducers.

Cell variants with an impaired ability to differentiate have not only been successfully exploited in the Friend erythroleukemia cell system (Eisen et al., 1978; Marks and Rifkind, 1978) but also in the myeloid leukemia cell system developed by Sachs et al. (1978). In the latter cell system variants have been isolated which resemble some of the variant HD3 sublines described here in that they have lost their inducibility by the physiological inducer (in this case a macrophage-granulocyte-specific factor) but are still partially inducible by treatment with certain chemicals (Sachs, 1980). In addition, these authors found that after addition of the physiological inducer some cell variants appeared to be capable of reaching only an incomplete stage of maturation (Sachs et al., 1978). In this regard, it has to be pointed out that the phenotypes assigned to the different HD3 sublines in this communication represent only a preliminary, very incomplete description of their actual properties. For example, some of our variants (such as 28 C1), when shifted to 42° in differentiation medium, became benzidine positive while they remained negative if kept in normal growth medium. We have also observed several variants that appear to represent intermediate phenotypes and that remain to be fully characterized (H.B., unpublished data). In this regard it needs to be pointed out that it is unclear whether the variants described represent true cellular mutants and whether they have undergone chromosome aberrations.

Despite of some of the difficulties still inherent to the *ts*-AEV-erythroblast cell line system described, it is one of the few cell systems in which differentiation can be studied in a homogenous population using a physiological trigger to induce differentiation. In addition, it offers the advantage that any results that are obtained with the continuous cell line can be checked on a smaller scale by using freshly transformed AEV-erythroblast strains (Beug et al., 1982). Artifacts due to possible aberrations inherent to many established cell lines can thus be avoided. In particular, our system may be useful to study the regulation of hemoglobin expression at the molecular level, determining not only which hemoglobin chains are turned on and in what sequence, but also how different variants might be affected in this regulation.

## ACKNOWLEDGMENTS

We thank Drs. Michael J. Hayman, Norman Iscove, and Jean F. Conscience for helpful suggestions and comments, Sigrid Grieser for excellent technical assistance, and Birgit Blanasch for typing.

## LITERATURE CITED

Beug, H., v.Kirchbach, A., Döderlein, G., Conscience, J.F.and Graf, T. (1979) Chicken hematopoietic cells transformed by seven strains of defective avian leukemia viruses display three distinct phenotypes of differentiation. Cell, *18:*375–390.

Beug, H., Kitchener, G., Döderlein, G., Graf, T. and Hayman M.J. (1980) Mutant of avian erythroblastosis virus defective for erythroblast transformation: Deletion in the *erb* portion of p75 suggests function of the protein in leukemogenesis. Proc. Natl. Acad. Sci. U.S.A., *77:*6683–6686.

Beug, H., Graf, T. and Hayman, M.J. (1981a) Production and characterization of antisera specific for the *erb*-portion of p75, the presumptive transforming protein of avian erythroblastosis virus. Virology, *111:*201–210.

Beug, H., Müller, H., Grieser, S., Döderlein, G. and Graf, T. (1981b) Hematopoietic cells transformed in vitro by $REV_T$ avian reticuloendotheliosis virus express characteristics of very immature lymphoid cells. Virology *115:*295–309.

Beug, H., Palmieri, S., Freudenstein, Ch., Zentgraf, H.W., and Graf, T. (1982) Hormone dependent terminal differentiation *in vitro* of chicken erythroleukemia cells transformed by *ts* mutants of avian erythroblastosis virus. Cell (in press).

Bruns, G.A.P. and Ingram, V.M. (1973) The erythroid cells and haemoglobins of the chick embryo. Philos. Trans. R. Soc. Lond. (Biol.), *266:*225–305.

Conscience, J.F., Miller, R.A., Henry, I. and Ruddle, H. (1977) Acetylcholinesterase, carbonic anhydrase and catalase activity in Friend erythroleukemia cells and non-erythroid mouse cell lines. Exp. Cell Res., *105:*401–412.

Dexter, T.M., Allen, T.D. and Lajtha (1977) Conditions controlling the proliferation of haemopoietic stem cells in vitro. J. Cell Physiol., *91:*335–344.

Eisen, H., Keppel-Ballivet, F., Georgopoulos, C.P., Sassa, S., Granick, J., Pragnell, I., and Ostertag, W. (1978) Biochemical and genetical analysis of erythroid differentiation in Friend virus-transformed murine erythroleukemia cells. In: Differentiation of normal and neoplastic hematopoietic cells. Clarkson, B., Marks, P.A. and Till, J.E., eds. Cold Spring Harbor Conferences on Cell Proliferation. Cold Spring Harbor Laboratory, Vol. 5, pp. 277–294.

Gazzolo, L., Samarut, J., Bonabdelli, M. and Blanchet, J.P. (1980) Early precursors in the erythroid lineage are the specific target cells of avian erythroblastosis virus in vitro. Cell, *22:*683–691.

Graf, T. (1973) Two types of target cells for transformation with avian myelocytomatosis virus. Virology, *54:*398–413.

Graf, T., Royer-Pokora, B. and Beug, H. (1976) Evidence for the multiple oncogenic potential of cloned leukemia virus: in vitro and in vivo studies with avian erythroblastosis virus. Virology, *71:*423–433.

Graf, T., Ade, N. and Beug, H. (1978a) Temperature-sensitive mutant of avian erythroblastosis virus suggests a block of differentiation as mechanism of leukaemogenesis. Nature, *257:*496–501.

Graf, T., Beug, H., Royer-Pokora, B. and Meyer-Glauner, W. (1978b) In vitro transformation of hematopoietic cells by avian erythroid and myeloid leukemia viruses: a model system for the differentiation of normal and neoplastic cells. In: Differentiation of Normal and Neoplastic Hematopoietic Cells. Clarkson, B., Marks, P.A. and Till, J.E., eds. Cold Spring Harbor Conferences on Cell Proliferation. Cold Spring Harbor Laboratory, pp. 625–639.

Graf, T., v.Kirchbach, A. and Beug, H. (1981) Characterization of the hematopoietic target cells of AEV, MC29, and AMV avian leukemia viruses. Exp. Cell Res., *131:*331–343.

Hankins, W.D., Kost, T.A., Koury, M.J. and Krantz, S.B. (1978) Erythroid bursts produced by Friend leukemia virus in vitro. Nature, *276:*506–508.

Hayman, M.D., Beug, H. and Savin, K.W. (1982) Changes in the expression of membrane antigens during the differentiation of chicken erythroblasts. J. Supramol. Struct. (in press).

Housman, D., Gusella, R., Geller, R., Levenson, R. and Weil, S. (1978) Differentiation of murine erythroleukemia cells: the central role of the committment event. In: Differentiation of normal and neoplastic hematopoietic cells. Clarkson, B., Marks, P.A. and Till, I.E., eds. Cold Spring Harbor Conferences on Cell Proliferation. Cold Spring Harbor Laboratory, Vol. *5*, pp. 193–207.

Housman, D. and Volloch, V. (1980) Stabilization of MEL cells at terminal stages of differentiation. In: Cellular and molecular regulation of hemoglobin switching. Stamatoyannopoulow, J. and Nienhuis, eds. (in press).

v.Kirchbach, A. (1981) Characterisierung der Zielzellen defekter Hühner leukämieviren mit Hilfe spezifischer Antiseren, Ph.D. Thesis, Tübingen, FRG.

Lewis, R.B., McClure, I., Rup, B., Niesel, D.W., Garry, R.F., Hoelzer, I.D., Nazerian, K. and Bose, H.R. (1981) Avian reticuloendotheliosis virus: Identification of the hematopoietic target cell for transformation. Cell, *25:*421–431.

McLeod, D.L., Shreeve, M.M. and Axelrad, A.A. (1974) Improved plasma culture system for production of erythrocytic colonies in vitro: Quantitative assay method for CFU-E. Blood, *44:*517–534.

Marks, P. and Rifkind, R. (1978) Erythroleukemic differentiation. Annu. Rev. Bioch., *47:*419–448.

Metcalf, D. and Moore, M.A.S. (1971) Haemopoietic cells. North-Holland Publishing Co., Amsterdam.

Orkin, S.H., Harosi, F.I. and Leder, P. (1975) Differentiation in erythroleukemic cells and their somatic hybrids. Proc. Natl. Acad. Sci. U.S.A., *72:*98–102.

Ross, I. and Sautner, D. (1976) Induction of globin mRNA accumulation by hemin in cultured erythroleukemic cells. Cell, *6:*513–520.

Sachs, L. (1978) Regulation of normal cells differentiation and malignancy in myeloid leukemia. In: Differentiation of normal and neoplastic hematopoietic cells. Clarkson, G., Marks, P.A. and Till, I.E., eds. Cold Spring Harbor Conferences on Cell Proliferation. Cold Spring Harbor Laboratory, Vol. *5*, pp. 411–426.

Sachs, L. (1980) Constitutive uncoupling of pathways of gene expression that control growth and differentiation in myeloid leukemia: A model for the origin and progression of malignancy. Proc. Natl. Acad. Sci. U.S.A., *77:*6152–6156.

Samarut, I., and Nigon, V. (1976) In vitro development of chicken erythropoietin sensitive cells. Exp. Cell Res., *100:*245–248.

Samarut, I. and Bouabdelli, M. (1980) In vitro development of CFU-E and BFU-E in cultures of embryonic and postembryonic chicken hematopoietic cells. J. Cell Physiol., *105:*553–563.

Till, I.E. and McCulloch, E.A. (1980) Hematopoietic stem cell differentiation. Biochim. Biophys. Acta, *605:*431–459.

Witter, R.L. (1979) A proposed method for designating avian cell lines and transplantable tumors. Avian Pathol. *8:*487–498.

JOURNAL OF CELLULAR PHYSIOLOGY SUPPLEMENT 1:209–212 (1982)

# Transforming Genes of Chicken Bursal Lymphomas

GEOFFREY M. COOPER*
*Sidney Farber Cancer Institute and Department of Pathology, Harvard Medical School, Boston, Massachusetts 02115*

Bursal lymphomas of chickens infected with lymphoid leukosis viruses contain at least two distinct activated transforming genes: 1) a c-*myc* gene activated by integration of viral DNA and 2) a cellular gene, not linked to viral DNA, which efficiently induces transformation of NIH 3T3 cells. Since lymphomagenesis by lymphoid leukosis viruses appears to be a multistep process, it is possible that activation of these two different genes occurs at different stages of the disease. The bursal lymphoma transforming gene detected by transfection has been isolated by molecular cloning. Hybridization analysis indicates that this gene is homologous to normal chicken DNA sequences.

Transfection by cellular DNAs has recently been used to investigate transforming genes of both normal and neoplastic cells. DNA fragments of normal cells have a low transforming activity (~0.003 transformants/μg DNA), which is thought to represent activation of potential transforming genes as a consequence of DNA rearrangements during donor DNA integration (Cooper et al., 1980). Consistent with this hypothesis is the observation that DNAs extracted from cells transformed by normal cell DNA fragments induce transformation with high efficiencies (0.1–1 transformants/μg DNA) in secondary transfection assays, indicating that these transformed cells contain activated transmissible transforming genes (Cooper et al., 1980).

DNAs of a variety of neoplastic cells induce transformation with high efficiency indicating that, in some transformed cells, carcinogenesis has involved DNA alterations, either by mutation or DNA rearrangement, leading to activation of cellular transforming genes. Neoplastic cells in which transforming gene activation has been reported include chemically transformed mouse fibroblasts (Shih et al., 1979); chicken bursal lymphomas (Cooper and Neiman, 1980); a chicken nephroblastoma (Cooper and Neiman, 1980); human, rabbit, and mouse bladder carcinomas (Krontiris and Cooper, 1981; Shih et al., 1981); mouse and human lung carcinomas (Shih et al., 1981; Krontiris and Cooper, unpublished observations); mouse and human mammary carcinomas (Lane et al., 1981); mouse and rat neuroblastomas and gliomas (Shih et al., 1981); and mouse and human lymphoid neoplasms (Lane et al., in preparation). Transforming genes activated in several different types of neoplasms and in several different vertebrate species have thus been detected by their transforming activity in transfection assays. In this article, I will discuss the transforming genes activated in chicken bursal lymphomas, the isolation of such a transforming gene by molecular cloning, and the specificity of transforming genes activated in neoplasms of different types of differentiated cells.

## TRANSFORMING GENES OF CHICKEN BURSAL LYMPHOMAS

Bursal lymphomas are induced in chickens by infection with lymphoid leukosis viruses (LLVs). LLVs are representative of a class of retroviruses that do not appear to contain specific viral transforming genes, do not transform cells in vitro, and require long latent periods before formation of neoplasms in vivo. In contrast, sarcoma and acute leukemia viruses contain specific genes responsible for oncogenicity, efficiently transform cells in vitro, and induce neoplasms with short latent periods in vivo. Oncogenesis by LLVs thus appears to be indirect in comparison to sarcoma and acute leukemia viruses and more closely resembles chemically induced and naturally occurring oncogenesis.

*To whom correspondence/reprint requests should be addressed.

Transfection assays of DNAs of LLV-induced bursal lymphomas indicated that these neoplasms contained transforming genes that induced transformation of NIH 3T3 mouse cells with efficiencies of approximately 0.1 transformants/μg DNA (Cooper and Neiman, 1980). In contrast, DNAs of LLV-infected non-neoplastic bursal tissues lacked transforming activity. LLV-induced lymphomagenesis thus appeared to involve genetic alterations leading to activation of transforming genes detectable in the transfection assay.

To determine whether the transforming genes detected by transfection of these lymphoma DNAs were linked to viral DNA sequences, transformed NIH cells were analyzed for the presence of LLV DNA by Southern (1975) blot hybridization. No viral DNA sequences were detected in NIH cells transformed by LLV-induced lymphoma DNAs using probes either representative of the entire LLV genome or specific for the LLV long terminal redundancy (LTR), which contains the viral transcriptional promoter (Cooper and Neiman, 1980). Therefore, the transforming genes of bursal lymphomas detected by transfection were not linked to viral DNA, indicating that lymphomagenesis by LLVs involved indirect activation of cellular transforming genes at some stage of the disease process.

Hayward et al. (1981) have reported that approximately 80% of LLV-induced bursal lymphomas contain viral LTR sequences integrated in the vicinity of c-*myc*, the cellular homolog of the presumptive transforming gene of the acute leukemia virus MC29. Integration of viral LTR sequences apparently results in increased transcription of c-*myc*, implicating c-*myc* activation as a direct consequence of LLV infection in lymphomagenesis (Hayward et al., 1981).

The LLV-induced lymphomas that were used as DNA donors in transfection assays (Cooper and Neiman, 1980) also contain viral LTR sequences integrated near c-*myc* (Cooper and Neiman, 1981). However, further analysis of NIH cells transformed by these lymphoma DNAs indicates that transformation was not mediated by transfer of these activated c-*myc* sequences (Cooper and Neiman, 1981). Bursal lymphomas induced by LLV thus appear to contain at least two different candidate transforming genes: 1) a c-*myc* gene activated by integration of viral LTR sequences and 2) a distinct cellular gene, not linked to LLV DNA, which efficiently induces transformation of NIH 3T3 cells. Activation of cellular transforming genes that are not linked to viral DNA has also been detected in mammary carcinomas induced by mouse mammary tumor virus that, like LLV, does not appear to contain a specific viral transforming gene and induces tumors with long latent periods (Lane et al., 1981).

The long latent period and pathogenesis of bursal lymphomas induced by LLV suggests a process that includes several preneoplastic and neoplastic stages and that might be expected to involve more than a single transformation event. Preneoplastic transformed follicles are observed in the bursas of most birds (10–100 transformed follicles per bursa) within approximately 40 days after LLV infection (Neiman et al., 1980). The majority of these preneoplastic follicles regress, but 1–2% appear to develop into neoplastic nodules that are the presumed precursors of metastatic lymphomas (Neiman et al., 1980). Proliferation of preneoplastic transformed follicles thus appears to be an early event in the disease process that is sometimes followed by progression to neoplasia. Viral activation of c-*myc* may thus be involved in a stage of lymphomagenesis that precedes or complements activation of transforming genes detected by transfection. For example, activation of c-*myc* might be an early event in lymphomagenesis resulting in proliferation of preneoplastic transformed follicles. Progression to neoplasia might then involve mutations or gene rearrangements within cells of this preneoplastic proliferative population resulting in activation of transforming genes detected by transfection. Neoplastic nodules and metastatic lymphomas contain both activated c-*myc* genes and activated transforming genes detectable in the transfection assay (Cooper and Neiman, 1980, 1981). However, the activity of either of these genes in preneoplastic transformed follicles is not known. Further work will thus be required to elucidate the roles of these two transforming genes in the disease process.

## MOLECULAR CLONING OF A CELLULAR TRANSFORMING GENE ACTIVATED IN BURSAL LYMPHOMAS

The cellular transforming gene activated in a bursal lymphoma cell line (RP-9) has recently been isolated by molecular cloning (Goubin et al., in preparation). DNA of NIH cells transformed by RP-9 DNA was subjected to partial digestion with the four-base recognition restriction endonuclease Mbo I and cloned in λ Charon 30 to produce a library of approximately 200,000 phage. This library was screened by sib-selection (Cavalli-Sforza and Lederberg, 1956) using transfection of NIH 3T3 cells to identify recombinant phage containing a biologically active transforming gene. The library was initially divided into 10 pools of 20,000 phage that were amplified and assayed by transfection. One pool, which was positive in this screen, was then divided into pools of 4,000 phage. A positive pool from this screen was divided into pools of 250 phage and a positive pool was used for isolation of individual phage plaques. A single positive phage plaque was then isolated and purified. This clone, which is designated λ ChBlym-1, contains an insert of approximately 4.5 Kb and induces transformation of NIH 3T3 cells with an efficiency corresponding to approximately 2,000 transformants/μg cell DNA. Restriction fragments of normal chicken DNA are detected in blot hybridizations with λ ChBlym-1 probe indicating that the λ ChBlym-1 transforming gene is homologous to normal chicken sequences (Goubin et al., in preparation). Blot hybridization analysis also indicates that the gene homologous to λ ChBlym-1 is conserved in mammalian species, including man.

## SPECIFICITY OF TRANSFORMING GENES ACTIVATED IN NEOPLASMS OF DIFFERENTIATED CELLS

Restriction endonuclease analysis of transforming genes activated in mammary carcinomas indicated that the same or closely related transforming genes were activated in six independent mouse mammary carcinomas, induced by either viral or chemical carcinogens, and in a human mammary carcinoma cell line (Lane et al., 1981). The susceptibility of mammary carcinoma transforming genes to restriction endonuclease cleavage differed from those of transforming genes activated in chicken bursal lymphomas (Cooper et al., 1982), human bladder carcinomas (Krontiris and Cooper, 1981), NIH cells transformed by fragments of normal chicken and human DNA (Cooper et al., 1980, 1982), chemically transformed mouse fibroblasts (Shilo and Weinberg, 1981), and mouse and human lymphoid neoplasms (Lane et al., in preparation). Analysis of the transforming genes of B and T cell neoplasms of mouse and human origin further indicates that specific transforming genes are activated in neoplasms representative of specific stages of lymphocyte differentiation (Lane et al., in preparation).

These results thus indicate that specific transforming genes are activated in neoplasms of particular differentiated cell types and suggest the hypothesis that neoplasia may result from abnormal expression of genes that are normally involved in differentiation-specific growth control. One possibility is that, in some differentiated cell types, only one or a few of the total number of potential transforming genes are susceptible targets for activation by mutations or DNA rearrangements. Alternatively, some differentiated cells might be phenotypically sensitive to the effects of only one or a few potential transforming genes.

Molecular cloning and analysis of the chicken bursal lymphoma transforming gene indicates that these sequences are highly conserved in evolution. It will thus be of interest to investigate the role of these genes in normal lymphocytes and in mouse and human neoplasms corresponding to different stages of lymphocyte differentiation.

## LITERATURE CITED

Cavalli-Sforza, L.L., and J. Lederberg (1956). Isolation of preadaptive mutants in bacteria by sib-selection. Genetics, *41*:367–381.

Cooper, G.M., M.A. Lane, T.G. Krontiris, and G. Goubin (1981). Analysis of cellular transforming genes by transfection. In Klein, G., ed.: Advances in Viral Oncology. New York: Raven Press, Vol. I, in press.

Cooper, G.M., and P.E. Neiman (1980). Transforming genes of neoplasms induced by avian lymphoid leukosis viruses. Nature, *287*:656–659.

Cooper, G.M., and P.E. Neiman (1981) Two distinct candidate transforming genes of lymphoid leukosis virus-induced neoplasms. Nature, *292*:857–858.

Cooper, G.M., S. Okenquist, and L. Silverman (1980). Transforming activity of DNA of chemically transformed and normal cells. Nature, *284*:418–421.

Goubin, G., P.E. Neiman, J. Luce, and G.M. Cooper Molecular cloning and analysis of a cellular transforming gene activated in avian bursal lymphomas. (in preparation).

Hayward, W.S., B.G. Neel, and S.M. Astrin (1981). Activation of a cellular *onc* gene by promoter insertion in ALV-induced lymphoid leukosis. Nature, *290*:475–480.

Krontiris, T.G., and G.M. Cooper (1981). Transforming activity of human tumor DNAs. Proc. Natl. Acad. Sci. U.S.A., *78*:1181–1184.

Lane, M.A., A. Sainten, and G.M. Cooper (1981). Activation of related transforming genes in mouse and human mammary carcinomas. Proc. Natl. Acad. Sci. U.S.A., *78*:5185–5189.

Lane, M.A., A. Sainten, and G.M. Cooper Transforming genes of T- and B-lymphocyte neoplasms of mouse and human origin. (in preparation).

Neiman, P.E., L.N. Payne, L. Jordan, and R.A. Weiss (1980). Malignant lymphoma of the bursa of Fabricius: Analysis of early transformation. In Essex, M., Todaro, G. eds.: Viruses in Naturally Occurring Cancer. Cold Spring Harbor Conference on Cell Proliferation, Vol. 7, pp. 519–528.

Shih, C., L.C. Padhy, M. Murray, and R.A. Weinberg (1981). Transforming genes of carcinomas and neuroblastomas introduced into mouse fibroblasts. Nature, *290:*261–264.

Shih, C., B.Z. Shilo, M.P. Goldfarb, A. Dannenberg, and R.A. Weinberg (1979). Passage of phenotypes of chemically transformed cells via transfection of DNA and chromatin. Proc. Natl. Acad. Sci. U.S.A., *76:*5714–5718.

Shilo, B.Z., and R.A. Weinberg (1981). Unique transforming gene in carcinogen-transformed mouse cells. Nature, *289:*607–609.

Southern, E.M. (1975). Detection of specific sequences among DNA fragments separated by gel electrophoresis. J. Mol. Biol., *98:*503–517.

JOURNAL OF CELLULAR PHYSIOLOGY SUPPLEMENT 1:213–217 (1982)

# Insertion of Drug Resistance Genes in Animals

MENASHE BAR-ELI, KAREN E. MERCOLA, DENNIS J. SLAMON, NILS MAURITZSON, HOWARD D. STANG, AND MARTIN J. CLINE*
*Department of Medicine, Center for Health Sciences, University of California at Los Angeles, Los Angeles, California 90024*

## TECHNIQUES FOR INSERTING NEW GENETIC INFORMATION INTO ANIMAL CELLS

A variety of techniques have been developed for introducing new genetic information into mammalian cells growing in tissue culture. All the techniques have in common the feature that selective conditions are necessary to favor the growth of the "transformed" cells that have acquired the new information.

Cell fusion techniques involve the physical fusion of cell membranes. After cell fusion and subsequent nuclear fusion between paired cells, a hybrid cell containing chromosomes from both parents is formed. Somatic cell hybridization has been used to map over 200 mammalian gene loci, including human genes (1, 2). In fusions between cells of different species, chromosome loss generally occurs, such that the proliferating hybrid cell contains only a subset of the genetic material of one of the parents (3).

Polyethylene glycol and inactivated Sendai virus have been most widely used as agents facilitating cell fusion (4). A variation of this approach has utilized artificially constructed lipid vesicles (liposomes) (5, 6) or erythrocytes (7), which are first loaded with nucleic acid and then fused with target cells. However, because of molecular size restrictions, these techniques have utilized RNA more often than DNA.

## CHROMOSOME-MEDIATED TRANSFER OF GENETIC INFORMATION

The technique of chromosome-mediated gene transfer utilizes isolated metaphase chromosomes as vectors for the transfer of genetic information (8–12). This process typically results in transfer of subchromosomal amounts of genetic material in a range of sizes up to 1% of the human haploid genome (13). In chromosome-mediated transfer the "transgenome" has been shown to be associated with several recipient chromosomes.

The technique of chromosome-mediated transfer makes it possible to determine gene order on a given chromosome. For example, Klobutcher and Ruddle (13) have prepared a number of stable subclones from the chromosomal transfer of linked human genes for thymidine kinase, galactokinase, and procollagen Type I. Relative sizes of human chromosome fragments in the clones were determined from metaphase spreads and measurement of enzymatic activity. Based on the assumption of a single break and fusion of a chromosomal fragment in the deletion analysis, a gene order of centromere galactokinase (thymidine kinase-procollagen Type I) on human chromosome 17 was determined.

## MICROCELL-MEDIATED TRANSFER

Microcells are single chromosomes or small groups of chromosomes contained within a nuclear membrane. They are prepared as follows: Appropriate tissue culture cells containing a potentially useful gene are arrested in mitosis for 24 to 95 hours with colchicine. The chromosomes accumulate in grape-like clusters of microcells on the cell surface. The microcells can be recovered free of cell cytoplasm by ultracentrifugation in the presence of cytochalasin B (14). The cytoplasm sticks to the surface of plastic on which the cells are grown, while the separated microcells and nuclei and some cell debris are pulled away under high centrifugal force. The microcells can then be physically separated from contaminating cells and nuclei because of their small size. Techniques also exist for isolating microcells from cells that grow in suspension culture rather than adherent to a surface (15). Fournier and Ruddle have successfully transferred small numbers of murine chromosomes into tissue culture cells of the mouse, Chinese hamster, and man (4).

## DNA-MEDIATED TRANSFORMATION

A method utilizing cellular endocytosis of calcium microprecipitates of DNA has been used to insert highly purified DNA sequences containing specific genes (16). These proce-

*To whom reprint requests/correspondence should be addressed.

0021-9541/82/SUPPL. 1-0213$02.00 

dures transform only a very small proportion of a recipient cell population, perhaps only one cell in a million. It is for this reason that selective techniques for favoring the growth of cells that have taken up the foreign DNA become very important. For example, the transformation of mouse cells lacking thymidine kinase ($Tk^-$ cells) have been utilized extensively as a model since cells transformed to $Tk^+$ phenotype are readily selected in medium containing hypoxanthine, aminopterin, and thymidine (HAT medium). $Tk^+$ cell clones have been produced in this manner by transformation of $Tk^-$ cells with purified TK genes from Herpes virus. High molecular weight DNA from a variety of species has also been used to transfer Tk activity to $Tk^-$ cells. In each instance the Tk produced is that of the DNA donor (21).

The DNA-mediated transformation technique can also be used to simultaneously insert two or more genes into cells—a process known as cotransformation. For example, mouse L cells have been simultaneously transformed with Herpes virus Tk and rabbit β globin (22). In this experiment the two genes were physically linked together. In other experiments two separate genes are coprecipitated with calcium phosphate and exposed to target cells. In general, cells competent to incorporate exogenous DNA will take up several varieties simultaneously.

## DIRECT INJECTION OF NEW GENES

Genetic material can be injected directly into the nucleus of cells. In some instances this material is integrated and expressed. Drosophila histone genes have been transcribed in Xenopus oocytes after injection of these genes together with SV-40 DNA (23). In 1974 Jaenisch and Mintz injected DNA from SV-40 into mouse embryos at the blastocyst (32-64 cell) stage and reimplanted the blastocysts into foster mothers. Some of the embryos reached full development and carried SV-40 DNA sequences one year later (24).

## USE OF VIRAL VECTORS TO INSERT NEW GENES

Modified SV-40 has been used as a vector to introduce cloned genes into mammalian cells. In combining prokaryotic or eukaryotic genes with SV-40 it is often necessary to delete genes that are essential for growth and replication of the virus (25, 26). The defective recombinant virus must be propagated in the presence of a helper virus. Using this approach, a variety of genes have been introduced into monkey kidney cells. Important questions concerning transcription and processing of RNA in eukaryotes have been approached using this system (31–33), but applicability of these methods in living animals is limited due to the infectious nature and transforming properties of the SV-40 virus.

Recently, discrete portions of the SV-40 viral genome have been covalently linked to specific prokaryotic and eukaryotic genes and used in DNA-mediated transformation experiments. The early gene promoter, which controls transcription of early region genes in SV-40, is active in a wide variety of host cells. A recombinant molecule made from covalent linkage of viral promoter and a gene such as β globin may facilitate active transcription of the globin gene in host cells (31–33).

## INSERTION OF NEW GENETIC INFORMATION INTO ADULT ANIMALS

Recently we described experiments aimed at inserting new genes into the bone marrow cells of living mice (36–38) based on previous studies in tissue culture (17–22, 28–35). Our strategy was based upon the possibility of introducing into hematopoietic stem cells new genes conveying resistance to the anticancer drug methotrexate (Mtx) and the administration of Mtx to the recipient animals to exert selective pressure favoring proliferation of cells incorporating and expressing the new genes. The DNA-mediated transformation technique was used. The marrow cells were subsequently returned to irradiated recipient animals where they proliferated and reconstituted hematopoiesis. Two sources of drug resistance genes were used: 1) high molecular weight DNA from a mouse cell line highly resistant to methotrexate and containing reiterated sequences of dihydrofolate reductase (DHFR, the enzyme inhibited by Mtx) [this DHFR was also unusual in that it had a reduced affinity for Mtx (38)] and 2) a purified gene from Herpes simplex virus coding for thymidine kinase (HSVTk).

## METHOTREXATE-RESISTANT DHFR

The commonly used folic acid antagonist, methotrexate, binds to and inhibits the mammalian enzyme, dihydrofolate reductase. Cell lines resistant to Mtx have been selected by mutagenesis and/or growth in medium in which Mtx concentration is increased in a stepwise fashion. Three mechanisms of resistance have been identified: mutations affecting intracellular transport of drug, mutations that alter the affinity of DHFR for Mtx, and reduplication of a presumably normal DHFR gene resulting in increased rates of enzyme synthesis and of cellular DHFR concentration. Gene duplication is apparently the commonest mechanism of acquired Mtx resistance. In most resistant cell lines the number of copies of the DHFR gene and the rate of mRNA synthesis are directly related (39).

Kellems et al. (40) described a Mtx-resistant mouse fibroblast cell line designated 3T6R1, which had an approximate 30-fold increase in the number of DHFR genes, of mRNA and enzyme synthesis above wild-type Mtx-sensitive fibroblasts. We extracted high molecular weight DNA from Mtx-resistant 3T6R1 cells and used it to transform mouse bone marrow cells in vitro utilizing the $CaPO_4$ technique. Transformed and control marrow cells were mixed and injected into irradiated syngeneic mice. During the period required to reestablish hematopoiesis from the injected marrow, Mtx was administered to recipient animals to favor proliferation of those stem cells that incorporated and expressed additional copies of the DHFR genes. A chromosomal marker (T6T6) was used to identify the progeny of transformed cells. A predominance of marrow cells with the karyotype of the transformed cell population was observed in most Mtx-treated recipient animals (36). Mtx-resistance and elevated levels of DHFR in hematopoietic cells were also noted (36).

These observations suggest successful introduction of new functional DHFR genes into marrow stem cells.

It is not known what increase above normal levels of DHFR are necessary to endow a stem cell with a proliferative advantage in an Mtx-treated mouse. If the required increase is more than twofold the normal levels as suggested by earlier observations, then several explanations of the predominance of the transformed marrow are possible: integration and expression of two or more normal DHFR genes simultaneously in the same stem cell, integration of a single normal gene which subsequently reduplicates in the cell in vivo to produce multiple functional copies of the DHFR gene, and integration of one or a few copies of a mutant gene producing a highly Mtx-resistant enzyme. In order to explore these alternatives we examined the DHFR of Mtx-sensitive 3T6 cells and Mtx-resistant 3T6R1 cells. We found that 3T6R1 DHFR has different properties from the normal enzyme and is extremely resistant to the inhibitory effect of Mtx (38).

## INHIBITION BY MTX

Equal amounts of cellular extract from either Mtx-sensitive 3T6 cells or Mtx-resistant 3T6R1 cells were incubated with increasing quantities of methotrexate and assayed for residual DHFR activity. From the titration curves shown in Figure 1, it is evident that the mutant and the wild-type enzymes differ in their sensitivity to inhibition by this folate antagonist. Mtx at $10^{-5}$ M inhibited 85–90% of the wild-type enzyme but only 25–30% of the DHFR activity of 3T6R1 cells. Assuming that the binding of Mtx to the reductases and inhibition of enzyme occurs in a stoichiometric manner (41), the increased resistance of 3T6R1 enzyme to Mtx inhibition probably results from the reduced binding of drug to a mutant enzyme.

## POLYACRYLAMIDE GEL ANALYSIS

Crude extracts from either wild-type or 3T6R1 cells were subjected to electrophoresis on native 7.5% polyacrylamide gels. Using a stain for DHFR enzymatic activity (42), the wild-type cells showed two minor bands and one major band while the Mtx-resistant cells demonstrated one minor and one major band, which differ in their electrophoretic mobility from the wild-type major band. Since each protein band corresponds to an area of enzymatic staining, all are dihydrofolate reductase. This conclusion is supported by the results of eluting the proteins from the gels and assaying each fraction for DHFR activity. The gels were sliced into 3 mm segments, eluted overnight, and then assayed for DHFR activity (Fig. 2). It is evident that the major band of DHFR activity differs in its electrophoretic mobility in wild-type and Mtx-resistant 3T6 cells. The mutant enzyme has a major band of DHFR activity migrating in advance of the major band of wild-type enzyme. The mutant enzyme

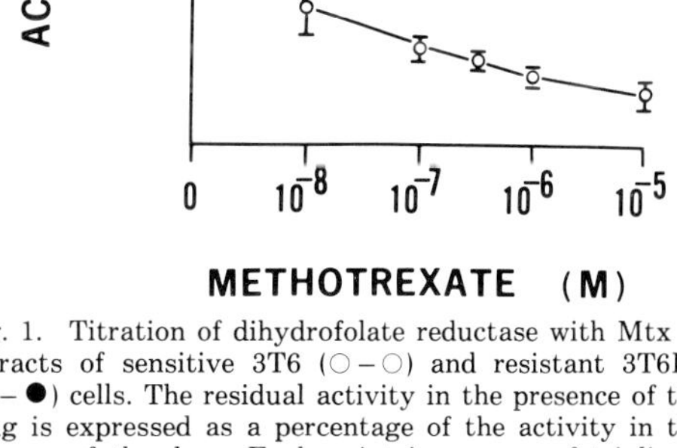


Fig. 1. Titration of dihydrofolate reductase with Mtx in extracts of sensitive 3T6 (○ – ○) and resistant 3T6R1 (● – ●) cells. The residual activity in the presence of the drug is expressed as a percentage of the activity in the absence of the drug. Each point is a mean of triplicate reaction.

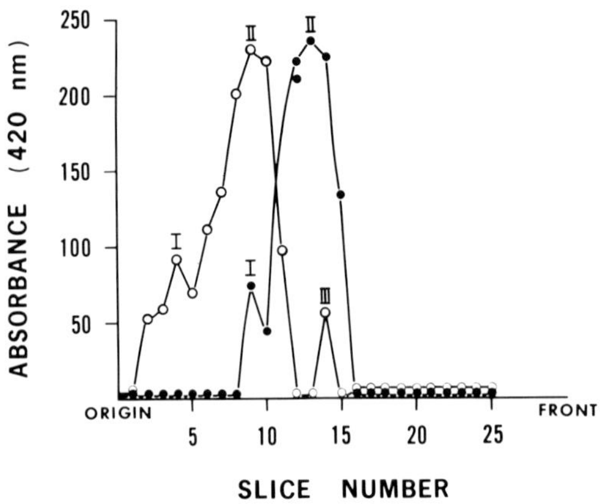


Fig. 2. Dihydrofolate reductase activity of wild-type 3T6 (○—○) and mutant 3T6R1 (●—●) extracts eluted from slices after polyacrylamide gel electrophoresis. Gels were sliced into 3 mm segments, eluted overnight in Tris buffer pH 7.5, and then assayed for DHFR activity.

also has a minor band with mobility similar to that of the wild-type major band. Wild-type DHFR has a minor band that migrated close to the mutant major band, suggesting that the origin of the altered mutant enzyme might be from band III of the wild-type DHFR.

Our observation that the DHFR of Mtx-resistant 3T6R1 differs in several properties from the enzyme of the Mtx-sensitive cell line suggests that 3T6R1 cells contain a structurally altered DHFR. The major features distinguishing the 3T6R1 enzyme are resistance to inhibition by Mtx, heat stability (data not shown), and an altered pattern of electrophoretic mobility. Since Mtx is a stoichiometric inhibitor of reductase, it is likely that the enzyme from resistant cells has an altered affinity for the drug.

A high proportion of Mtx-treated mice receiving bone marrow cells exposed to 3T6R1 DNA demonstrate evidence of successful transformation including predominance of the karyotype of the transformed marrow population and relative resistance to Mtx of hematopoietic tissues. However, we have been unsuccessful in obtaining evidence for successful transformation of bone marrow when we have used DNA from other Mtx-resistance cell lines. These cell lines are thought to contain reiterated copies of normal mouse DHFR ($Mtx^R$ L121O cells) or reiterated copies of a mutant hamster DHFR (CHO $Pro^-$ Mtx, Flintoff, W.F., et al. 1976, Somat. Cell Genet. 2:245–261). The mutant hamster DHFR is, however, far less resistant to Mtx than the reductase from 3T6R1 cells. These observations suggest that the successful introduction of new functional DHFR genes into marrow stem cells is most likely due to integration of one or a few copies of a mutant gene producing a highly Mtx-resistant dihydrofolate reductase.

## LITERATURE CITED

1. Ruddle, F.H., and Creagan, R.P. Parasexual approaches to the genetics of man. Annu. Rev. Genet., *9:*407–486, 1975.
2. McKusick, V.A., and Ruddle, F.H. The status of the gene map of the human chromosomes. Science, *177:*390–405, 1977.
3. Fournier, R.E.K., and Ruddle, F.H. Microcell-mediated chromosome transfer. In Molecular Human Cytogenetics, ICN-UCLA Symposia on Molecular and Cell Biology. *7:*189–199, 1977.
4. Fournier, R.E.K., and Ruddle, F.H. Microcell-mediated transfer of murine chromosomes into mouse, Chinese hamster, and human somatic cells. Proc. Natl. Acad. Sci. U.S.A., *74:*319–323, 1977.
5. Ostro, M.J., Giacomoni, D., Lavelle, D., Paxton, W., and Dray, S. Evidence for translation of rabbit globin messenger RNA after liposome-mediated insertion into a human cell line. Nature, *274:*921–923, 1978.
6. Dimitriadis, G.J. Translation of rabbit globin in RNA introduced by liposomes into mouse lymphocytes. Nature, *274:*923–924, 1978.
7. Rechsteiner, M. Red-cell-mediated microinjection. Natl. Cancer Inst. Monogr., *48:*57–64, 1978.
8. McBride, O.W., and Ozer, H.L. Transfer of genetic information by purified metaphase chromosomes. Proc. Natl. Acad. Sci. U.S.A., *70:*1258–1262, 1973.
9. McBride, O.W., Burch, J.W., and Ruddle, F.H. Cotransfer of thymidine kinase and galactokinase genes by chromosome-mediated gene transfer. Proc. Natl. Acad. Sci. U.S.A., *75:*914–918, 1978.
10. Degnen, G.E., Miller, I.L., Adelberg, E.A., and Eisenstadt, J.M. Chromosome-mediated gene transfer between closely related strains of cultured mouse cells. Proc. Natl. Acad. Sci. U.S.A., *73:*2838–2842, 1976.
11. Willecke, K., Lange, R., Krüger, A., and Reber, T. Cotransfer of two linked genes into cultured mouse cells. Proc. Natl. Acad. Sci. U.S.A., *73:*1274–1278, 1976.
12. Wullem, G.J., van der Horst, and J., Bootsma, D. Incorporation of isolated chromosomes and induction of hypoxanthine phosphoribosyltransferase in Chinese hamster cells. Somat. Cell Genet., *1:*137–152, 1975.
13. Klobutcher, L.A., and Ruddle, F.H. Phenotype stabilization and integration of transferred material in chromosome-mediated gene transfer. Nature, *280:*657–660, 1979.
14. Ege, T., and Ringertz, N.R. Preparation of microcells by enucleation of micronucleate cells. Exp. Cell Res., *87:*378–382, 1974.
15. Wigler, M.H., Neugut, A.I., and Weinstein, I.B. Enucleation of mammalian cells in suspension. Meth. Cell Biol., *14:*87–93, 1976.
16. Graham, F.L., and van der Eb, A.J. A new technique for the assay of infectivity of human adenovirus 5 DNA. Virology, *52:*456–567, 1973.
17. Wigler, M., Sweet, R., Sim, G.K., et al. Transformation of mammalian cells with genes from prokaryotes and eukaryotes. Cell, *16:*777–785, 1979.
18. Wigler, M., Silverstein, S., Lee, L-S, et al. Transfer of purified herpes virus thymidine kinase gene to cultured mouse cells. Cell, *11:*223–232, 1977.

19. Pellicer, A., Wigler, M., Axel, R., and Silverstein, S. The transfer and stable integration of the HSV thymidine kinase gene into mouse cells. Cell, *14:*133–141, 1978.
20. Bacchetti, S., and Graham, F.L. Transfer of gene for thymidine kinase-deficient human cells by purified herpes simplex viral DNA. Proc. Natl. Acad. Sci. U.S.A., *74:*1590–1594, 1977.
21. Wigler, M., Pellicer, A., Silverstein, S., and Axel, R. Biochemical transfer of single-copy eukaryotic genes using total cellular DNA donor. Cell, *14:*725–731, 1978.
22. Mantei, N., Boll, W., and Weissman, C. Rabbit β-globin messenger-RNA production in mouse L-cells transformed with cloned rabbit β-globin chromosomal DNA. Nature, *281:*40–46, 1979.
23. Mertz, J.E., and Gurdon, J.B. Purified DNAs are transcribed after microinjection into xenopus oocytes. Proc. Natl. Acad. Sci U.S.A., *74:*1502–1506, 1977.
24. Jaenisch, R., and Mintz, B. Simian virus 40 DNA sequences in DNA of healthy adult mice derived from preimplantation blastocysts injected with viral DNA. Proc. Natl. Acad. Sci. U.S.A., *71:*1250–1254, 1974.
25. Cole, C.N., Landers, T., Goff, S.P., et al. Physical and genetic characterization of deletion mutants of simian virus-40 constructed in vitro. J. Virol., *24:*277–294, 1977.
26. Lai, C.J., and Nathans, D. Mapping temperature-sensitive mutants of simian virus 40: rescue of mutants by fragments of viral DNA. Virology, *60:*466–475, 1974.
27. Peterson, J.L., and McBride, O.W. Cotransfer of linked eukaryotic genes and efficient transfer of hypoxanthine phosphoribosyl transferase by DNA-mediated gene transfer. Proc. Natl. Acad. Sci. U.S.A., *77:*1583–1587, 1980.
28. Wold, B., Wigler, M., Lacy, E., Maniatis, T., et al. Introduction and expression of a rabbit β-globin gene in mouse fibroblasts. Proc. Natl. Acad. Sci. U.S.A., *76:*5684–5688, 1979.
29. Breathnack, R., Mantei, N., and Chambon, P. Correct splicing of a chicken ovalbumin gene transcript in mouse L-cells. Proc. Natl. Acad. Sci. U.S.A., *77:*740–744, 1980.
30. Wigler, M., Pellicer, A., Silverstein, S., et al. DNA-mediated transfer of the adenine phosphoribosyl transferase by locus into mammalian cells. Proc. Natl. Acad. Sci. U.S.A., *76:*1373–1376, 1979.
31. Mulligan, R.C., Howard, B.H., and Berg, P. Synthesis of rabbit β-globin in cultured monkey kidney cells following infection with a SV 40 β-globin recombinant genome. Nature, *277:*108–114, 1979.
32. Hamer, D.H., and Leder, P. SV 40 recombinants carrying a functional RNA splice junction and polyadenylation site from the chromosomal mouse β maj-globin gene. Cell, *17:*737–747, 1979.
33. Hamer, D.H., and Leder, P. Expression of the chromosomal mouse β maj-globin gene cloned in SV 40. Nature, *281:*35–40, 1979.
34. Upcroft, P., Skolnik, H., Upcroft, J.A., et al. Transduction of a bacterial gene into mammalian cells. Proc. Natl. Acad. Sci. U.S.A., *75:*2117–2121, 1978.
35. Capecchi, M. Cell *22:*479–488, 1980.
36. Cline, M.J., Stang, H., Mercola, K., Morse, L., Ruprecht, R., Browne, J., and Salser, W. Gene transfer in intact animals. Nature, *284:*422–425, 1980.
37. Mercola, K.E., Stang, H.D., Browne, J., Salser, W., and Cline, M.J. Insertion of a new gene of viral origin into bone marrow cells of mice. Science, *208:*1033–1035, 1980.
38. Bar-Eli, M., and Cline, M.J. Cold Spring Harbor Symposium on Production and Insertion of Mammalian Vectors. (in press), 1981.
39. Schimke, R.T., Kaufman, R.J., Alt, F.W., and Kellems, R.E. Gene amplification and drug resistance in cultured murine cells. Science, *202:*1051, 1978.
40. Kellems, R.E., Morhenn, V.B., Pfendt, E.V., et al. Polyoma virus and cyclic AMP-mediated control of dihydrofolate reductase mRNA abundance in methotrexate-resistant mouse fibroblasts. J. Biol. Chem., *254:*309, 1979.
41. Blakley, R.L. The Biochemistry of Folic Acid and Related Pteridines. Amsterdam: North-Holland Publishing Co., 1969, pp. 139–187.
42. Hiebert, M., Gauldie, J., and Hillcoat, B.L. Multiple enzyme forms from protein-bromophenol blue interaction during gel electrophoresis. Anal. Biochem., *46:*433, 1972.

JOURNAL OF CELLULAR PHYSIOLOGY SUPPLEMENT 1:219–226 (1982)

# Gene Transfer Into the Mouse Germ-Line

**FRANKLIN COSTANTINI* AND ELIZABETH LACY**
*Department of Zoology, University of Oxford, Oxford OX1 3PS, United Kingdom*

A cloned rabbit DNA fragment containing the adult β-globin gene has been introduced into the mouse germ-line by DNA microinjection into the pronuclei of fertilized mouse eggs. Twenty-four adult mice derived from eggs injected with the cloned DNA were screened for the presence of the rabbit gene in their liver DNA by Southern blot hybridization. Nine of these mice contain the rabbit gene in liver and at least six of them transmit it through the germ-line to a fraction of their offspring. Thus, it should be possible to derive permanent strains of mice carrying the foreign gene. The estimated numbers of copies of the rabbit gene per diploid mouse genome range from one or two up to 20 or more. The multiple copies of the rabbit DNA clone appear to be arranged in tandem head-to-tail arrays that are integrated into mouse chromosomal DNA. Chromosomal integration has been directly demonstrated for one mouse by in situ hybridization to metaphase chromosomes using a radioiodinated probe. In this mouse, most if not all the copies of the rabbit DNA fragment are integrated into one homologue of mouse chromosome 1. The possible use of this technique for the study of globin gene regulation is discussed.

The transfer of cloned genes into the tissues of living mice is potentially a useful method for studying the mechanisms that regulate gene expression during animal development. Gene transfer into the bone marrow cells of mice has been accomplished by introducing cloned genes into cultured erythroid cells that were subsequently transplanted into suitable recipient mice (Mercola et al., 1980; Cline et al., 1980). This approach may provide extremely useful animal models for gene therapy in particular tissues, but the new genes thus introduced are not inherited and can only be propagated to future generations by repeated bone marrow transplants. For many purposes it would be desirable to transfer a gene into the germ-line so that it would be inherited and carried in all the cells of future generations of mice throughout their development. Embryonal carcinoma cells have been proposed as a vehicle for gene transfer into mice (Mintz, 1977) because of their ability to contribute to mouse tissues (including the germ-line, in rare instances) after injection into embryos (Mintz and Illmensee, 1975; Papaioannou et al., 1978). However, it remains to be demonstrated that embryonal carcinoma cell lines can colonize the germ-line after transformation by foreign genes.

An alternative approach is the direct introduction of cloned genes into mice by DNA microinjection into eggs or early embryos. Several previous observations suggested that this approach might be feasible. First it was shown by Jaenisch and Mintz (1974) that SV40 DNA sequences persisted in the somatic tissues of adult mice derived from blastocysts that had been injected with SV40 DNA. Second, Moloney leukemia viral genomes could be introduced into mouse chromosomes and transmitted through the germ-line after infection of early embryos with the virus, despite the fact that mouse embryos are nonpermissive for replication of the virus (Jaenisch, 1976). Third, cloned genes carried in bacterial plasmids could be introduced into cultured cells at a high frequency by direct microinjection into cell nuclei (Capecchi, 1980; Anderson et al., 1980). Recently, Gordon et al. (1980) reported the introduction of a cloned herpes simplex virus-1 thymidine kinase gene into the tissues of a newborn mouse by injection of the cloned gene into the pronuclei of fertilized mouse eggs. Although the frequency of gene transfer was low and partial deletion or rearrangement of the injected DNA appeared to occur, these results indicated that cloned DNA molecules, and not only whole viral genomes, could be

*To whom correspondence/reprint requests should be addressed.

introduced into somatic tissues of the mouse by this method.

This paper describes our attempts to introduce a cloned rabbit β-globin gene into mice by microinjection of fertilized eggs. The DNA we have used (Fig. 1) consists of a 19 kilobase (kb) rabbit DNA fragment containing the adult β-globin gene and a β-like pseudogene, which is cloned in the bacteriophage lambda vector CH4A (λRβG2; Lacy et al., 1979; Hardison et al., 1979; Lacy and Maniatis, 1980). As we have recently reported (Costantini and Lacy, 1981), the rabbit β-globin clone has been introduced at a high frequency into the mouse germ-line

## MATERIALS AND METHODS

### Microinjection of fertilized mouse eggs

Fertilized one-cell eggs were recovered from (C57BL/6 × CBA/H)$F_1$ female mice after natural matings with (C57BL/6 × CBA/H)$F_1$ males. Eggs were recovered from the oviduct 3 to 7 hr before first cleavage, cultured in M16 medium (Whittingham, 1971) in a 5% $CO_2$/95% air atmosphere for up to 2 hr before injection. For injection, eggs were transferred in batches of 15–20 to a drop of PB1 medium (Whittingham and Wales, 1969) on a plastic Petri dish and overlaid with parafin oil. Injection was performed under brightfield illumination using a Leitz Diavert microscope fitted with Leitz micromanipulators. Injection pipettes were pulled from "kwik-fil" thin-walled glass tubing (Clark Electromedical Instrument CG100-TF) and had a tip diameter of less than one micron. After the pipette tips were filled with the DNA solution by capillary action, the barrel was filled with 3MM Fluorinert FC77 (Gordon et al., 1980), and the pipette was placed in a Leitz instrument holder connected to an air-filled 50 ml ground glass syringe (Graessmann et al., 1980).

Eggs were held by suction with a blunt-ended pipette and one pronucleus, either male or female, was injected with DNA by gentle pressure on the injection syringe until a slight pronuclear swelling was visible. The volume of DNA solution injected could not be accurately measured, but we estimate that it was between 0.5 and 2 pl. The λRβG2 DNA was isolated from phage particles as described by Maniatis et al (1978). The λRβG2 DNA used for injection was dissolved in 10 mM Tris-HCl pH 7.4, 0.1 mM EDTA at concentrations that varied from 5 μg/ml to 50 μg/ml for different batches of eggs. Thus the number of λRβG2 molecules injected was about 50–200 at the lowest concentration and 500–2000 at the highest concentration. Fifty percent of the eggs survived injection, and these were cultured for 30 min to 2 hr in M16 medium before being transferred in batches of 20–30 to the oviducts of pseudopregnant (C57BL/6 × CBA/H)$F_1$ female mice. Fifteen percent of these eggs developed to term.

### Screening mice for λRβG2 sequences

Total DNA was extracted from liver biopsy samples or whole newborn mice and screened for λRβG2 sequences by Southern blot hybridization (Southern, 1975) as previously described (Costantini and Lacy, 1981).

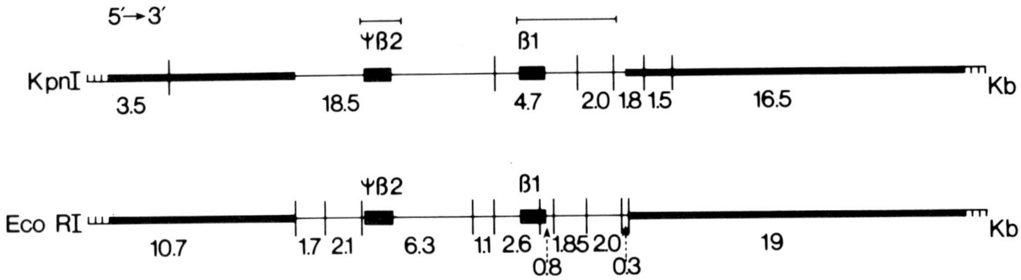


Fig. 1. Structure of clone λRβG2. Thick lines represent the bacteriophage vector and thin lines represent the rabbit chromosomal DNA insert. The solid boxes, labeled β1 and ψβ2, are the adult β-globin gene and a β-like pseudogene, respectively. The symbols ⊔⊔ and ⊓⊓ represent the single-standard cohesive ends of the bacteriophage lambda vector. Vertical lines show the location of KpnI (top) and EcoRI (bottom) restriction sites. The bars above the KpnI map are the DNA sequences used as probes in the experiment shown in Figure 2. Figures 1, 2, and 3 are reproduced from Nature by permission.

### In situ hybridization to metaphase chromosomes

Metaphase chromosome spreads were prepared from peripheral blood and Geimsa banded by Ted Evans and Mike Burtenshaw of Dunn School of Pathology, Oxford. In situ hybridization was performed as described by Robins et al. (1981). Slides were dipped for autoradiography in Ilford L4 emulsion and exposed for six days.

## RESULTS

### Detection of λRβG2 sequences in mouse liver DNA samples

Fifty-one mice were obtained from eggs that had been injected with a mixture of linear monomeric λRβG2 DNA (80–90%) and concatemeric λRβG2 DNA (10–20%). Twenty-four of these mice have been screened for the presence of λRβG2 sequences by Southern blot hybridization to total DNA extracted from liver biopsy samples using probes for the rabbit β-globin genes β1 and ψβ2 (see Fig. 1). Nine of these 24 mice (38%) have been shown to contain copies of λRβG2 in liver DNA (Costantini and Lacy, 1981). Figure 2 shows an example of the screening procedure in which the mouse DNA samples were digested with the restriction endonuclease KpnI, fractionated on agarose gels, transferred to nitrocellulose, and hybridized to the $^{32}$P-labeled rabbit β-globin probe. Of the nine mouse DNA samples on this gel, five (mice 3,4,7,9, and 13) show a hybridization pattern characteristic of λRβG2 DNA (lanes k and l), while the other four show no hybridization to the rabbit β-globin probe. The number of copies of λRβG2 sequences per average liver cell in these mice can be estimated from the intensity of hybridization relative to the hybridization with known amounts of λRβG2 DNA in lanes k and l. These numbers range from one to two copies per cell (mouse 7) up to 20 or more copies per cell (mouse 9).

### Transmission of λRβG2 through the mouse germ-line

To determine if the injected rabbit β-globin genes had become incorporated into germ-line cells, progeny from matings between the positive mice and normal mice were screened for the presence of λRβG2 sequences. The DNA for Southern blot hybridizations was prepared from either whole newborn mice or from liver biopsy samples from live progeny. All the mice tested so far (four males and two females) transmit the rabbit DNA fragments to a fraction of their offspring. (The frequencies of transmission so far observed are listed in Table 1). Figure 3 demonstrates the transmission of λRβG2 sequences by mouse 13 to two out of six of its progeny.

### The structure of λRβG2 copies in mouse DNAs

All the expected restriction fragments representing the rabbit DNA insert of λRβG2 are visible in KpnI digests (Fig. 2) and EcoRI digests (e.g., Fig. 3) of the mouse DNA samples. Thus a majority of the λRβG2 copies in each mouse have not undergone deletions or sequence rearrangements. When the mouse DNAs were digested with EcoRI (which cleaves the 19 kb and 10.7 kb lambda vector arms from the rabbit DNA insert) and probed with $^{32}$P-labeled λRβG2, a 30 kb band was seen in every case instead of the 19 kb and 10.7 kb bands. indicated that left and right arms of the lambda vector were covalently joined in the mouse cells. Such a result would be obtained if the λRβG2 molecules were replicating as circular episomes, or if they were integrated into the mouse DNA as head-to-tail linear concatemers. In addition to the above-mentioned fragments, most of the mouse DNAs show a few (usually one to three) fainter new EcoRI bands that hybridize with the λRβG2 probe. These bands are of different molecular weights

TABLE 1. Frequencies of transmission of λRβG2 sequences to offspring[1]

| Parent mouse number | Sex | Number of positive offspring/ total number tested |
|---|---|---|
| 4 | Male | 1/5 |
| 7 | Male | 2/4 |
| 13 | Male | 7/16 |
| 23 | Male | 4/10 |
| 37 | Female | 1/4 |
| 38 | Female | 1/4 |

[1]Parent mice are the mice which developed from eggs that were injected with λRβG2 and that are known to contain λRβG2 sequences in their liver DNA. The offspring were screened by Southern blot hybridization to DNA prepared from either liver biopsy samples or the whole newborn mouse.

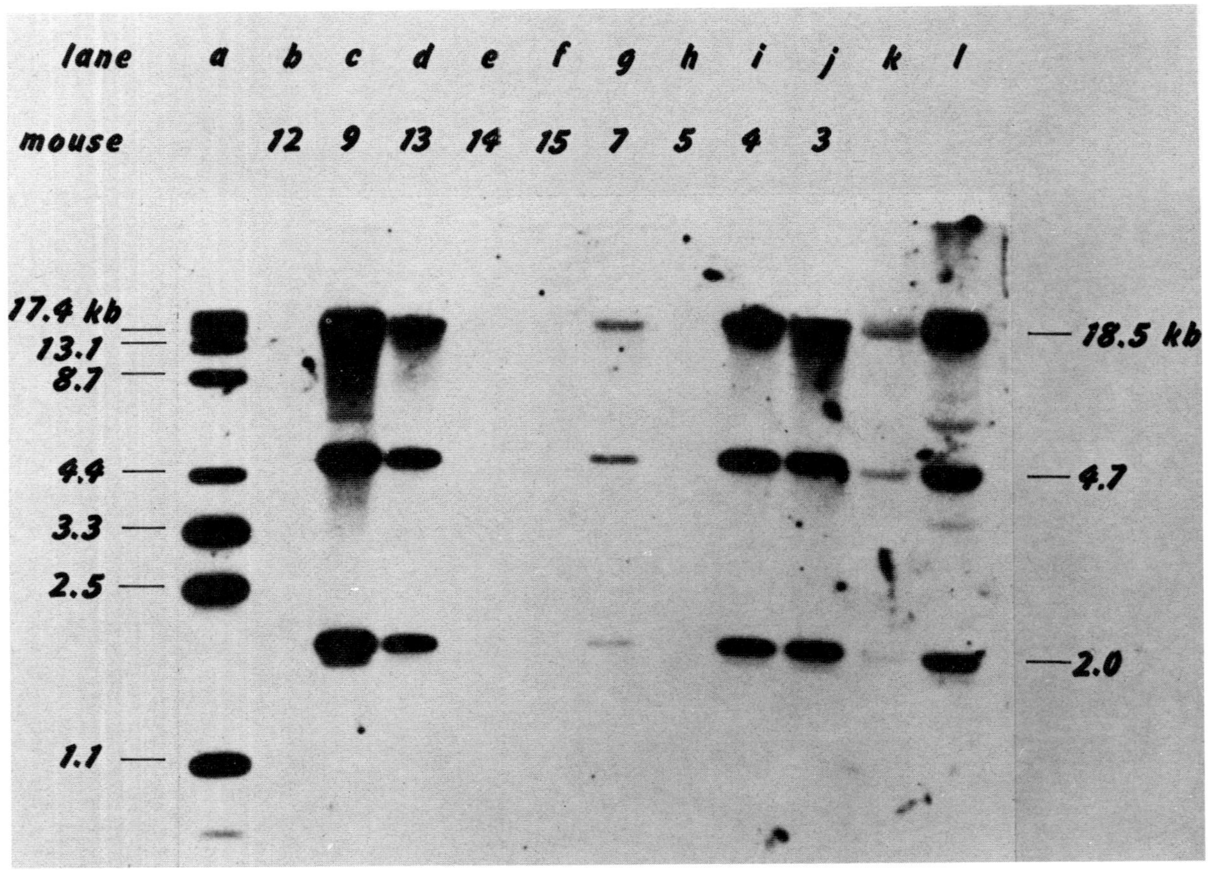


Fig. 2. Southern blot hybridization to mouse liver DNA samples. Mouse liver DNA samples were purified, digested with KpnI, fractionated on a 0.75% agarose gel, blotted to nitrocellulose, and hybridized to a $^{32}P$-labeled probe as described previously (Costantini and Lacy, 1981). The hybridization probe was a mixture of two recombinant plasmids containing the sequences from λRβG2 shown by the horizontal bars in Figure 1. Lane a contains molecular weight markers. Lanes b-j each contain 10 μg of KpnI digested liver DNA from mice derived from injected eggs. Lanes k and l contain, respectively, 80 and 800 pg of KpnI digested λRβG2 DNA as well as 10 μg of salmon sperm carrier DNA. These lanes represent the amount of hybridization expected for mouse DNAs containing one and ten copies of λRβG2 per diploid genome. The 18.5, 4.7, and 2.0 kb fragments of λRβG2 hybridize intensely, while the 1.8 kb fragment hybridizes only weakly with the probes used. The additional bands visible on the digest of cloned λRβG2 DNA are partial digestion products.

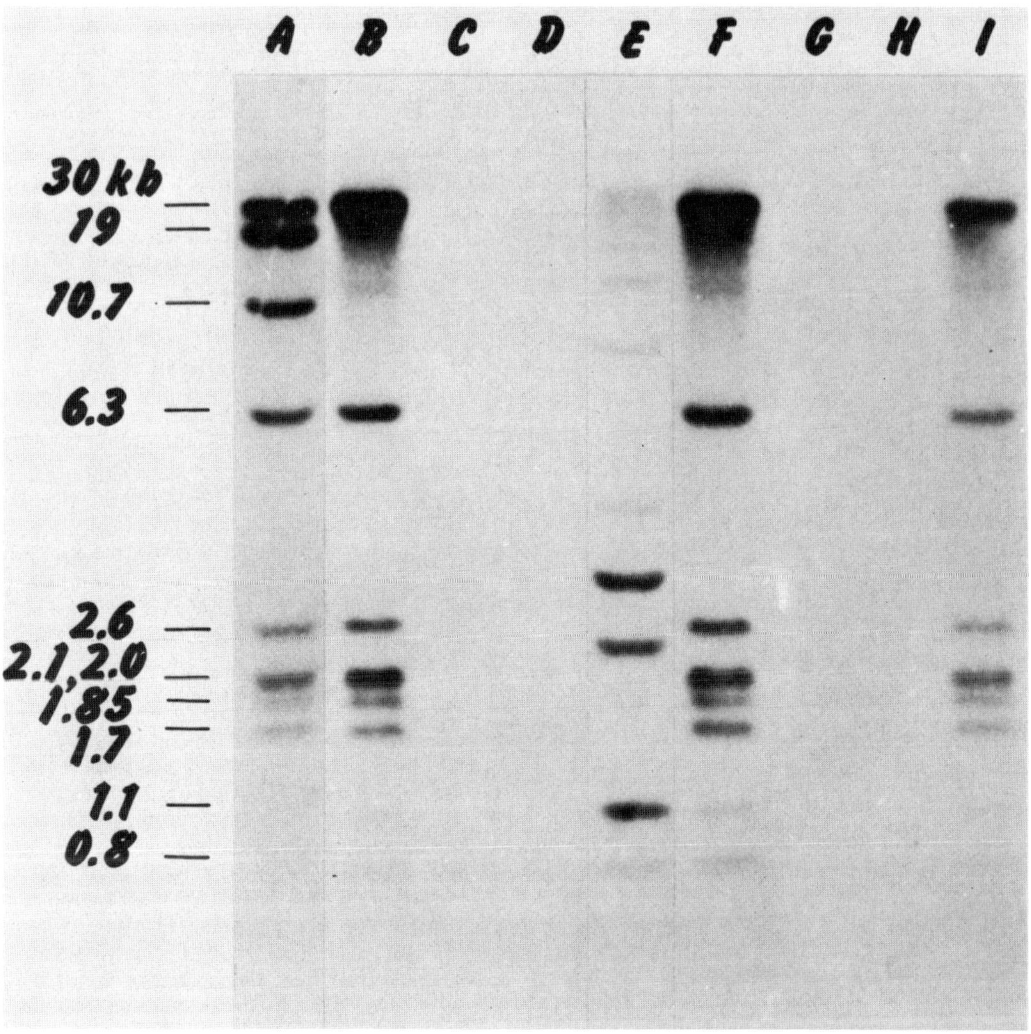


Fig. 3. Germ-line transmission of λRβG2 sequences by mouse 13. Mouse 13, a male, was mated to a normal CBA/H female and total DNA was isolated from six whole newborn offspring. Ten micrograms of liver DNA from mouse 13 (lane I) and 10 μg of DNA from each of six offspring (lanes B,C,D,F,G, and H) were digested with EcoRI and heated for 5 min at 65°C in 0.01 M Tris, pH 8. DNAs were fractionated on an agarose gel, Southern blotted, and hybridized to $^{32}$P-labeled λRβG2 DNA, as previously described (Costantini and Lacy, 1981). Also included was a mixture of linear and concatermeric λRβG2 DNA, digested with EcoRI (lane A), and molecular weight markers (lane E). The 19 and 10.7 kb bands in lane A represent the lambda vector arms of linear monomeric λRβG2, while the 30 kb band derives from the concatmers and consists of the two lambda arms joined by the lambda cohesive ends. The bands of between 0.8 and 6.3 kb are the EcoRI cleavage products of the rabbit DNA insert of λRβG2.

in each of the original positive mice and are inherited by the progeny (Costantini and Lacy, 1981). For example, the liver DNA of mouse 13 (lane I, Fig. 3) shows an EcoRI fragment of about 12 kb, and this new faint band is inherited by both of the positive progeny shown (lanes B and F) as well as by every other positive progeny so far identified (data not shown). As discussed below, we favor the interpretation that the λRβG2 copies are integrated into mouse chromosomes in tandem arrays, and that the new bands represent the EcoRI fragments generated at sites of recombination between λRβG2 and mouse chromosomal DNA.

Definitive evidence for chromosomal integration has so far been obtained for one parent mouse (mouse 23) by in situ hybridization to metaphase chromosome spreads prepared from peripheral blood. The chromosomes were identified by Geimsa banding, and hybridized to

$^{125}$I-labeled λRβG2 DNA (specific activity $10^9$ cpm/μg). In every metaphase spread examined (19 out of 19), a cluster of grains was observed over one homologue of chromosome 1, and no other consistent sites of labeling were apparent (data not shown). Although we cannot rule out the possibility that a minority of the λRβG2 copies in this mouse are located at other chromosomal sites, it appears that most of the copies are integrated into chromosome 1, presumably at a single site.

## DISCUSSION

We have shown that a large cloned eucaryotic DNA fragment may be introduced at a high frequency into the germ-line DNA of mice by microinjection into the pronuclei of fertilized eggs. Integration of the foreign DNA into a mouse chromosome has been demonstrated in one mouse by in situ hybridization. Therefore, the ability of a DNA fragment to become integrated into embryonic DNA and to be propagated in somatic and germ-line cells is not an exclusive property of viral genomes or viral DNA sequences. While some work has suggested that the frequency of gene transfer by microinjection into cultured cells is increased by linkage to SV40 DNA fragments containing the viral origin of replication (Capecchi, 1980), such procedures are apparently not necessary to achieve a high frequency of gene transfer into the mouse germ-line. It remains possible, of course, that viral vectors will be advantageous for some purposes, such as promoting the expression of a foreign gene in the mouse (Oskarsson et al., 1980) or introducing a gene into specific target tissues of an adult mouse (Jaenisch, 1976; Jaenisch, 1980).

The genomic Southern blotting data described above indicate that most of the λRβG2 copies in the mouse DNAs preserve the original structure of the rabbit DNA clone and have not undergone gross rearrangements. This contrasts with the results reported by Gordon et al. (1980), as well as other cases involving microinjection of recombinant DNAs into cells (Anderson et al, 1981), where deletions or rearrangements have been observed. The organization of the λRβG2 copies in the DNAs of our positive mice cannot be easily addressed by genomic Southern blotting, due to the large size of λRβG2 (48.5 kb) and the presence of multiple copies per diploid genome. The available data, however, are most consistent with the model that λRβG2 is integrated into mouse chromosomes in tandem, multiple arrays. First, a 30 kb EcoRI fragment derived from the covalent joining of left and right arms of lambda is seen in every positive mouse DNA. Second, the relatively weak hybridization to the new EcoRI bands (Fig. 3) indicates that these new fragments are present in fewer copies per cell than the fragments derived from the intact λRβG2 copies. Finally, a different set of new bands is characteristic of each of the original positive mice, and the characteristic set of bands is inherited by the offspring of each mouse. These are the results predicted if the new bands represent junction fragments between mouse chromosomal DNA and the ends of multiple λRβG2 arrays. The strongest evidence for chromosomal integration, however, is the intense hybridization of radioiodinated λRβG2 DNA to a single chromosome, chromosome 1, in metaphase spreads prepared from the blood of mouse 23. This mouse contains more than 20 copies of λRβG2 per diploid genome (at least in liver cells; see Fig. 2), and the in situ hybridization result indicates that most of them are integrated into a single chromosome. We do not, however, discount the possibility that a fraction of the λRβG2 sequences in this mouse might exist as free circles or might be integrated as individual copies at separate chromosomal locations.

It has been observed that linear DNA molecules become ligated into high molecular weight concatemers following injection into Xenopus eggs (Bendig, 1981) and also following introduction into cultured cells by calcium phosphate precipitation (Perucho et al., 1980; Robins et al., 1981). Possibly, the tandem arrays of λRβG2 in the mouse DNAs have been generated by similar processes in the mouse eggs. However, the concatemers of λRβG2 might derive directly from the concatemeric forms present in the injected DNA sample.

The high frequency of gene transfer that we have observed using λRβG2 indicates that this technique can be used routinely for the production of mice carrying different cloned foreign genes or altered forms of a gene generated by in vitro mutagenesis. With some practice it is possible to inject over 100 eggs in a few hours, and at the rate we have observed (see Materials and Methods) these eggs should yield several mice carrying the injected DNA fragment in germ line DNA. While the transmission of foreign DNA sequences through multiple generations remains to be demonstrated, the stability we have observed during two generations suggests that it will be possible to derive permanent lines of mice carrying copies of the rabbit β-globin gene (or other genes) in different chromosomal locations. An important question that can now be addressed concerns the ability of newly introduced genes to be expressed in the recipient mice. Experiments are currently in progress in our laboratory to determine whether the rabbit gene is expressed in the positive mice and, if so, whether the expression is tissue specific. Since chromosomal integration will presumably occur at random locations in the mouse genome, the regulation of a newly introduced gene may be affected by its unusual chromosomal environment. In this case its expression might be repressed or might be inappropriate in tissue specificity or developmental timing. Alternately, if all the essential cis-acting regulatory sequences are closely linked to the gene and contained on the transferred DNA fragment, it could be expressed normally despite an abnormal chromosomal location. This would also require that the regulatory mechanisms for the gene of interest are not entirely species specific. In the case of globin genes, there is evidence from studies on human thalassemic patients with deletions in the β-globin cluster (Fritsch et al., 1979; Van der Ploeg et al., 1980) that suggests that the expression of globin genes may be regulated by sequences at considerable distances (20 kb or more) from the affected genes. Thus it may be necessary to transfer extremely long DNA fragments in order for normal globin gene switching mechanisms to operate in a recipient mouse. Fragments up to 45 kb can be isolated using cosmid vectors, and our results indicate that it will be possible to transfer fragments of this length into the mouse genome. It may therefore be feasible to identify DNA sequences involved in globin gene switching by the transfer into mice of a series of large DNA fragments containing different portions of the rabbit (or another) β-globin gene cluster. While these questions can also be addressed by the transfer of globin genes into cultured erythroid cells, a potential advantage of the system we have described is that a foreign globin gene inherited through the mouse germ-line will be present in the hemopoietic stem cells and thus exposed to all the normal events of erythroid cell development, which may be essential for its normal expression.

## ACKNOWLEDGMENTS

We thank Chris Graham for helpful discussions and Sally Roberts and Ivor Mason for technical assistance. F.C. and E.L. were recipients of postdoctoral fellowships from the NIH (HDO5893-02) and the American Cancer Society (PF-1759), respectively. This research was supported by a grant to C.F. Graham from the Cancer Research Campaign.

## LITERATURE CITED

Anderson, W.F., L. Killos, L. Sanders-Haigh, P.J. Kretschmer, and E.G. Diacumakos (1980) Replication and expression of thymidine kinase and human globin genes microinjected into mouse fibroblasts. Proc.Natl.Acad.Sci. U.S.A., 77:5399–5403.

Anderson, W.F., P.J. Kretschmer, L. Sanders-Haigh, L. Killos, and E.G. Diacumakos (1981) Gene transfer and in vitro expression. In Stamatoyannopoulos, G., Nienhuis, A.W. eds.: Organization and Expression of Globin Genes. New York: Alan R. Liss, Inc., pp. 301–312.

Bendig, M.M. (1981) Persistence and expression of histone genes injected into Xenopus eggs in early development. Nature, 292:65–67.

Capecchi, M.R. (1980) High efficiency transformation by direct microinjection of DNA into cultured mammalian cells. Cell, 22:479–488.

Cline, M.J., H. Stang, K. Mercola, L. Morse, R. Ruprecht, J. Browne, and W. Salser (1980) Gene transfer in intact animals. Nature, 284:422–425.

Costantini, F., and E. Lacy (1981) Introduction of a rabbit β-globin gene into the mouse germ line. Nature, 294:92–94.

Fritsch, E.F., R. M. Lawn, and T. Manatis (1979) Characterization of deletions which affect the expression of fetal globin genes in man. Nature, 279:598–603.

Gordon, J.W., G.A. Scangos, D.J. Plotkin, J.A. Barbosa, and F.H. Ruddle (1980) Genetic transformation of mouse embryos by microinjection of purified DNA. Proc.Natl.Acad.Sci.U.S.A., 77:7380–7384.

Graessmann, A., M. Graessmann, and C. Mueller (1980) Microinjection of early SV40 DNA fragments and T antigen. In Methods in Enzymology. New York: Academic Press, Vol. 65, pp. 816–825.

Hardison, R.C. E.T. Butler, III, E. Lacy, T. Maniatis, N. Rosenthal, and A. Efstratiadis (1979) The structure and transcription of four linked rabbit β-like globin genes. Cell, 18:1285–1297.

Jaenisch, R. (1976) Germ-line integration and Mendelian transmission of the exogenous Moloney leukemia virus. Proc.Natl.Acad.Sci.U.S.A., 73:1260–1264.

Jaenisch, R. (1980) Retroviruses and embryogenesis: Microinjection of Moloney leukemia virus into midgestation mouse embryos. Cell, 19:181–188.

Jaenisch, R., and B. Mintz (1974) Simian virus 40 DNA sequences in DNA of healthy adult mice derived from preimplantation blastocysts injected with viral DNA. Proc.Natl.Acad.Sci.U.S.A., 71:1250–1254.

Lacy, E., R.C. Hardison, D. Quon, and T. Maniatis (1979) The linkage arrangement of four rabbit β-like globin genes. Cell, 18:1273–1283.

Lacy, E., and T. Maniatis (1980) The nucleotide sequence of a rabbit β-globin pseudogene. Cell, 21:545–553.

Maniatis, T., R.C. Hardison, E. Lacy, J. Lauer, C. O'Connell, D. Quon, G.K. Sim, and A. Efstratiadis (1978) The isolation of structural genes from libraries of eucaryotic DNA. Cell, 15:687–701.

Mercola, K.E., H.D. Stang, J. Browne, W. Salser, and M.J. Cline (1980) Insertion of a new gene of viral origin into bone marrow cells of mice. Science, 208:1033–1035.

Mintz, B. (1977) Teratocarcinoma cells as vehicles for mutant and foreign genes. Brookhaven Symp.Biol 29:82–95.

Mintz, B., and K. Illmensee (1975) Normal genetically mosaic mice produced from malignant teratocarcinoma cells. Proc.Natl.Acad.Sci.U.S.A., 72:3585–3589.

Oskarsson, M., W.L. McClements, D.G. Blair, J.V. Maizel, and G.F. Vande Woude (1980) Properties of a normal mouse cell DNA sequence (sarc) homologous to the Src sequence of Moloney sarcoma virus. Science, 207:1222–1224.

Papaioannou, V.E., R.L. Gardner, M.W. McBurney, C. Babinet, and M.J. Evans (1978) Participation of cultured teratocarcinoma cells in mouse embryogenesis. J.Embryol.Exp.Morphol., 44:93–104.

Perucho, M., D. Hanahan, and M. Wigler (1980) Genetic and physical linkage of exogenous sequences in transformed cells. Cell, 22:309–317.

Robins, D.M., S. Ripley, A.S. Henderson, and R. Axel (1981) Transforming DNA integrates into the host chromosome. Cell, 23:29–39.

Southern, E.M. (1975) Detection of specific sequences among DNA fragments separated by gel electrophoresis. J.Mol.Biol., 980:503–517.

Van der Ploeg, L.H.T., A. Konings, M. Oort, D. Roos, L. Bennini, and R.A. Flavell (1980) γ-β thalassemia: Deletion of the γ- and δ- genes influences β-globin gene expression in man. Nature, 283:637–642.

Whittingham, D.G. (1971) Culture of mouse ova. J.Reprod.Fertil. (Suppl.), 14:7–21.

Whittingham, D.G., and R.G. Wales (1969) Storage of two-cell mouse embryos in vitro. Aust.J.Biol.Sci., 22:1065–1068.

# Index